The Cathode-Ray Tube

THE CATHODE-RAY TUBE

Technology, History and Applications

Peter A. Keller

Palisades Press

An Imprint of
Palisades Institute for Research Services, Inc.
New York, New York

Palisades Institute for Research Services, Inc.
201 Varick Street, Suite 1140, New York, NY 10014

Printed in the United States of America

Library of Congress Cataloging in Publication Data

Keller, Peter A., 1937–
The cathode-ray tube: technology, history, and applications / Peter A. Keller
320 p. 19 × 23.5 cm.
Includes bibliographical references and index.
ISBN: 0-9631559-0-3
1. Cathode ray tubes. I. Title.
TK7871.73.K46 1991
621.3815′42—dc20 91-68267
CIP

Consulting Editor: Kenneth I. Werner
Editor: Holly S. McLean-Aldis
Book Design: Nutmeg Publishing Consultants

Contents

Acknowledgements

The author wishes to express his gratitude to the following individuals for their many helpful suggestions, interest and contributions. The Tektronix Corporate Library staff—Mary Bagly, Derr Kleintob and Georgia George—provided invaluable leads and references without flinching at the most obscure of publications. Also, thanks to Julianne Williams for her indexing, and Jim Richardson and Dick Compton for photographic assistance. A special thank you to Irene for her research, editorial suggestions, encouragement, help in searching out vintage CRTs in dusty attics and warehouses, and her tolerance during the prolonged conversion of the family dining room to a photography studio.

Bill Beran	Tektronix (formerly of Westinghouse)
Jake Brain	Thomas Electronics
Bill Brown	Tektronix
Ken Compton	Clinton Electronics
Michael Cross	Historical Electronics Museum/ Westinghouse
Chris Curtin	Formerly of Tektronix
Tom Goldsmith	Retired (formerly of Du Mont)
Joe Griffith	Formerly of Tektronix
Pat Green	Tektronix
Joe Hallett	Consultant (formerly of Sylvania and Tektronix)
Jim Hawes	Tektronix
Ken Hawken	Tektronix
Bob Hohlbauch	FAA
Vern Issac	Formerly of Tektronix
Bruce Kelley	Antique Wireless Association
Lionel Kreps	Formerly of Tektronix
Phil Krzyzkowski	U.S. Army, Ft. Monmouth
Bob Matthew	Corning Glass Works
Ken McGhee	Electronic Industries Association
John McQueen	Southwest Vacuum Devices (formerly of Du Mont)
Walt Melnick	Wright-Patterson Air Force Base
Dick Momberger	Philips (formerly of Sylvania)

Bailey Neal	Philips (formerly of Sylvania)
Carol Parcels	Hewlett-Packard
Norm Patrick	Litton (formerly of RCA Lancaster)
Ron Petersen	Formerly of Tektronix
Ed Powers	Raytheon
Jacob Ruiter	Ruiter Associates (formerly of Du Mont)
Jack Sauerborn	Hughes Aircraft
Peter Seats	Thomson Electron Tubes and Devices
Ludwell Sibley	Antique Wireless Association
Elliot Sivowitch	Smithsonian Institution
John Sonneborn	Tektronix
Larry Tannas	Society for Information Display
Jim Weager	Panasonic
Jim Wurtz	Litton Electron Devices

Additional acknowledgements accompany illustrations throughout the text.

Preface

This book culminates the author's longtime fascination with the cathode-ray tube and its ability to convert a stream of invisible electrons into patterns capable of revealing the innermost workings (or nonworkings as is often the case) of an electronic device or displaying a picture rivaling a photograph in quality.

For a device whose demise has been predicted for years due to the advent of solid-state devices, the CRT is remarkably healthy; indeed, it is stronger than ever. The cathode-ray tube antedates the receiving vacuum tube and is now at its zenith, some 20 years following the vacuum tube's demise.

The relative low cost, extreme versatility, ability to display full-color, high-resolution images, and a design that lends itself to continuing improvement has allowed the CRT to maintain a strong lead over competing technologies. One need only look around to see the all-pervasiveness of the cathode-ray tube. Television, personal computers, video games, electronic test and measurement instruments, computer-aided workstations, airline reservation systems, cockpit flight instrument displays, pay-telephones, cash registers, automatic bank teller machines, and even beverage can sorting/counting/smashing machines are all examples of our daily contact with the ubiquitous cathode-ray tube.

Just when it seemed that every possible use for a CRT display had been developed, the author stumbled on the "video scale" at a rest stop along the New York State Thruway while researching material for this volume. Of course, it requires considerably more than a penny for your weight to be displayed on a color, shadow-mask, cathode-ray tube.

This book traces the evolutionary development of the cathode-ray tube from the earliest experiments with fluorescence and electrical discharges at reduced atmospheric pressure through the latest full-color, high-resolution tubes for high-technology computer displays. The text is presented with a balance between historical and technical advancements and will be of interest and value to designers, users, historians, collectors and students. Much of the information presented has heretofore been available only from scattered sources. Other information has been in danger of being lost forever as early researchers have retired and corporate files have been purged of old records. Wherever possible, early diagrams are included for historical flavor. An extensive list of references is included to allow the reader further study of topics of interest.

No attempt has been made to include information on the many spin-offs of the cathode-ray tube developed for nonviewable applications such as X-ray, scan conversion, television pickup, etc. Likewise, equipment utilizing cathode-ray tubes and the circuitry required to drive them is discussed only as necessary to understand evolutionary trends and tube performance demands. These demands are both those placed on the tube by the equipment performance specifications and those imposed on the equipment by the tube operating requirements.

Generally, tube type numbers are provided up to the late 1950s when the quantity of new tubes with small variations and "house-numbers" instead of industry-registered numbers began to grow exponentially. It would be impossible to cover them all, therefore only those considered to be milestone designs, typical of a classification or a particularly widely used type, are identified by type number. Much of the information on the tubes themselves is from the JEDEC registration files of the Electronic Industries Association in Washington, D.C., as well as individual manufacturers' technical data sheets. Where the tube was supplied with several different phosphors, the phosphor designation number often has been replaced by a hyphen in the text, for example "16AMP-" to simplify the description. In the case of tubes with house numbers, such as "CK-1352," "22M30" or "T4220," the phosphor designation usually has been omitted for clarity since some of these designations can become quite unwieldy.

In a few places there is repetition of information from other chapters. The purpose of such repetition is to make the chapters complete in themselves, rather than requiring the book to be read cover-to-cover and eliminates constant referral to other chapters.

In many cases it is difficult to identify the "true" inventor of a concept. Differences in defining the concept, time lags in publishing results, inadequate records (and recollections), geography and whether or not a working model was demonstrated all influence the determination of the person to whom an invention is attributed. In many cases, almost simultaneous discovery occurred on opposite sides of the Atlantic Ocean. I. E. Mouromtseff summed it up well in a 1950 article in the *Proceedings of the IRE* by commenting, "And yet, one can hardly think of a single important invention for which the full credit would be unequivocally given by vox populi—and off the record, even by experts—to a single person, or to the same person. Moreover, the favored person is by far not always the official patentee. A good illustration of this apparently strange phenomenon is the case of radio."*

If Mouromtseff thought radio was bad, he should study the evolution of radar. In its present form radar is often more or less officially attributed to Robert Watson-Watt. However, it also has been attributed to or claimed in varying degrees by, Hertz, Marconi, Hulsmeyer, Fessenden, Baird, Du Mont, Taylor and Young, Breit and Tuve, and others. Just ask, "Who made radar practical?"

*Mouromtseff, I. E. "Who Is the True Inventor?", *Proc. IRE,* Vol. 38, No. 6, pp. 609–611, 1950.

and the list of names probably will double. Ask the same questions in the USSR, Germany or France, or ask, "Which company developed radar?" and you better get a long sheet of paper for your list! Bear in mind that most developments described in this book actually were the work of several individuals, especially in the years since 1940. Wherever possible, as many individuals as could be identified are mentioned or included in the references.

Some omissions are bound to occur in a subject as complex and lengthy as the development of the cathode-ray tube, and the author regrets any such oversights. It should be noted that the conclusions drawn as to firsts, dates, and individuals are those of the author and were derived in large part from papers, patents and other published accounts. Despite extensive research, inaccuracies may exist due to unavailability of published material for security or competitive reasons or mere oversight.

Lastly, the author welcomes corrections and additions to the material presented so that this book's future editions will represent the most accurate picture of the many design improvements and the long history behind the modern cathode-ray tube. It is hoped that this book will help to uncover other historical facts from previously untapped sources. If interest and additional facts warrant, the author would like to publish a revised and expanded second edition for the 100-year anniversary of the Braun tube in 1997.

Peter A. Keller
March 1, 1990

1

Cathode-Ray Tube Fundamentals

1.1 General form

In order to understand the importance of the evolutionary changes in cathode-ray technology and their impact on equipment using them it is helpful to have an understanding of the fundamental properties and operation of the cathode-ray tube (CRT).

Literally thousands of different CRT designs exist today with each one developed for a particular need. It is a tribute to the versatility of the CRT that so many variations are possible without economics becoming too much of a hindrance. CRT design involves a considerable amount of mix and match juggling. New tube types often are created by taking a gun from one tube, sealing it on a bulb from a second tube, and selecting an appropriate phosphor. This should not imply that CRT design is that simple. Design of new, high-performance, electron guns optimized for certain characteristics, or shadow-mask color tubes, involve the most sophisticated of computer simulations for determining electron trajectories with many variables requiring definition. Design compromises are required at every step with such opposing characteristics as brightness, resolution, size and cost. Wherever possible, parts of existing designs are utilized to minimize tooling costs and shorten development cycles.

All cathode-ray tubes (CRTs) are divided into two major classifications depending on the means used to position or deflect the electron beam to a desired location on the tube's screen. The first, and currently most important, is the *electromagnetically deflected CRT* used almost universally in television, computer displays and radar applications. It is used wherever fixed scan format, short tube length, and relatively low cost are required. The other classification, the *electrostatically deflected CRT,* is used primarily for oscilloscopes where extreme speed is required to position the electron beam to a desired location on the screen in order to follow rapidly changing waveforms. Although these two deflection methods greatly affect the tube's mechanical form, as shown in Figures 1.1 and 1.2, the fundamentals of operation are similar in most other respects.

One specialized form of CRT that will be discussed in Chapters 3, 4, 7 and 8 is the direct-view storage tube. The storage tube not only displays information, it can also retain

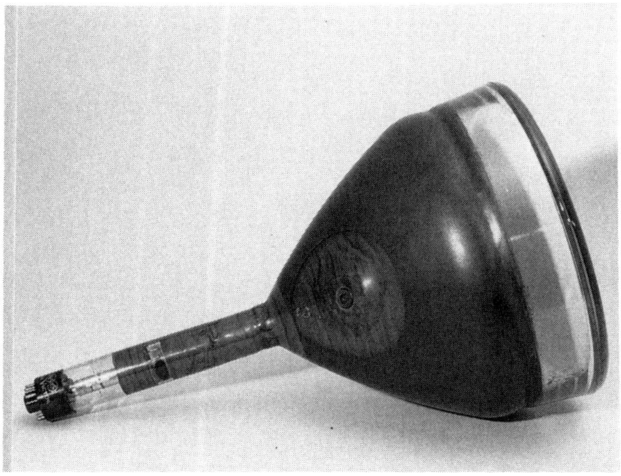

Figure 1.1 RCA Victor 10BP4 electromagnetic-deflection and focus cathode-ray tube, circa 1946. (From the author's collection.)

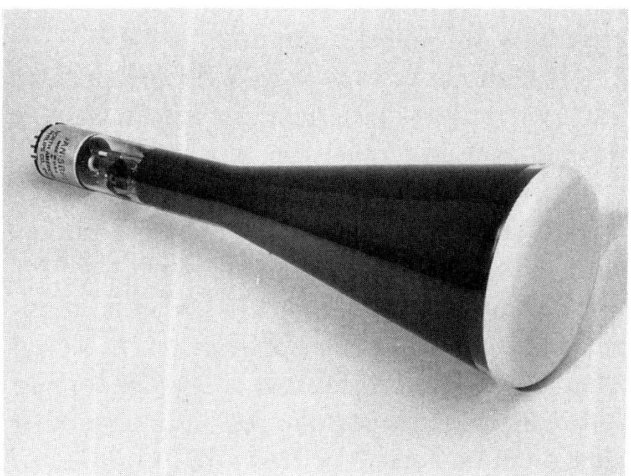

Figure 1.2 Norelco 5BP1 electrostatic-deflection and focus cathode-ray tube, circa 1945. (From the author's collection.)

it for a period of time without the need for continually refreshing the display by the associated driving electronics.

Most color CRTs are similar to other electromagnetically deflected CRTs except that they contain the equivalent of three conventional electron guns in one envelope along with a considerably more complex screen structure. (The color picture tube will be discussed in greater detail in Chapter 6.)

The basic configuration of electromagnetically and electrostatically deflected CRTs[1–8] is shown in Figures 1.3 and 1.4, respectively. The six basic functions of the modern cathode-ray tube were firmly in place by the late 1930s, yet much refinement and innovation occurred during the next 50 years as

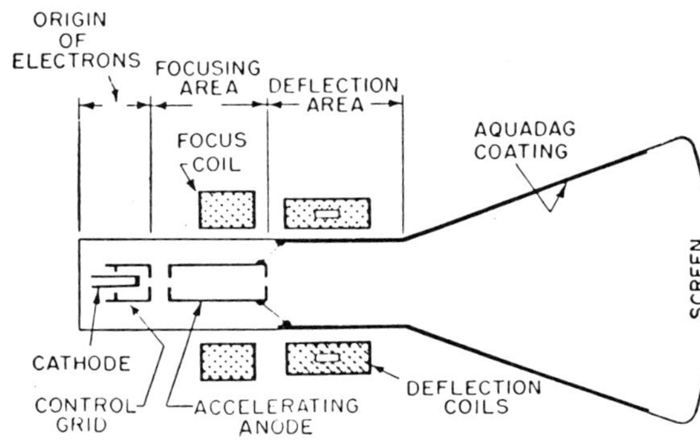

Figure 1.3 Electromagnetic-deflection cathode-ray tube schematic.[4]

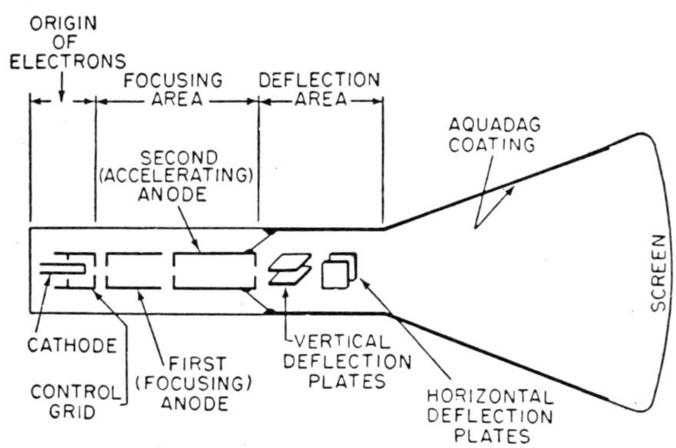

Figure 1.4 Electrostatic-deflection cathode-ray tube schematic.[4]

the cathode-ray tube developed to become a major part of almost all high-technology electronic products. The six basic functions are:

1. The *electron beam source* and *beam intensity control section.*
2. One or more *acceleration electrodes* which increase the electron beam's velocity for increased light output from the screen.
3. *Focusing section* which brings the electron beam to a sharp focus at the screen.
4. *Deflection system* positions the beam to a desired location on the screen or is used to scan the beam in a repetitive pattern.
5. *Phosphor screen* converts the invisible electron beam to visible light.
6. A mechanical structure known as the *envelope* maintains a high vacuum, provides electrical connections to the various electrodes and insulates them from each other.

These six basic functions are discussed in Sections 1.2 through 1.7. Evolutionary steps prior to about 1940 are covered in Chapter 2. Developments after 1940 branched out in several fields of application and will be discussed in the chapters covering the respective major CRT applications.

The assembly of electrodes or elements mounted within the neck of the CRT is commonly known as the *electron gun* (Figures 1.5 and 1.6). This is an appropriate analogy because it is the function of the electron gun to "shoot" a beam of electrons toward the screen or target. The velocity of the electron beam is a function of the overall accelerating voltage applied to the tube. For a CRT operating at an accelerating voltage of 20,000 volts, the electron velocity at the screen is approximately 250,000,000 miles per hour (mph) or about 37 percent of the velocity of light. To carry the ballistics analogy a step further, this translates to 370,000,000 feet per second (fps) compared to a typical high-power rifle bullet velocity of 3,000 fps. Although the velocity of the electrons is extremely high, the mass is very small, thus, the result normally is luminescence of the phosphor screen where

Figure 1.5 Electromagnetic-deflection and focus-triode electron gun from a National Union 7BP7, circa 1943.

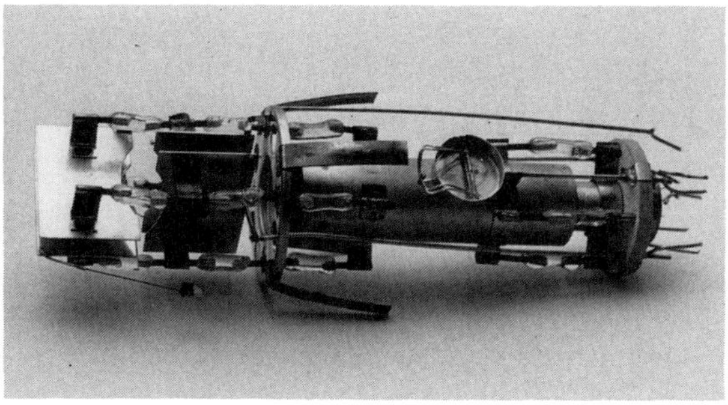

Figure 1.6 Electrostatic-deflection electron gun from a Sylvania 3AP1, circa 1944.

struck by the beam. If the beam power (accelerating voltage times beam current) is sufficiently high, however, intense localized heating may occur with a resultant phosphor burn or glass damage.

The earliest tubes used to investigate the properties of cathode rays consisted of only two electrodes in a glass envelope. These electrodes were the *cathode,* which was the source of electrons, and the *anode,* which accelerated the electrons to a high velocity. No phosphor screen was used and the cathode rays were made visible by fluorescence resulting from the collision of the electrons with residual gas atoms or the tube's glass walls. Chapter 2 traces the evolution as the focusing, deflection and screen functions were added.

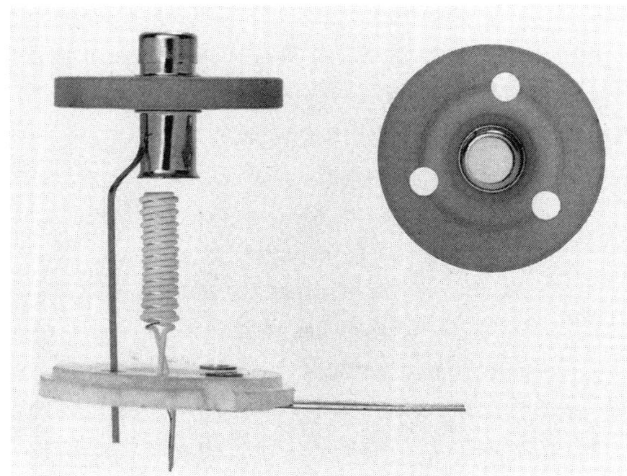

Figure 1.7 Parts of a heater-cathode assembly. The emitting surface of the cathode cup is in the upper right. (Courtesy of Tektronix, Inc.)

1.2 THE ELECTRON SOURCE

The cathode is the source of electrons for the electron beam (cathode rays). It may be a cold-cathode, in which electrons are pulled from it through application of extremely high anode voltages of 50,000 to 100,000 volts (as were used in the earliest cathode-ray tubes), or it may be a hot-cathode (as used in modern tubes) with anode voltages of 1,000 to 30,000 volts. Hot-cathode tubes are divided into two types: (1) the directly heated type, in which a hot tungsten filament, or one coated with thorium, emits electrons directly from the filament; and (2) the indirectly heated type, in which an oxide-coated metal cup containing a tungsten heater (as shown in Figure 1.7) emits the electrons from the oxide coating. The latter is used universally today.

Heater–The heater is operated at approximately 1,000 to 1,100 degrees Celsius in order to warm the cathode to a sufficient temperature for electron emission to occur, usually about 800 degrees Celsius. The correct cathode temperature is crucial to maintaining stable operation and long life. The heater power consumption is typically between one and six watts at 6.3 volts although 2.5 volts at 2.1 amps was the most common rating during the 1930s. Either AC or DC may be used for the heater's power supply.

Failure mechanisms for the heater include burn-out, which is usually the result of excessive heater voltage and electrical leakage or shorts to the closely spaced cathode.

Cathode–The hot-cathode was originally developed by Artur R. B. Wehnelt in 1903 to 1904. It was a key element in the advancement of the modern cathode-ray tube (see Chapter 2). More accurately, Wehnelt "discovered" the hot-cathode when he observed the emission of cathode rays from some areas of a platinum wire heated to only 800 to 1000° C in a vacuum. Since electron emission will not occur from a metal at such a low temperature, Wehnelt correctly surmised that the emission was a result of oxides on the platinum surface.

The oxide-cathode for cathode-ray tubes evolved into a cup-shaped element containing the heater (refer again to Figure 1.7). The end of the cathode cup, usually made of nickel with controlled impurities, is coated with barium, strontium and sometimes calcium carbonate to form an electron-emitting surface.[9–11] The cathode is operated at a high negative voltage relative to the screen which is at the anode voltage. The negative electrons emitted by the cathode are attracted toward the more positive screen. All other voltages applied to the CRT are referenced relative to the cathode which is considered to be at 0 volts. In reality, either the cathode or the anode may be near ground potential as long as the correct voltages for the other electrodes are maintained relative to them. In some equipment, such as oscilloscopes, the actual ground point usually is somewhere between the cathode and screen.

Cathode loading, or the current density per unit of cathode area, is an important design factor in obtaining small spot size, high beam current for good brightness and long cathode life. Cathode loading is a function of grid-to-cathode spacing, grid aperture size, and drive voltage applied between the cathode and grid No. 2 or anode No. 1 voltage. Normally, peak loadings of less than one amp/cm^2 are used although up to 30 amp/cm^2 may be used for special purposes where duty cycles are limited or short life is acceptable.

Average beam current usually is within the range of one microampere to one milliampere at the screen. Beam current reaching the screen often is less than the cathode current due to interception of electrons by beam-shaping or beam-limiting apertures farther down the electron gun. High-resolution guns may have beam currents of only 10 percent or less of the initial cathode current reaching the screen. Still, the actual quantity of electrons is enormous. A beam current of 100 microamperes, typical of many CRT applications, represents a quantity of 6.24×10^{14} (624 trillion) electrons per second! Figured over a 20,000 hour life expectancy, the total number of electrons delivered by a cathode is on the order of 5×10^{22} (50,000,000,000,000,000,000,000).

Cathode activation is the process used to bake out the nitrocellulose binder, which secures the cathode coating to the cathode cup, and to convert the carbonate cathode coating into oxides capable of maximum emission. The process consists of operating the newly pumped CRT at higher than normal heater voltage to increase the cathode temperature, and drawing high current from the cathode during part of the cycle. Each manufacturer has their own favorite cycle profile of heater voltage and drive voltage versus time.

The cathode is one of the key items for long CRT life which currently is often in the range of 10,000 to 20,000 hours. Pure oxide coatings, clean metal cathode cups, proper activation, controlled operating temperature, reasonable loading, proper spacing to the control grid and a good vacuum must all be maintained for maximum life. Failure modes may involve depletion of the oxide coating causing low emission, damage of the coating by positive ion bombardment created by the electron beam colliding with residual gas atoms, shorts to the closely spaced control grid and physical damage resulting from high-voltage arcs within the tube.

Control grid–Grid No. 1, or the control grid, is the remaining element of the electron source region of the CRT. The term *grid* is a carryover from the receiving tube days when a grid was actually a spiral-wound wire grid surrounding the cathode. The grid's function was to control the number of electrons reaching the tube's positive "plate." By applying a negative voltage to the grid, the negative electrons were repelled back toward the cathode. If the grid was made less negative, i.e., closer to the cathode potential, more electrons were allowed to reach the plate and the plate current increased. A small change in grid voltage caused a large change in plate current and produced signal amplification or gain. The function of the grid is similar in a CRT, with only the physical structure being different. The CRT grid, usually identified as G1, consists of another metal cup concentric with the cathode cup as shown in Figure 1.8. The grid cup has a small hole or aperture in the center which allows electrons to exit the cathode in the direction of the screen and simultaneously restricts the size of the electron beam. Applying a negative voltage to the grid reduces the number of electrons in the beam, thus controlling the brightness at the screen. At a sufficiently high negative voltage, known as the *cutoff* voltage, all electrons are prevented from reaching the screen. Typically, the cutoff voltage is near 60 volts. The cutoff voltage is a function of the spacing between grid and cathode, size of the grid aperture and the voltage applied to the first anode or G2. A low cutoff voltage is desirable to allow the grid to be controlled with smaller voltages, however trade-offs in beam current, spot size, beam current stability and close manufacturing tolerances required for the grid spacing tend to limit the practical sensitivity obtainable. With too-close spacing, the distance may vary with temperature as the cathode heats up causing a brightness change. It

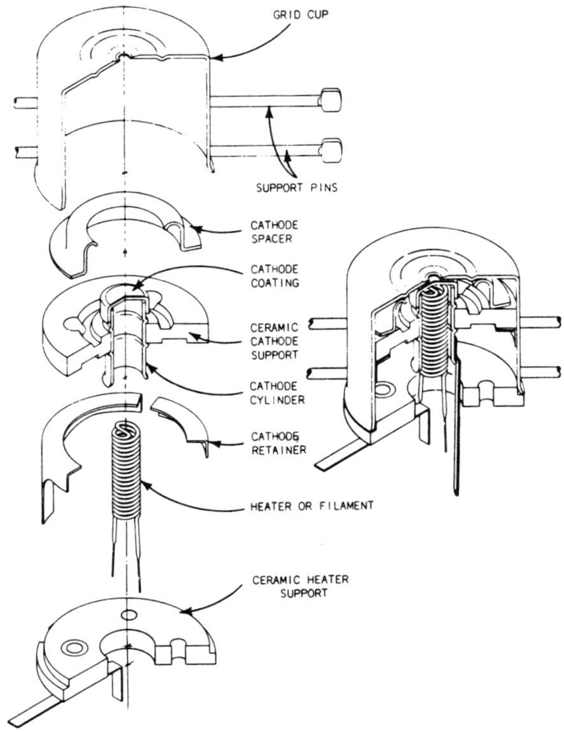

Figure 1.8 Details of a grid-cathode assembly.[7] (Courtesy of Tektronix, Inc.)

might even create a short circuit between the two elements, thus causing the beam to turn on to maximum current with no control over it. This occasional CRT failure mode is referred to as a G-K short.

The grid must not be driven positive with respect to the cathode or damage may occur to the cathode coating due to overheating from drawing excessive cathode current.

1.3 ACCELERATION SYSTEM

There will always be one or more accelerating electrodes to increase the beam velocity before reaching the screen. The accelerators are referred to by a variety of names depending on the tube's manufacturer and its exact location and function. These designations range from grid No. 2 to grid No. 5, anode No. 1 to anode No. 3 and various combinations of the words *post-* and *accelerator,* or *ultor*. Regardless of what one calls them, the effect is basically the same, accelerating electrodes add energy to the electron beam to produce more light from the screen.

Anode No. 1–The accelerating electrode closest to the grid/cathode assembly usually is called grid No. 2 (G2) in electromagnetically deflected tubes and the first anode (A1) or accelerator in electrostatically deflected CRTs (see Figures 1.3 and 1.4), although as mentioned the designation varies.

This electrode is also usually in the form of a small cup with a central beam-limiting aperture. The function of the accelerating electrode is to form a positive field to pull the electrons away from the cathode, accelerate them toward the screen and limit the beam's diameter by intercepting the more divergent electrons from the beam bundle. The first accelerator usually operates between 250 and 500 volts in magnetic CRTs and 1,000 to 4,000 volts in electrostatic tubes. Some radar tubes during World War II did not use a gun-mounted accelerating electrode, but instead utilized the high-voltage anode for the same function. The result was a very simple but reportedly excellent-performing CRT.

Ion trap–Externally mounted ion traps in the G2/anode region frequently were used on magnetically deflected television picture tubes in the years immediately following World War II. Residual gas atoms within the tube cause ions (charged atoms) to be produced when struck by electrons. Negative ions originate at or near the cathode, follow the same path as electrons in an electrostatic field but are only slightly affected by magnetic fields.[12–14] Ions are considerably larger and heavier than electrons and the ions are capable of causing damage to the phosphor screen where they strike it. Since the magnetic-deflection field has much less effect on ions than on electrons, the ions tend to impact the screen near the center forming a darkened area known as an *ion spot*.

Several methods of ion trapping were developed to prevent these effects. The earliest was the bent-neck CRT introduced by R. M. Bowie.[15,16] The electron gun's axis was aligned with the edge of the screen so that the ion damage did not affect the center of the picture, but occurred at the edge where it was less noticeable. Pattern distortion would have been severe with this arrangement. The next approach was to have only a portion of the CRT's neck bent with an electromagnet at the bend to align the beam with the tube's centerline while the ions continued going straight into the neck wall (Figure 1.9). Obviously this design would not be adaptable to high volume manufacturing. Later approaches used conventional CRT envelopes combined with modified electron guns. These modifications were in the G2 and anode electrode region and were in conjunction with externally mounted electromagnets or permanent magnets located in that region. Three major methods were used extensively in the U.S.—the slashed-field, offset and bent electron guns. Both single- and double-magnet ion traps were used (Figure 1.10).

The slashed-field electron gun (Figure 1.11) used G2 and anode elements which were cut diagonally to form a slant electron lens with an offset axis for the electrons and ions which leave the cathode region on the same path. The double-magnet ion trap restores the elec-

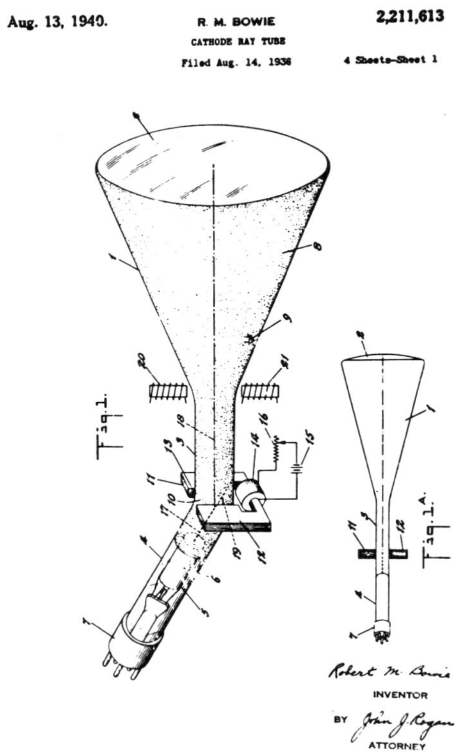

Figure 1.9 Bent-neck ion trap patent drawing.[15]

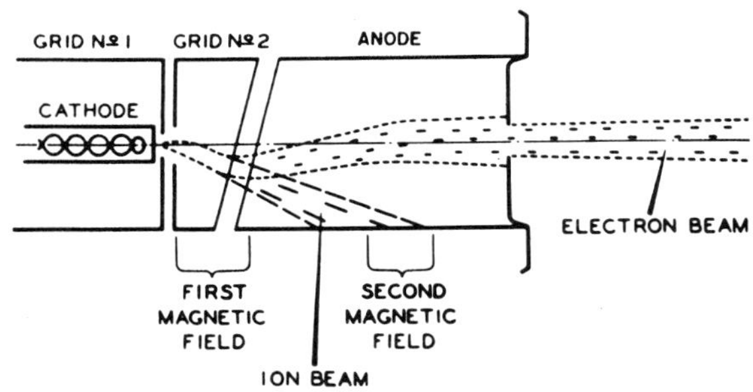

Figure 1.11 Slashed-field electron gun ion trap.[2]

trons to the normal tube axis while the ions are unaffected and trapped within the anode barrel. Some distortion of the electron beam to an elliptical spot was associated with this approach. This was a variation of a method originally proposed by R. M. Bowie. His method used an additional electrode to cause the electrostatic deflection of the beam instead of the slashed-field lens.

The offset electron gun's (Figure 1.12) construction was similar to the slashed-field gun except that the entire gun was tilted several degrees from the tube's center axis. Again, the electrons and ions were directed toward the anode wall by the slanted lens formed between the two elements. A single-

Figure 1.10 Single permanent magnet, double permanent magnet and double electromagnet ion traps. (Courtesy of James Richardson.)

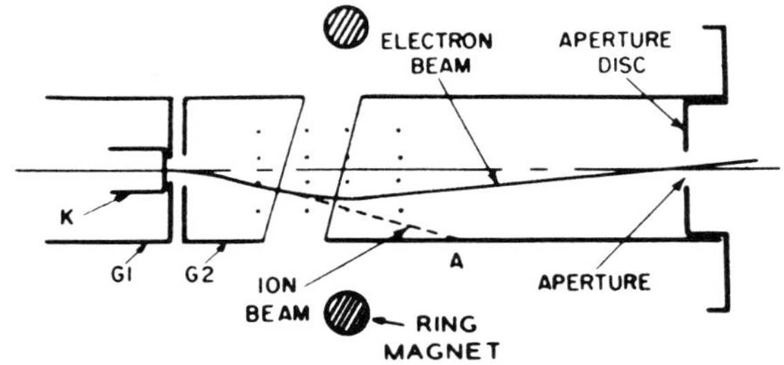

Figure 1.12 Offset electron gun ion trap.[2]

magnet ion trap then restored the electron path to the tube's centerline while leaving the ion path unchanged. Benefits of this method included minimizing of beam ellipticity and the use of a lower cost single-magnet ion trap.

The bent gun (Figure 1.13) was invented at Allen B. Du Mont Laboratories[17] and used a physically bent anode barrel. The cathode, control grid and G2 were inclined to the tube center axis while the bent anode barrel and single-magnet ion trap restored the electron beam path to coincide with the center axis, as in the offset gun.

Other methods were developed by L. A. Woodbridge[14] and H. Branson.[18] Woodbridge's method was employed in England by Electronic Tubes Limited. It utilized an offset cathode and grid aperture with a magnetic field to align the electron beam with the tube's axis while the ions were trapped in the anode barrel.

The mass production of aluminized screens in the early 1950s led to the elimination of the need for ion traps since the aluminum layer prevented the physically large ions from penetrating it while allowing penetration of electrons (see Section 1.6). Electrostatically deflected CRTs have no need for ion traps because the ions and electrons are deflected equally by electrostatic fields, thus spreading the ions over a large enough area to diminish their effects and to make them indistinguishable from normal screen aging.

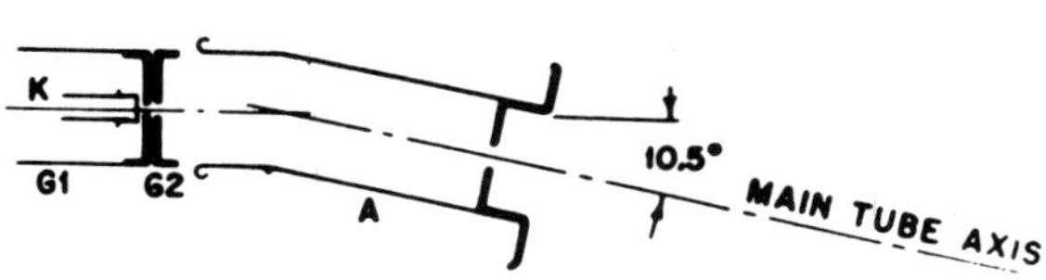

Figure 1.13 Bent-gun ion trap.[2]

Anode–This is another electrode that is known by several names. Often it is called the anode, but may be called grid No. 4 (G4), second anode (A2) or the ultor (as RCA referred to it). Usually its function is to provide increased acceleration for greater screen brightness, but in electrostatic CRTs it is used as part of the focusing lens and operates at or near the voltage applied to the first anode. In magnetic-deflection CRTs it often is operated at up to 30,000 volts and for some theater projection tubes may operate as high as 80,000 volts. The anode includes the entire screen end of the tube in magnetically deflected CRTs and monoaccelerator electrostatic CRTs. It may also be connected to a gun-mounted anode wafer or barrel through spring contacts.

Post-Accelerator–Also known as anode No. 3 (A3), intensifier, post-deflection accelerator (PDA) and post-ultor, this electrode is used only in electrostatic deflection CRTs and consists of a conductive band painted or evaporated on the inside of the CRT glass walls near the screen (Figure 1.14). Since it is located beyond the deflection system, it allows deflection of a lower voltage beam which is more sensitive while retaining an overall high-voltage for high brightness. Voltages of from 2,000 to 25,000 volts usually are used for the post-accelerator. The post-accelerator was described in a patent application by D. E. Howes of Westinghouse, filed in 1924.[19] It was also suggested by W. F. G. Swann in 1918, W. Rogowski in 1920 and A. B. Wood in 1923. Operating post-deflection accelerator CRTs were described by E. Sommerfeld in 1928 and F. K. Harris in 1934.[103]

1.4 FOCUSING SYSTEM

Two methods are used in modern cathode-ray tubes to bring the electron beam to a sharp

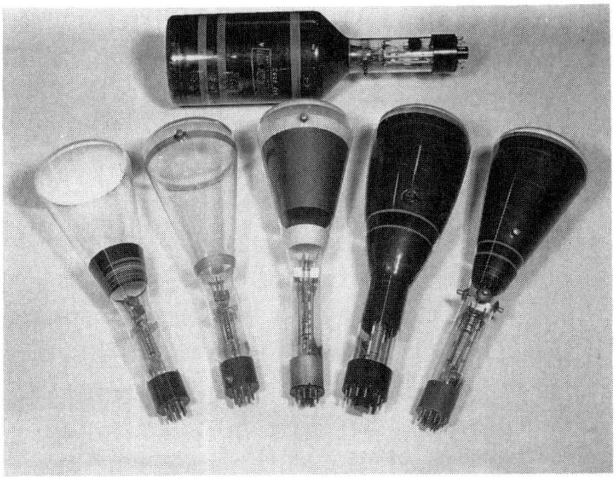

Figure 1.14 Post-deflection accelerator cathode-ray tubes, circa 1940–1945. (Courtesy of Jacob Ruiter.)

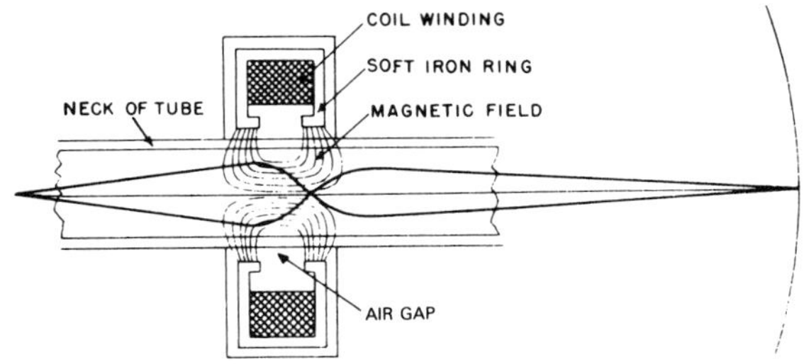

Figure 1.15 Magnetic focus coil and its effect on the electron-beam divergence.[3]

focus at the screen: magnetic and electrostatic focusing. A third method, gas focusing, was used in the 1920s and early 1930s as described in Chapter 2. Gas focusing, however, had several shortcomings and became obsolete as magnetic and electrostatic focusing were developed.

Magnetic focus–During the 1940s, magnetic focus was used in most electromagnetically deflected cathode-ray tubes for television and radar. Both electromagnetic focus coils and permanent magnet focus assemblies were commonly used. Focus coils or assemblies were mounted over the neck of the tube in the first anode region. The electron beam, which normally diverges in that region, was altered to converge to a focus at the screen through the action of the magnetic lens effect as shown in Figure 1.15 Around the time of the Korean conflict, magnetic focusing was replaced with electrostatic focus in order to conserve strategic materials, such as the cobalt used in Alnico permanent magnets, and for weight and cost savings.[20] Today, magnetic focusing is used only for some projection CRTs and the highest resolution film recording applications. Its advantage for these applications is that it may be made with much greater diameter than is possible with an electrostatic lens, which must be smaller than the CRT's neck diameter. Larger diameter focus lenses have less spherical aberration, consequently a sharper defined spot on the CRT screen.

Electrostatic focusing–This consists of one or more electrodes added to the electron gun to form an electrostatic lens. The term lens is used in the electron-optics sense whereby electrons may be refracted by electric fields in a manner similar to that of light by lenses and glass surfaces. The lens is formed by the voltage gradient field between adjacent electrodes at differing voltages. The voltage on the focus electrode is adjusted in order to converge the electron beam to a focus at the screen.

Cathode-ray tubes of the 1930s used a simple bi-potential electrostatic focus method.[100] In recent years, bi-potential focus also has been used extensively in color and projection CRT designs because of its advantages in tubes requiring high beam current.

Bi-potential focus was superseded late in World War II by the *einzel* focusing system for electrostatically deflected tubes as shown in Figure 1.16. Einzel translates from German as "one," which refers to the two outermost elements (of the three) being at the same or one voltage.[21] The einzel focus system, also known as unipotential focus, has two advantages. (1) No interception of the electron beam by the focus electrode and (2) fixed voltage between the first anode and grid/cathode assembly. These advantages prevented current from being drawn in the focusing circuit and changes in beam current with focus adjustment.[22] When applied to magnetically deflected CRTs, low-focus voltages were achieved, thus simplifying focus control circuitry.

Dynamic focusing may be provided with electrostatic focusing systems for wide angle deflection CRTs to correct for the spot defocus at the screen's corners caused by the longer "throw" distance from the deflection

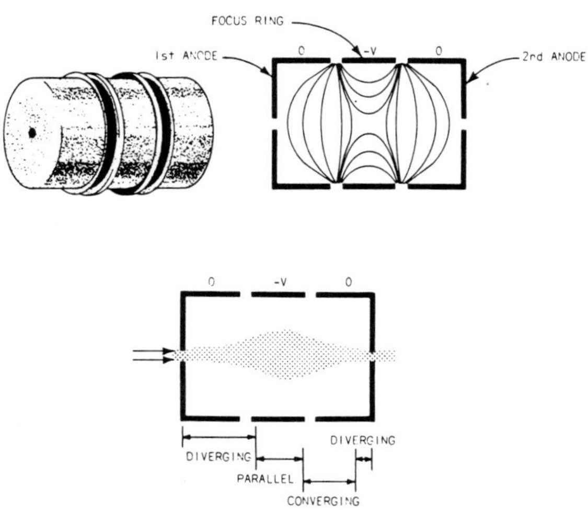

Figure 1.16 Einzel electrostatic focusing lens.[7] *(Courtesy of Tektronix, Inc.)*

yoke to the screen corners. Parabolic waveforms derived from the deflection amplifiers are superimposed on the focus voltage and their amplitude is adjusted to refocus the beam in the corners.

1.5 DEFLECTION SYSTEM

As with focusing, the same two methods (electromagnetic and electrostatic) exist for deflection of the electron beam to any desired location on the screen. The choice of methods is dictated primarily by the scanning frequencies and tube length requirements as described in the following paragraphs and Section 1.9.

Electromagnetic deflection–The most prevalent deflection method today is electromagnetic. The observation that cathode rays could be deflected by a magnetic field dates back to 1859 and the work of Julius Plücker (see Chapter 2). To be useful, the electron beam must be deflected with an alternating current electromagnetic field rather than a permanent magnet in order to sweep the beam repetitively across the CRT screen. This is accomplished with two pairs of series-connected coils mounted around the neck of the tube as shown in Figure 1.17. The two sets of coils are arranged 90 degrees from each other in order to provide magnetic fields in both the vertical and horizontal axes for deflecting the beam to any point on the screen.

Magnetic deflection has several advantages. The most important is the ability to deflect the beam at wide angles, thus allowing relatively short tube length. Forty-degree deflection angles were used in the late 1930s and 50 to 55 degrees was the range of prevalent angles used through the 1940s. Deflection angles increased to 70 degrees in 1950. As screen sizes grew ever larger deflection

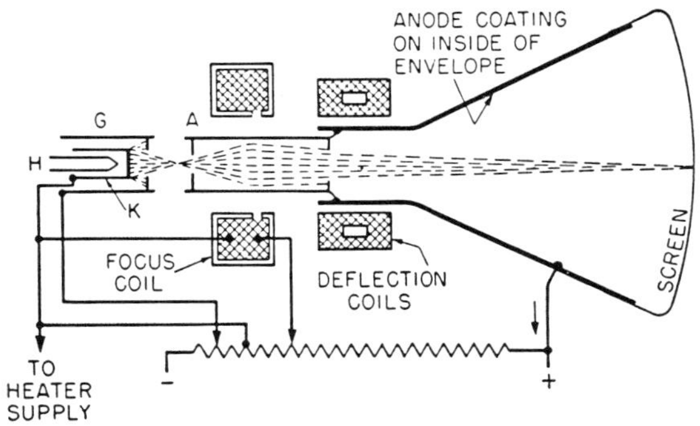

Figure 1.17 Electromagnetic deflection system.[4]

angles increased up to 114 degrees. An advantage of magnetic deflection is that tubes may be operated at higher accelerating voltages since the deflection sensitivity only decreases by the square root of the accelerating voltage instead of linearly, as in an electrostatic CRT, due to increased beam "stiffness" that results from higher voltages.

Magnetic deflection is particularly useful for television raster scan applications where the frequencies used to drive the deflection yoke are relatively low. Deflection frequencies above 100 kilohertz are difficult to work with due to the time-constants of the inductive coils which limit the rate at which the electron beam may be moved across the screen (slew rate). Special low inductance yokes requiring high driving current are used for high scan frequencies.

Electrostatic deflection–When the electron beam must be moved rapidly across the screen (as in an oscilloscope), electrostatic deflection is dictated. Deflection frequencies of 1,000+ megahertz have been achieved in commercial products in recent years, although in 1940 the practical limit was a few hundred kilohertz. Philips N.V. has produced an oscilloscope CRT capable of seven GHz using a helical deflection transmission line.[23]

Electrostatic deflection utilizes two pairs of metal deflection plates with the pairs mounted 90 degrees apart as shown in Figure 1.18. The electron beam passes between each pair on its way to the screen. When voltages are applied to any of the deflection plates, the beam of negative electrons is either attracted toward the plate if it is positive, or repelled if it is negative. Typically, each pair of deflection plates is driven "push-pull" to minimize pattern distortion, i.e., one plate is driven positive at the same time the other is driven negative an equal amount.

The frequency limitations for electrostatic deflection are much less severe than with a magnetic deflection yoke. Each pair of deflection plates forms the plates of a capacitor having only a few picofarads of capacitance and, therefore, little loss of high frequency signals. The deflection amplifier used to drive the deflection plates usually is the bandwidth limiting factor.

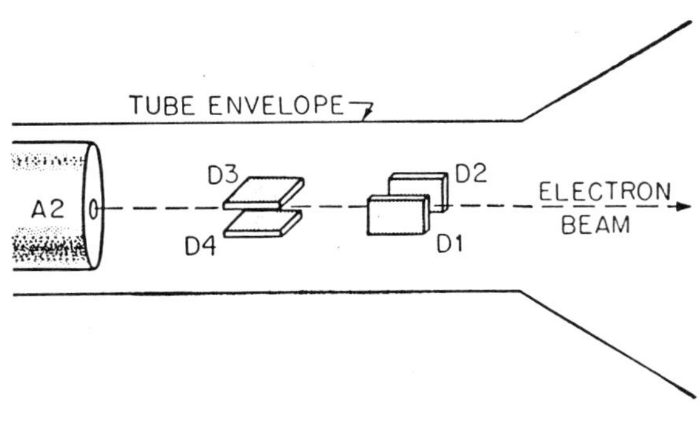

Figure 1.18 Electrostatic deflection system.[4] *Usually the plates closest to the screen, D1–D2, are used for the less critical requirements of horizontal deflection since they have lower sensitivity.*

Electrostatic CRTs are being replaced by electromagnetically deflected CRTs with raster scanning as digital electronics allow high frequency signals to be stored and converted to the raster format with its more leisurely deflection speeds.

Combination magnetic and electrostatic deflection–A few cathode-ray tubes have been made using both electromagnetic and electrostatic deflection. Some CRTs built around the beginning of the twentieth century had two sets of deflection plates, but also could be used with deflection coils. Others, such as the RCA 904 of the mid-1930s and the RCA 9JP1 introduced in 1942, had one set of vertical deflection plates and were intended to be deflected in the horizontal axis by electromagnetic deflection only. In oscillographic applications, the horizontal deflection rates are sometimes at a low frequency with fixed scan rates suitable for an electromagnetic-deflection yoke.

1.6 PHOSPHOR SCREENS

To make the otherwise invisible electron beam visible to the human eye, a coating of a fluorescent material called a phosphor is applied to the screen end of the cathode-ray tube. The phosphor consists of an inorganic crystalline powder containing traces of a controlled impurity atom known as an *activator* or *dopant*. Zinc silicate with a manganese activator, known as phosphor No. 1 in the 1930s and later as P1, was the most common phosphor in early CRTs and is still used to some extent. Zinc silicate is an efficient green-emitting mineral that occurs in nature as willemite. Willemite was first found in Liege and was named for King William I of the Netherlands. It was mined in the now closed zinc mines at Sterling Hill and Franklin, New Jersey.[24] Today CRT phosphors are synthesized from high-purity chemicals in order to achieve the highest possible efficiency under electron beam excitation.[1–7]

Zinc sulfide and zinc-cadmium sulfide are widely utilized for television picture tubes and, depending on the activator used, can produce a wide range of colors from orange through blue. Thousands of materials are luminescent and approximately 100 have been registered as CRT phosphors that are used commercially (see Appendix 7–9 for some commonly used phosphors). The newest phosphors are the "rare-earth" phosphors used for color television and aircraft cockpit displays.[25–28]

Phosphors emit light due to the energy added to the electrons in the phosphor crystal's atoms when bombarded by the high-energy electron beam. The added energy causes the electrons to jump to a higher energy orbit where they can exist only for a finite time before returning to their normal "ground state" orbit. A portion of the added energy is released in the form of one or more photons of light. The photon's energy determines the wavelength or color of the emitted light. It is interesting to note that most phosphors do not emit light without the controlled impurity atoms added as activators. The activators distort the crystal lattice structure making them luminescent.

There are a number of characteristics that dictate the phosphor choice for a specific application, including color, persistence, efficiency, the ability to absorb high electron beam currents without damage due to burning or loss of efficiency by saturation, resolution and suitability to the processes used for tube fabrication.[7,25–40]

Color–CRT phosphors have been developed to produce almost any visible color (and a few that are not visible). The choice of phosphor color used in a given CRT is partially dictated by the color sensitivity of the detector that will be viewing the display. Detectors include the human eye, photographic film or a photomultiplier tube. (Matching of the detector to the phosphor will be discussed later in this section under Efficiency.)

The match of the phosphor to the detector for efficiency isn't the only criterion for color choice. Black and white television naturally looks best in black and white, not black and green or some other color. Well, that is almost true. In reality, viewers usually prefer a somewhat bluish appearance as it appears "crisper." This seems to be related to the preference for blueing added to clothes washing detergents to make the wash appear whiter. An attempt to market television receivers with sepia-toned screens during the early 1950s for easier-on-the eyes viewing was a commercial failure. Two approaches were used: National Video Corp. produced a 21-inch picture tube, the 21AMP23A, with a sepia-toned phosphor (P23) and Hoffman used standard picture tubes from 12 to 24-inch sizes combined with sepia filter glass under the trade name "Hoffman Easy-Vision."

In other applications, color becomes a matter of personal preference, as has become evident with personal computer displays where the merits of green, amber and white, and "paper white" screens is vigorously debated among dedicated enthusiasts. When computer users have multicolor displays with a choice of background, text and editing colors, one will find some very interesting color combinations, indeed.

Color television screens use three phosphors applied in a fine dot, stripe or line pattern. The three color phosphors are the three additive color primaries of red, green and blue. By exciting them individually and in varying combinations with three electron beams, it is possible to produce most visible colors.

Colors are measured quantitatively and expressed in a pair of coordinates that may be located on a chart known as a *chromaticity diagram* as shown in Figure 1.19. The diagram shown is the current 1976 CIE-UCS version with u′ and v′ coordinates which is replacing the 1931 version utilizing x and y coordinates depicted in Figure 1.20.[41] The 1976 version is similar in concept, but has been mathematically squeezed and stretched in various regions so that equal distances between different pairs of colors in any color region will be perceived as being approximately equal color differences by the human eye. For color picture tubes, any color within a triangle formed by the three phosphor primaries may be reproduced by varying the relative proportions of the three corresponding electron beam currents.

The color coordinates are computed from information contained in the spectral curve as measured with a spectroradiometer. Figure 1.21 is a curve of P1, a green oscilloscope phosphor having a simple bell-shaped curve. P4, which is the standard black and white television phosphor and is a blend of blue and yellow phosphors that produce white, is shown in Figure 1.22. A rare-earth phosphor, P45, which is used to reproduce medical images, is shown in Figure 1.23. Note the very narrow line emissions which characterize rare-earth phosphors.

Persistence–Persistence, or decay, is the term used to describe how long a phosphor continues to emit light after the electron beam

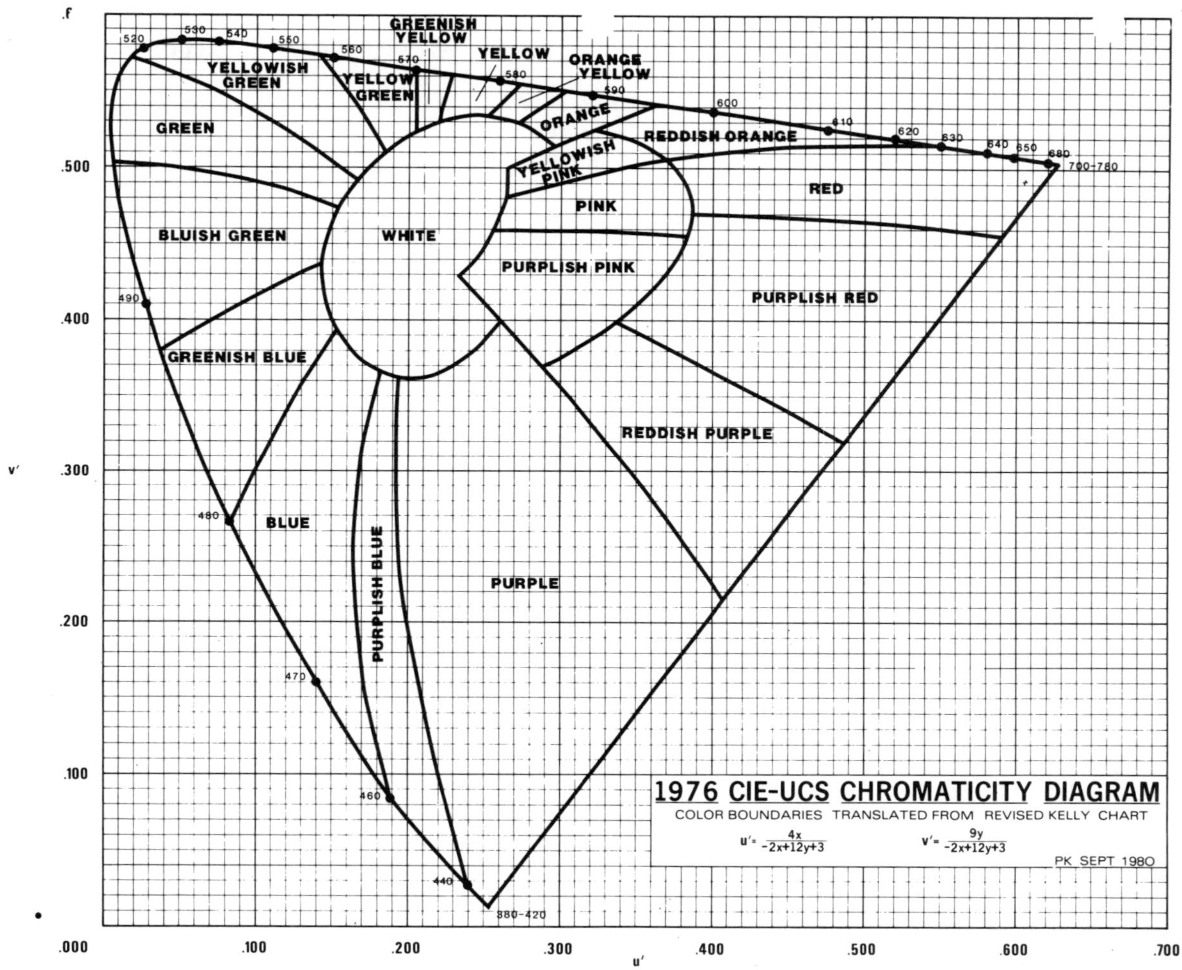

Figure 1.19 1976 CIE-UCS chromaticity chart.[41]

excitation is suddenly removed. Previously it was explained that the light is generated at some finite time after excitation by the return of the phosphor crystal's orbital electrons to their normal state. Each phosphor has a characteristic light output curve versus time after removal of excitation. Figure 1.24 shows how a phosphor responds to pulsed excitation with its own risetime and falltime. Figure 1.25 shows a typical persistence curve for P1 phosphor which has a medium persistence and has been used widely in low-voltage oscilloscopes. Phosphors may be chosen for a given application by persistence characteristics as well as color. For example, radar indicators require a long-persistence phosphor, such as P7, to provide image retention during the time between complete revolutions of the mechan-

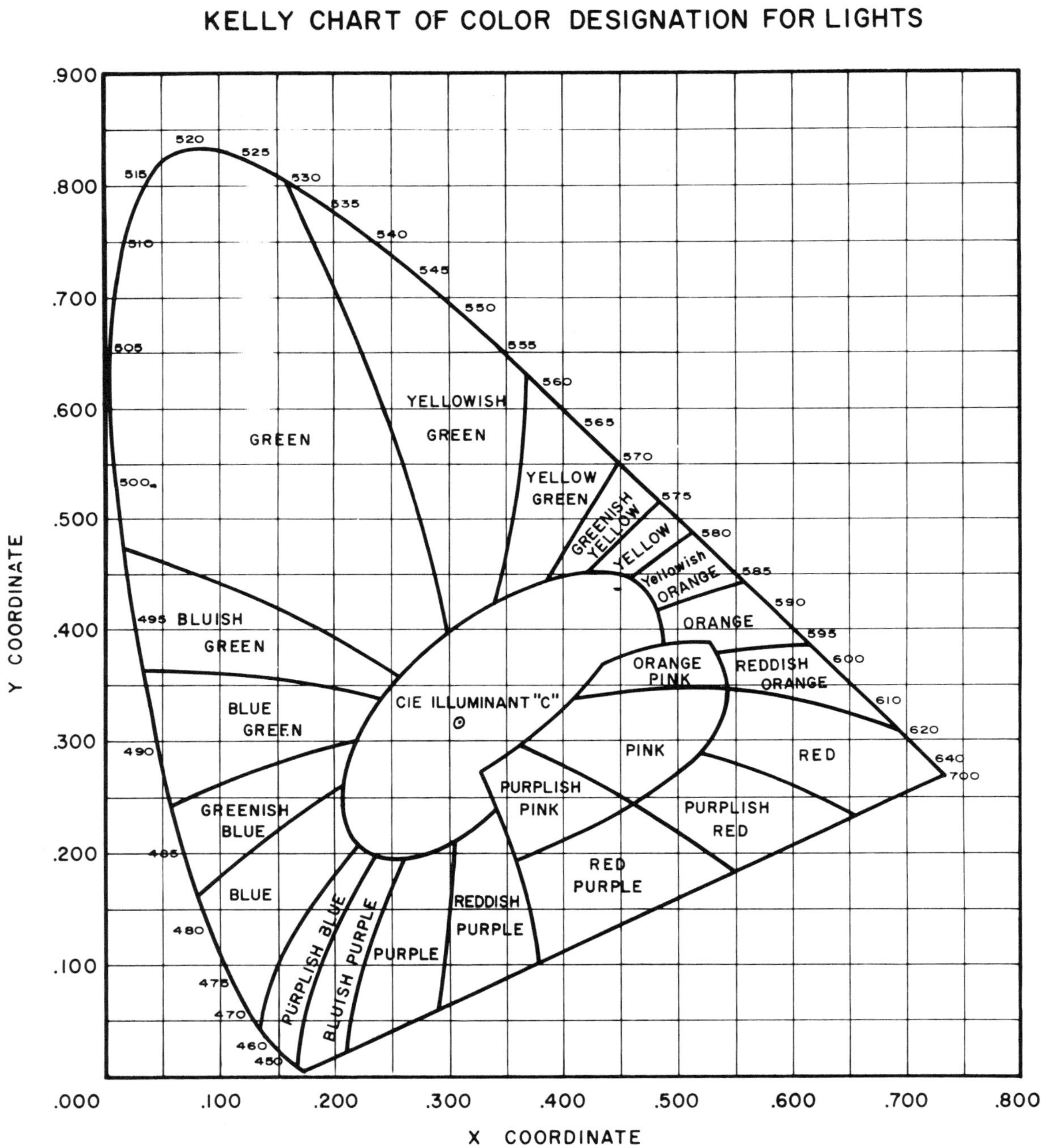

Figure 1.20 1931 CIE chromaticity chart.[41]

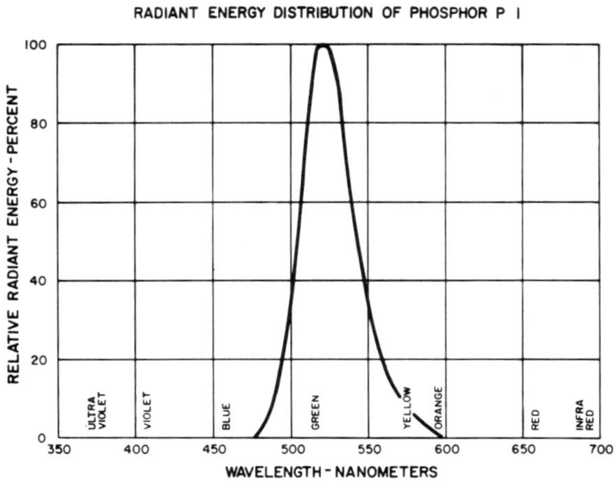

Figure 1.21 Spectral emission characteristics of P1 phosphor. (Courtesy of Tektronix, Inc.)

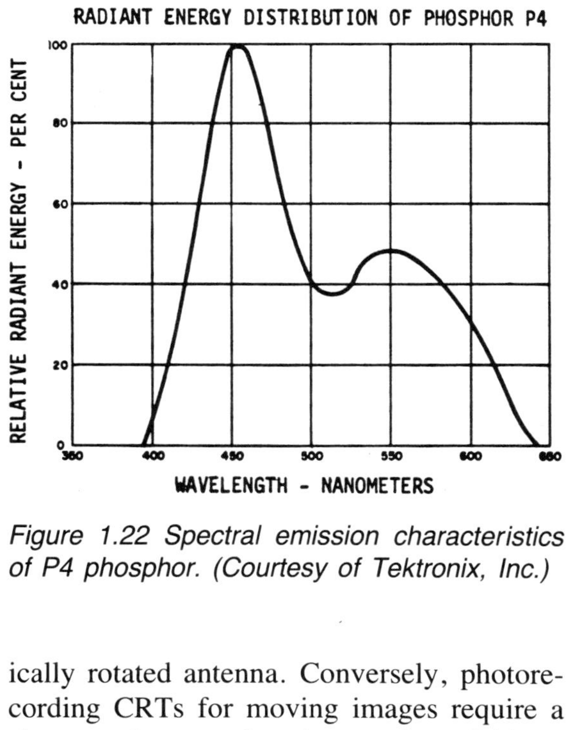

Figure 1.22 Spectral emission characteristics of P4 phosphor. (Courtesy of Tektronix, Inc.)

ically rotated antenna. Conversely, photorecording CRTs for moving images require a short persistence phosphor, such as P11, to prevent smearing of moving objects in the images. The range of persistences for commonly available phosphors range from nanoseconds to seconds. The persistence of sulfide phosphors is strongly dependent on beam current density and duration. Published persistence data therefore applies only to the operating conditions specified.

P7 is a particularly interesting phosphor that was developed by the British early in World War II for radar indicators. It consists of two layers of phosphors in what is known as a *cascade screen*. The phosphor that is deposited directly on the glass tube face is a long-persistence, yellow-green phosphor. Since this phosphor has longer persistence when excited by ultraviolet light than under electron beam bombardment, a second layer of near ultraviolet emitting phosphor is deposited on the first phosphor so that it is the phosphor struck by the electron beam. The near ultraviolet light emitted then excites the longer decay phosphor closest to the viewer. Some blue light is emitted by the near ultraviolet phosphor. Usually this is filtered out by placing an amber filter in front of the tube to block the "blue flash."

Another screen developed by Philco during World War II for radar was P10. P10 wasn't really a phosphor at all but a *scotophor*. By definition, a phosphor emits light when excited by some form of energy. P10 is instead a "dark trace" storage cathode-ray tube screen consisting of potassium chloride. Wherever struck by an electron beam, the normally white screen turns magenta. This coloration remains until "erased" by infrared radiation. Scotophors are also known as *Cathodochromic*. Chapter 3 discusses additional information on dark-trace storage tubes.

P34 is another unconventional phosphor and was developed by Ferranti, Ltd. As with many other phosphors, zinc sulfide is its primary

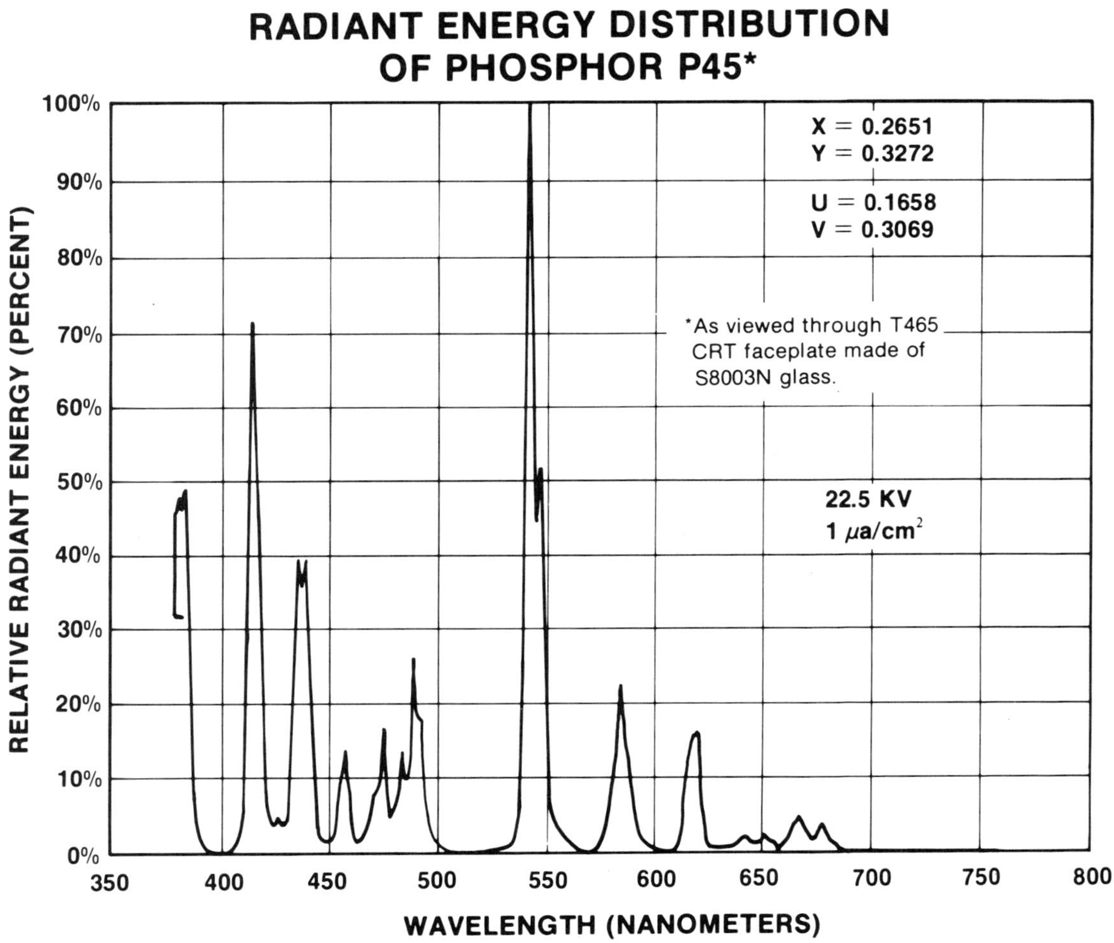

Figure 1.23 Spectral emission characteristics of P45 phosphor. (Courtesy of Tektronix, Inc.)

material with lead and copper acting as the activators. P34 emits light conventionally under electron beam excitation, but has the unique property of storing some of the beam energy within the phosphor due to the electrons changing orbits to levels that are traps. Additional energy applied to the screen from an infrared lamp causes the electrons to return to their rest or normal state with the consequential emission of light. This action may be triggered up to one minute or more after removal of the excitation by the beam.

Efficiency–As mentioned briefly in the discussion of phosphor color, color choice is important to provide an efficient match to the detector that will be viewing the display. More often than not, this will be the human eye which has highest sensitivity in the yellow-green region of the spectrum. Usually phosphors having most of their output in that color

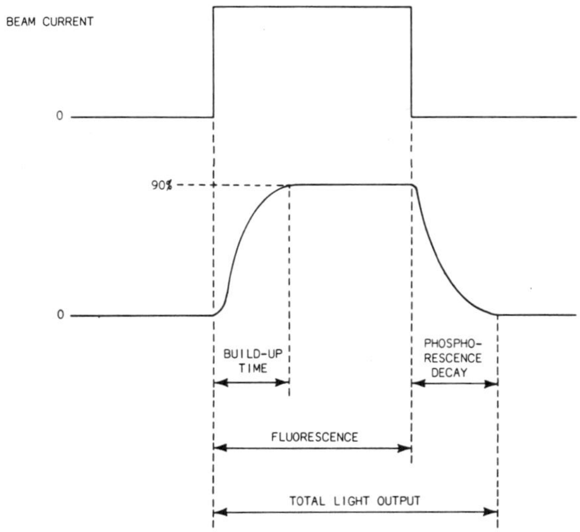

Figure 1.24 Typical phosphor time response to instantaneous changes of beam current.[7] *(Courtesy of Tektronix, Inc.)*

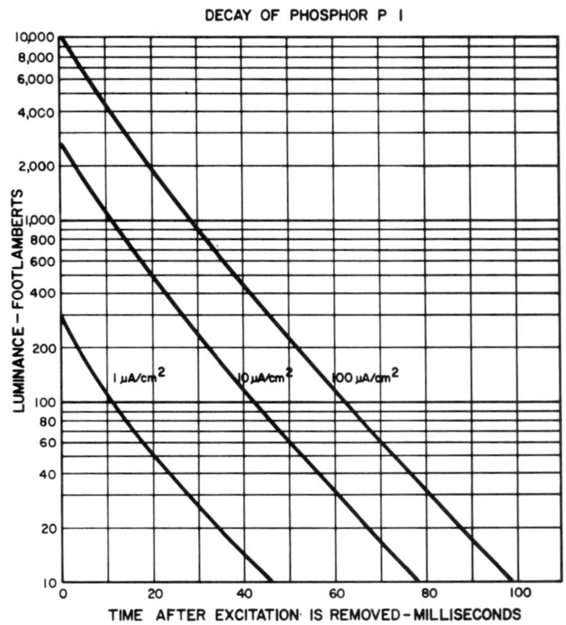

Figure 1.25 Persistence characteristics of P1 phosphor. (Courtesy of Tektronix, Inc.)

range will appear the brightest, thus minimizing the beam current required to excite the phosphor. Lower beam current results in longer cathode and phosphor life, smaller spot size and less likelihood of phosphor damage (burning). Photographic applications, and those using a photomultiplier tube for pickup, usually use a blue-emitting phosphor to match the peak sensitivity of the film or photomultiplier which is in the blue region. A number of other factors play a part in determining the efficiency of the phosphor screen. These include:

1. *Phosphor composition*–Some materials are inherently more efficient, e.g., the zinc sulfides. Generally speaking, very long and very short persistence phosphors sacrifice efficiency to obtain these extremes.
2. *Phosphor particle size*–Usually larger crystals are more efficient due to crystal fracture damage that occurs with the grinding required to obtain smaller particle size. Large particle size does not have very high resolution capability due to light being diffused throughout the entire phosphor crystal, therefore efficiency must be traded off in high-resolution tubes.
3. *Phosphor thickness (screen weight)*–Phosphor layers that are too thin waste beam energy by not totally absorbing it within the phosphor. The remainder penetrates into the glass and is dissipated as heat. If the phosphor is too thick, all of the light will be produced on the side closest to the electron gun and some of the light will be absorbed

and scattered as it passes through the extra phosphor on the way to the viewer. Again, there will be some loss of resolution in the latter situation.

4. *Accelerating voltage*–Higher accelerating voltages are more efficient due to deeper penetration into the phosphor crystals. The crystal surfaces are less efficient due to fracturing that occurs during synthesis of the phosphor and to surface contaminants on the phosphor crystals. Beyond a certain voltage known as the *sticking potential,* little additional brightness can be obtained. This is due to the phenomenon of secondary electron emission which increases with increasing voltage. Because the phosphor, which is an insulator, emits an increasing number of electrons as the beam voltage is raised, a point will be reached where as many will leave the screen as are received from the beam, thus no further brightness gain is realized.[101,102] This problem has been solved in recent years through applying conductive coatings to the screen, in the form of aluminizing, or a transparent tin oxide film on the glass faceplate.
5. *Beam current*–Brightness increases with increasing beam current density up to the point at which saturation or phosphor damage occurs. The actual efficiency or light output per watt of excitation power decreases, however, because as the current increases, the phosphor temperature also increases, which lowers the efficiency until a point is reached where the phosphor saturates and actually produces less light. Carried to the extreme, irreversible damage to the phosphor results due to thermal damage to the phosphor crystals. Another detrimental effect of high beam current is the increase in beam diameter due to mutual *space-charge* repulsion of the electrons. Cathode life also will be sacrificed.
6. *Aluminizing*–In 1939, the screen aluminizing process was developed by R. R. Law of RCA. It was later refined by Law and V. J. Schaefer of General Electric. Aluminizing significantly improved the screen's efficiency, eliminated sticking potential effects and prevented ion burn damage.[42–46] The process consisted of first coating the phosphor screen with a thin nitrocellulose lacquer layer to provide a smooth surface and then evaporating a thin film of aluminum onto the lacquer surface. Baking the screen removed the lacquer and allowed the smooth aluminum film to settle into contact with the phosphor. The mirrorlike film reflected light toward the viewer; light that previously was wasted to the rear of the tube. A gain in light output of about two times was obtained and contrast also improved. Without aluminizing, the light emitted rearward was partially reflected back to the screen with some of it striking black areas of the picture making them appear gray. The aluminum was allowed to contact the anode coating on the wall of the tube, thus providing a means to maintain the screen at anode potential. This eliminated screen charging and the sticking potential that limited the maximum anode voltage which

could be applied. Screen damage in the form of ion burns was eliminated since the heavy ions cannot penetrate the aluminum layer while electrons, being smaller, readily pass through as long as they have an accelerating voltage of at least 4,000 to 5,000 volts. Aluminized screens were called metal-backed or metalized screens, as well as manufacturers' trade names such as Silverama™ and Silver Screen 85™.

7. *Burn resistance*–A screen must be able to absorb the electron beam energy and consequent heating without damage at reasonable intensity levels in order to provide useful life. Most of the commonly used zinc silicate, zinc sulfide and rare-earth phosphors are quite burn resistant. Some very long persistence phosphors in the magnesium fluoride family, such as P19, that are used for slow radar sweeps, are easily damaged and equipment adjustments must be made carefully to avoid operating the tube at excessive beam currents. Some near ultraviolet phosphors also are prone to burning for a different reason. Most of the light that they emit is far removed from the green-yellow spectral region of maximum sensitivity of the human eye. The tendency is to turn up the beam current until enough light is produced to be seen easily, but this may occur at a current level in excess of what the phosphor can safely dissipate.
8. *Screen aging*–A long-term effect that occurs over the life of the cathode-ray tube is manifested as a loss of screen efficiency referred to as *aging*. Screen aging is due to two processes which occur in any area of the screen subjected to prolonged electron beam bombardment. The first process is *electron browning* created by a gradual reduction of oxides within the surface of the glass and phosphor by electron beam heating. The other process is *X-ray browning*. X-rays are produced whenever high-velocity electrons are stopped suddenly by a solid material such as the phosphor or glass. X-rays having an energy proportional to the accelerating voltage are generated. These will be absorbed by the phosphor and glass with the formation of color centers due to a change in ionization states within the material. X-ray browning is reversible by heating, but electron browning results in permanent physical damage to the phosphor or glass. Both types of aging have the same end effect. The brown discoloration of either absorbs some of the light emitted from the phosphor, especially the shorter (bluer) wavelengths of light. Screen aging is proportional to the total charge (coulombs) deposited on the screen by the electron beam and usually is not a problem until the accumulation of thousands of hours of operation.
9. *Resolution*–The ability to display extremely fine detail is limited to some degree by the size of the phosphor-crystals and how densely packed they are. Large crystals trap light within the crystals which can exit the crystal at some finite distance from where the electron beam struck it. Light also is scattered from crystal to crystal making the illuminated area larger than the

beam's area. Thin screens of small particle size phosphor will have a higher resolution capability, but at the expense of efficiency loss. One technique dating to a 1934 patent by J. H. de Boer,[47] mentioned by Williams in 1947,[48] and extensively researched during the 1950s for ultra-high resolution CRTs by Studer and Cusano of General Electric[49–52] is the transparent or evaporated phosphor. A phosphor is evaporated at high temperature within the CRT bulb and allowed to condense in a continuous transparent film on the faceplate. This effectively results in a large diameter, thin single phosphor crystal without the diffusing nature of a conventional powder type screen. Other advantages of the transparent screen include high contrast in ambient light, and excellent resistance to burning even at high beam current since the phosphor is in intimate thermal contact with the glass faceplate. This is not without a substantial penalty in light output since light is trapped within the crystal and "light-piped" out the screen's edges. Despite promising results, the transparent phosphor screen has not achieved commercial success.

10. *Processing*–The choice of phosphors may be influenced by compatibility with the process used to apply the phosphor to the glass faceplate. Screen uniformity, contamination and phosphor adherence all depend on the phosphor and screening application method as well as the materials used. Settling from a water/phosphor slurry is the most common method, although photo-deposition, painting, decals and spraying all have been or are still used.

1.7 ENVELOPE

The cathode-ray tube envelope, also called the bulb, bottle and, less reverently, the jug, serves five important functions: (1) Permits light from the phosphor screen to be seen from the outside, (2) provides an enclosure capable of maintaining a high vacuum, (3) provides mechanical support for the internal electron gun and various conductive and insulating coatings, (4) allows electrical connection between the internal electrodes and the outside world and (5) insulates the electrodes and coatings from each other. Additionally, it is called on to withstand high processing temperatures, high applied voltages and atmospheric forces of up to several tons for large tubes. For obvious reasons, glass has been the predominant candidate for the envelope, although metal and ceramic have been used for the funnel-shaped portion for certain applications.

The envelope is divided into four major parts as indicated in Figure 1.26. These are the *faceplate* (referred to as the panel in the television industry) which contains the screen; the *neck* which contains the electron gun, the

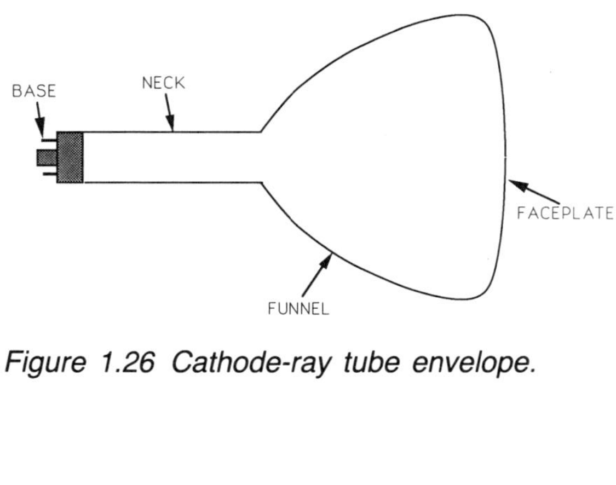

Figure 1.26 Cathode-ray tube envelope.

base which provides most of the electrical connections and the *funnel* which essentially contains nothing. Each will be discussed briefly.

Early CRT envelopes were one-piece, hand-blown glass bulbs and were similar to, and in a number of cases actually made from, the Erlenmeyer flask found in chemistry labs. These bulbs were made of hard glass (Pyrex™). During the early 1940s, separate pressed-glass funnels and faceplates of soft-glass flame-sealed together came into usage as flatter screens of better optical quality were developed for radar. As larger television screens were developed in the postwar years, lighter weight metal cone tubes were developed at RCA using a sagged plate-glass faceplate and glass to metal seals for the faceplate and neck.[53,54] (The metal cone picture tube will be discussed in greater detail in Chapter 5.)

The metal cone tube was soon superseded by large, pressed-glass faceplates and funnels sealed with the Electroseal™ process developed by Corning Glass Works. The Electroseal process solved the problem of sealing the thick glass parts required to withstand the tremendous atmospheric pressure on a large-screen picture tube. The glass to be sealed was first heated with a gas flame to a temperature at which it became electrically conductive. A radio frequency electric current applied to the heated region then effectively "welded" the two glass parts together.[55]

Today, pressed glass funnels and faceplates (Figure 1.27) with a *devitrifying frit* or *solderglass* seal are used extensively. Corning Glass Works developed the frit-seal process which was first introduced in 1958 by RCA for their all-glass 21CYP22 color picture tube.[56] The frit-sealing process consists of a toothpastelike material containing powdered glass/ceramic material which is applied to the junction of the parts to be sealed and fuses under pressure at high temperatures. The result is a glass-hard ceramic splice as shown in Figure 1.28. If necessary, the splice may be removed with nitric acid to reclaim the funnel and faceplate for reprocessing.

Figure 1.27 Color picture tube glass funnel and panel.

Modern color television picture tube glass

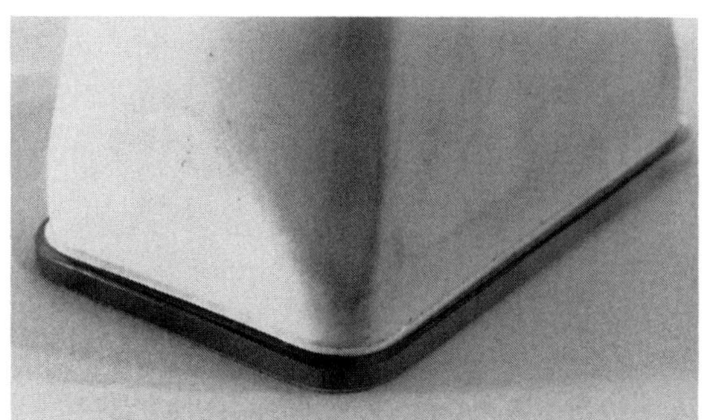

Figure 1.28 Frit seal of glass faceplate to ceramic funnel.

is produced in huge volume using computer-controlled processing.[57] Most monochrome glass bulbs and many color glass bulbs are now manufactured overseas.

Ceramic funnels were developed by Tektronix in 1963 and have been used for oscilloscope, computer display and film recording CRTs.[58] Ceramic funnels allow special funnel shapes to be more easily developed since the tooling consists mainly of a machined aluminum mandril. The process is well-suited for lower volume manufacturing scales than pressed glass allows. High tooling costs and glass melts measured in tons limit pressed glass to high-volume applications. The ceramic envelope also aids in the application of a silk-screened internal *graticule* (measuring scale) before the faceplate is sealed to the funnel. Plate glass or fiber-optic faceplates, as well as tubular glass neck tubing, are frit-sealed to the ceramic funnels. (See Chapter 4, Figure 4.69 which illustrates a ceramic funnel and a finished storage tube.) Ceramic envelopes also have been applied to avionics CRTs by Rank Electronic Tubes in England[59] and Tektronix in the United States.[60]

In all cases where two or more CRT parts are sealed together, it is necessary to match the thermal expansion characteristics (temperature coefficient) of the materials to prevent cracking or strain due to the differential expansion that the envelope is subjected to during processing.

To illustrate the extremes in materials that have been used for CRT envelopes, almost every CRT lab has at one time or another fabricated a "Coketron" made from that one glass envelope that always seemed to be present in any lab, the Coca-Cola™ bottle (Figure 1.29). A little phosphor settled in the bottom and an electron gun sealed to the neck results in an operable CRT that never ceases to amuse lab visitors.

Faceplate–CRT faceplates have undergone a series of evolutionary steps from the beginning when the faceplate was an integral part of the entire bulb and hand-blown of hard-glass (Pyrex™). Although a number of exceptions exist, faceplates usually were spherically shaped for strength up until about 1950. A crude handmade Farnsworth experimental Oscillite CRT with a flat faceplate (circa 1930) exists in the archives of the Smithsonian Institution (Figure 1.30). The first commercial flat-face instrument CRT faceplate was announced by von Ardenne in 1937[61] but did not come into wide usage until the mid 1940s when Du Mont began using them for high-performance CRTs.[62] As radar CRT faceplates became flatter in the early 1940s, the faceplate was shaped like a shallow, flat-bottomed cylinder and the transition to flame-sealed, separate faceplates and funnels began. These were of a "softer" glass such as soda-lime, lead or barium-silicate. Lead glass has an advantage of being a good absorber of the X-rays generated by the electron beam, but is more prone to browning with use. As accelerating voltages and beam currents increased for brighter displays, such as projection, cerium glass faceplates were developed to greatly reduce electron and X-ray browning.[63] Television picture tube screens became rectangular and sizes grew through the 1950s. To maintain strength against atmospheric pressure, the thickness also increased up to one half inch or more. Screen sizes of from one half inch to 45 inches have been produced.

Faceplate shapes were round until about 1949 when straight-sided rectangular screens were introduced. As screen sizes grew be-

Figure 1.29 "Coketron." (From the author's collection.) "Coke" and "Coca-Cola" are registered trademarks of the Coca-Cola Company.

Figure 1.30 Farnsworth experimental Oscillite with flat faceplate circa 1930. (Courtesy of National Museum of American History, Smithsonian Institution.)

yond 16 inches, the screen sides were curved for added strength. The curvature increased with the introduction of the wide-deflection angle monochrome picture tubes of the late 1950s. The current trend is back to straight sides with squarish corners and flatter screens. These screens are referred to as flat- or full-square tubes. Computer stress analysis has allowed the use of flatter surfaces while still meeting strength requirements.[64]

The color picture tubes introduced in the late 1950s and 1960s operated at high accelerating voltages of up to 25,000 volts or more for brighter pictures. X-rays are generated internally by all cathode-ray tubes, but can only penetrate the glass at the highest voltages. Concerns about possible harmful X-radiation emission at these voltages prompted the development of glass containing significant amounts of strontium, an X-ray absorber. This glass, as well as the appreciable thickness required for strength of large faceplates, has reduced the radiation emission to virtually unmeasurable levels. Recently, glasses with zirconium and barium have been investigated due to their better X-ray absorption characteristics.

CRTs used for oscilloscopes and photo-recording began to use plate-glass faceplates for optical flatness as the frit-sealing process became widespread. Tubes as large as 14 inches have been manufactured commercially using plate-glass.[65–67]

Faceplates have undergone a number of contrast enhancement improvements over the years to allow viewing under bright ambient illumination. Neutral gray glass faceplates were introduced in 1950 for television picture tubes. The contrast enhancement resulted because ambient light passes twice through the absorbing gray glass as it is reflected from the phosphor, while the light from the picture produced by the phosphor passes through it only once (Figure 1.31). A related method used in monochrome avionics displays that use green phosphor is the use of a narrow spectral transmission dark green glass faceplate which passes only the color of the phosphor.[68]

External spectrally selective filters for the enhancement of display contrast were described in a German patent application filed in 1936 and in U.S. Patent applications filed in 1937[69] and 1958.[70] Neodymium, praseodymium and a mixture of the two, didymium, glasses were used because of their unique spectral transmission characteristics. Didymium glass was investigated along with neodymium glass for color television picture tube filters by Zenith in 1969.[71] These glasses have a series of peaks and valleys across the spectrum which conveniently pass the colors of

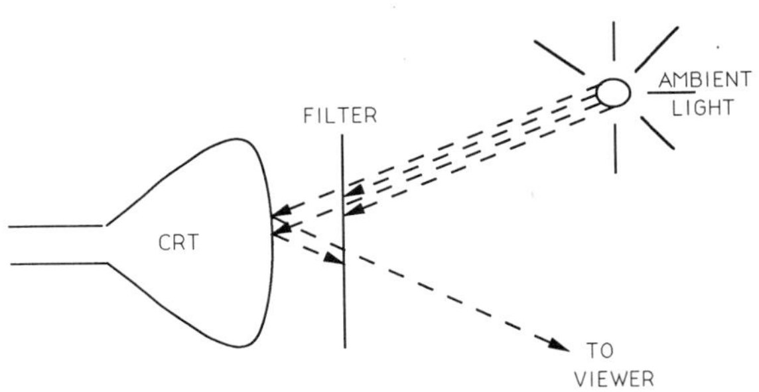

Figure 1.31 Contrast enhancement by use of a neutral gray filter in front of CRT. Note that light from the CRT passes through the filter only once while ambient light must make two passes through the filter, thus heavily attenuating it.

the three color phosphor primaries with little loss, while ambient light of other colors is heavily absorbed. Under ambient fluorescent light, they have a pleasant pale blue coloration. This approach was used later for color CRT panels by Mitsubishi[72] and General Electric, the latter using the name Neovision™. Nippon Electric Glass is a supplier of neodymium-doped, barium-strontium glass panels. Higher cost appears to be the main reason that it has not gained widespread acceptance.

Other contrast enhancement techniques use various surface treatment processes to reduce the mirrorlike reflections of about 4 percent that occur from a polished glass surface. The most common surface treatment is a surface etching that breaks up direct reflections from the front surface of the glass and scatters the light in random directions. Another technique is the evaporation of a magnesium-fluoride coating of one quarter wavelength thickness on the glass surface similar to that on a high-quality camera lens. This coating is called an anti-reflection, AR or HEA™ (High Efficiency Anti-Reflection) coating and serves to match the index of refraction of the glass to that of air, thus reducing surface reflections from 4 percent to less than 1 percent. The only drawbacks are higher manufacturing cost and the reflective fingerprint spots that become evident because the oils from a finger reduce the absorption of reflections. Thus, more frequent cleaning is required. Still other methods use external glass or plastic filters employing one or more of the above processes. These may be either laminated directly to the CRT panel or spaced at some distance from it. They may also provide protection against accidental implosion of the CRT. (Implosion is the reverse of explosion and is a result of the glass, which is under great atmospheric pressure, collapsing inward before being expelled outward.)

Funnel–The funnel usually contains one or more conductive internal coatings that form the anode(s). Electrical connection to the coating is made using a metal feed-through sealed into the glass with a glass-to-metal seal or through the base for lower voltage CRTs. The conductive coating, developed around 1934 by Acheson Colloids, is a gray or black colloidal graphite suspension (known as *aquadag* or *dag*) which is painted on the funnel walls and baked to dry. Where more than one anode is required, separate rings are painted and individual feed-throughs provided. Insulating coatings may be painted on the bare glass between anodes to prevent electrical charging of the glass. Aluminized CRTs often employ the aluminizing as the anode by allowing the aluminum to be deposited on most or all of the bulb's interior. Ceramic funnels may employ evaporated gold or other metals for anode coatings. Metal cone tubes, of course, use the metal cone itself as the anode and high-voltage insulation for the entire exterior of the cone usually is required for safety during equipment servicing to protect personnel from contacting the high voltage.

The anode connection originally consisted of just a wire sealed through the glass or a connection through the base in early CRTs. In the 1930s and early 1940s, receiving tube grid or plate caps were sometimes used (Figure 1.32). Radar CRTs of the 1940s used an anode connector called a small ball contact as shown in Figure 1.33. These were used extensively on oscilloscope CRTs and some early postwar television picture tubes, especially those of Du Mont. In 1946 the recessed

Figure 1.32 Medium cap anode connector for Research Enterprises, Ltd. 12FP7 similar to that used on receiving tubes. These were easily damaged when handling CRTs and were superseded by recessed type connectors during the early 1940s. (From the author's collection.)

Figure 1.33 Recessed small ball anode connector as used on Du Mont 5AFP1. (From the author's collection.)

small cavity cap, commonly referred to as the *anode button* or *button* (Figure 1.34) was introduced for television picture tubes and is still widely used today. The only other anode connection that is commonly used is the "flying-lead" permanently attached high-voltage lead, which is potted at the funnel connection to avoid corona discharge or arc-

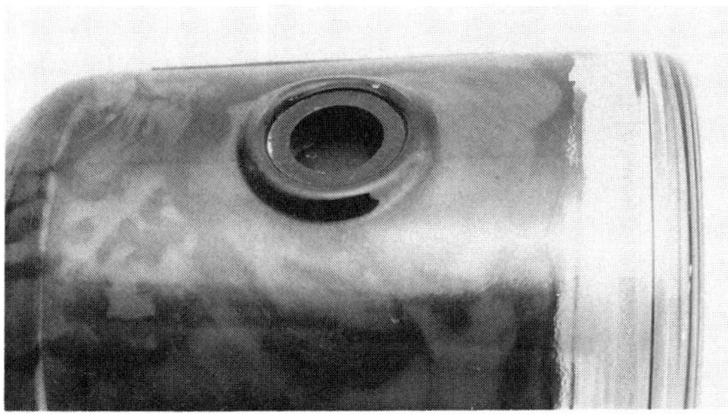

Figure 1.34 Recessed cavity anode connector as used on Waterman 2FP31. (From the author's collection.)

ing in humid environments and at high altitude. Figure 1.35 illustrates a flying-lead anode connection on an electrostatically deflected oscilloscope CRT.

Often an external conductive coating on the funnel is used for magnetic deflection picture tubes to provide extra filter capacitance for the high-voltage power supply or to reduce high voltage charging of the bulb's exterior.

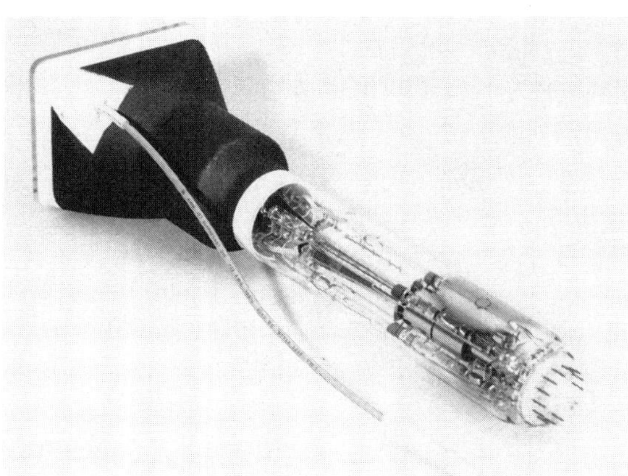

Figure 1.35 Flying-lead anode connector as used on Tektronix T4650. (From the author's collection.)

The conductive dag coating is somewhat similar to the internal anode coating except that it is lacquer-based and usually applied by spraying. A grounding spring contact is used to connect the dag to the chassis. An insulating coating also is often applied around the anode button to help prevent arc-over or electrical leakage in humid weather.

Neck–The neck is the extension of the bulb that contains the electron gun. The gun is positioned within the tubular glass neck by metal spring clips called *snubber springs* to allow movement due to thermal expansion during the processing bake cycles, and cushioning for shock and vibration during shipping and handling. The snubbers also may provide electrical contact between an anode element in the gun and the anode dag coating which extends partway down the neck tubing. Neck diameters range in size from less than one inch for data display tubes to three inches or more for multigun electrostatic tubes.

The neck also provides the means for mounting the deflection yoke, focus and other magnets such as the ion trap, centering magnets, color convergence assembly, etc., for electromagnetically deflected tubes. By having these components close to the electron beam central axis, lower deflection power and magnetic field strengths are required for higher efficiency. Larger neck diameters are sometimes preferable for electrostatically focused, high resolution data display CRTs to allow a larger diameter focus electrode to be used for reduced spherical aberration of the electron beam.

Many electrostatic deflection CRTs deflection plate connections feed through the neck rather than using longer connections down the length of the gun to the base. The short lead length greatly reduces the capacitance and allows much higher frequency signals to be displayed without attenuation. Tubes constructed with neck connections date back to the early 1930s and the work of Manfred von Ardenne when binding posts were used on tubular glass protuberances.[73] The methods for neck-mounted deflection plate connection progressed through receiving tube grid caps in the late 1930s (Figure 1.36), small ball caps in the late 1940s and early 1950s (similar to those used for anode connections) (Figure 1.37), 14- and 22-pin collars in the early 1950s for the many deflection plate connections required for multigun tubes (Figure 1.38), to small pieces of wire sealed in the glass in the late 1950s (Figure 1.39).

Base–Electrical connections to the electron gun are made through the tube base which mates with a matching socket. Early bases consisted of common radio tube bases of five to seven large diameter pins. A radial contact base of English origin was used in the late 1930s by Du Mont, but the common 11-pin "magnal" base developed specifically for CRTs became the standard for electrostatic deflection tubes during the early war years. The magnal base had enough pins to allow connection to all of the deflection plates without resorting to tying one of each pair of

Figure 1.36 Small cap deflection connectors as used on Du Mont 5SP7. (From the author's collection.)

Figure 1.37 Small ball deflection plate connectors as used on Du Mont 5RP11. (From the author's collection.)

Figure 1.38 Fourteen-pin collar (left) and 22-pin collar (right) deflection plate connectors as used on Du Mont 5AFP7 and Magnavox 7YP2, respectively. (From the author's collection.)

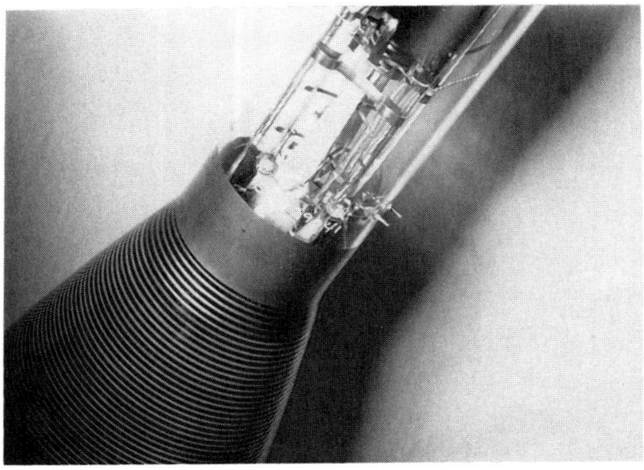

Figure 1.39 Wire feed-through deflection plate connectors as used on Tektronix T51. (From the author's collection.)

deflection plates to the anode to save pins—a practice common in early electrostatic-deflection CRTs. By driving the plates in "push-pull" rather than single-ended, deflection defocusing was reduced and sensitivity increased.

"Octal" eight-pin bases also were used for some one- and two-inch electrostatic CRTs, such as the RCA 913 and 902, respectively, that used the common internally connected deflection plate/anode connection. The standard receiving tube base at that time, octal bases were widely used for the 3HP7, 5FP7, 7BP7 and 12DP7 radar indicator CRTs during World War II because the electromagnetically deflected tubes required fewer base connections.

Larger diameter 14-pin diheptal bases with greater spacing between pins were developed in about 1943 to provide better high voltage insulation for electrostatically deflected CRTs operated at high altitude in aircraft.

Television contributed a base of its own in 1946 when the 12-pin duodecal base became the standard for television picture tubes. This was a rather curious situation since only five to seven of the pins were normally used and the others omitted. It was of some value for a few tubes which used high-voltage electrostatic focus when the pin that was used for focus was located on the opposite side of the socket from the five lower voltage pins. It also was an excellent base for some electrostatically deflected tubes, such as the 2BP1, 3RP1 and 5UP1 introduced in 1950. Figure 1.40 illustrates a number of the bases that were used commercially from about 1930 to 1960. The British refer to this style of basing as *overcapped*.

The last developments in this style of base occurred in the early 1950s when a 20-pin bidecal base (Figure 1.41) was used for the first three-gun color picture tube; the RCA 15GP22. A 25-pin pentaquintal base (Figure

Figure 1.40 Overcapped CRT bases used until the 1950s. From left to right: (1) Seven-pin receiving tube base as used on Sylvania 3AP1, (2) eight-pin octal base as used on RCA 902-A, (3) 11-pin magnal base as used on RCA 2AP1, (4) 12-pin duodecal base as used on RCA 2BP1 (also used on many television picture tubes prior to 1960 by omitting five to seven pins), (5) 14-pin diheptal base as used on Du Mont 3BP1. (From the author's collection.)

Figure 1.41 Twenty-pin bidecal base as used on the RCA 15GP22 and other early color picture tubes. Note that several pins are omitted on either side of the focus electrode pin to provide adequate insulation for the high focusing voltage. (Courtesy of C. E. "Sonny" Clutter.)

Figure 1.42 Twenty-five-pin pentaquintal base used for multigun electrostatic CRTs, such as this Magnavox 7YP2. (From the author's collection.)

1.42) came into limited use for multigun electrostatic CRTs; and a small diameter (1-1/8 inches) six-pin base (Figure 1.43) was used for early 110-degree deflection monochrome television picture tubes.

The late 1950s and early 1960s saw the adoption of the hard pin stem that eliminated the separate base and instead used heavier wire feed-throughs in the stem (utilized as pins directly). Inexpensive plastic caps are cemented to the header to provide a keyway for indexing of the mating socket. A large number of styles of these bases have been used since then with no universal type emerging. Figure 1.44 shows several hard pin bases.

Figure 1.43 Six-pin base used for television picture tubes with small neck diameters during the late 1950s. Tube shown is Sylvania 8YP4. (From the collection of Jerry Talbott.)

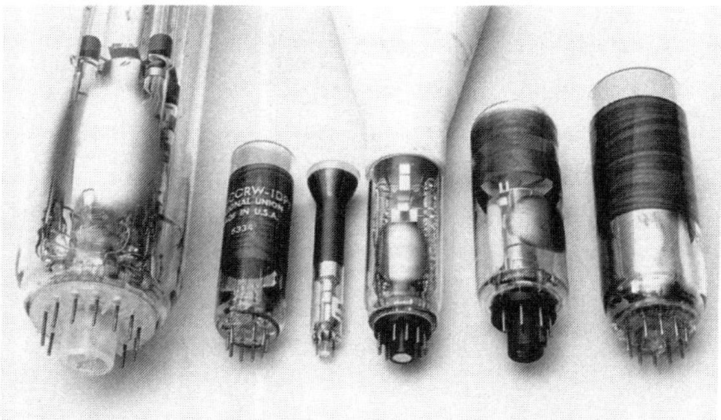

Figure 1.44 Hard pin bases as used from the mid-1950s to the present. Many variations in number of pins, diameter of pin circles, pin diameters and lengths, and plastic keyway configurations exist.

1.8 VACUUM PROCESSING

To achieve the longest life and best performance, a "clean" vacuum is necessary. Typically, pressures on the order of 1×10^{-6} to 1×10^{-8} torr (mm of mercury) are required for commercial cathode-ray tubes.[74,75] These pressures are attained through a combination of high vacuum pumps, parts cleaning, baking during pumping and "getters" which adsorb residual gases after the tube is pumped and sealed off. Poor vacuum cleanliness can result in gas discharge or arcs which can damage the circuits in an instrument using the CRT. Short cathode life also occurs due to positive ion bombardment.

Before pumping, all parts are thoroughly cleaned by washing with deionized water, acid rinsing, ultrasonic cleaning, vapor degreasing, etc., to remove surface contaminants. Usually pumping is done first using a mechanical "roughing" pump to remove a large percentage of the air from the tube until a satisfactorily low pressure is obtained for an oil-diffusion pump to take over for the final exhaust. During the pumping time, the entire tube is baked to a temperature of approximately 350 to 400 degrees Celsius to "outgas" occluded gases from the bulb's and coatings' surfaces. RF heating of the electron gun parts is used simultaneously. After the desired pressure is obtained, the tube is sealed off by fusing the exhaust glass tubulation through which the pumping system is connected. The tubulation is a part of the stem which contains the wire feed-throughs for electrical connection to the gun electrodes and usually is protected in the finished tube by the keyway on the base. Following seal-off, or tip-off, the bake temperature is reduced slowly to prevent glass strains or implosion due to thermal shock.

The final step in the exhaust process is to flash the *getter*. The getter usually consists of a metallic loop (Figure 1.45) containing barium and mounted on the electron gun or in the funnel. It is flashed through heating induced by a closely coupled coil placed near the neck or bulb which is excited by a high radio frequency electrical current. This leaves a silvery patch of metallic barium on the inside of the glass which is usually visible (Figure 1.46). The function of the getter is to re-

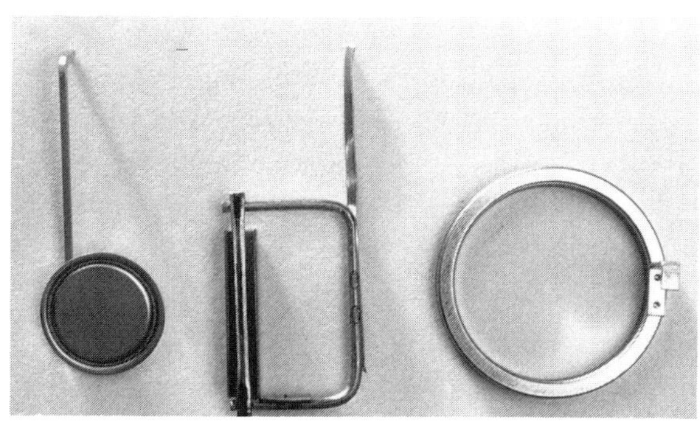

Figure 1.45 Commonly used getters for maintaining good vacuum for long CRT life.

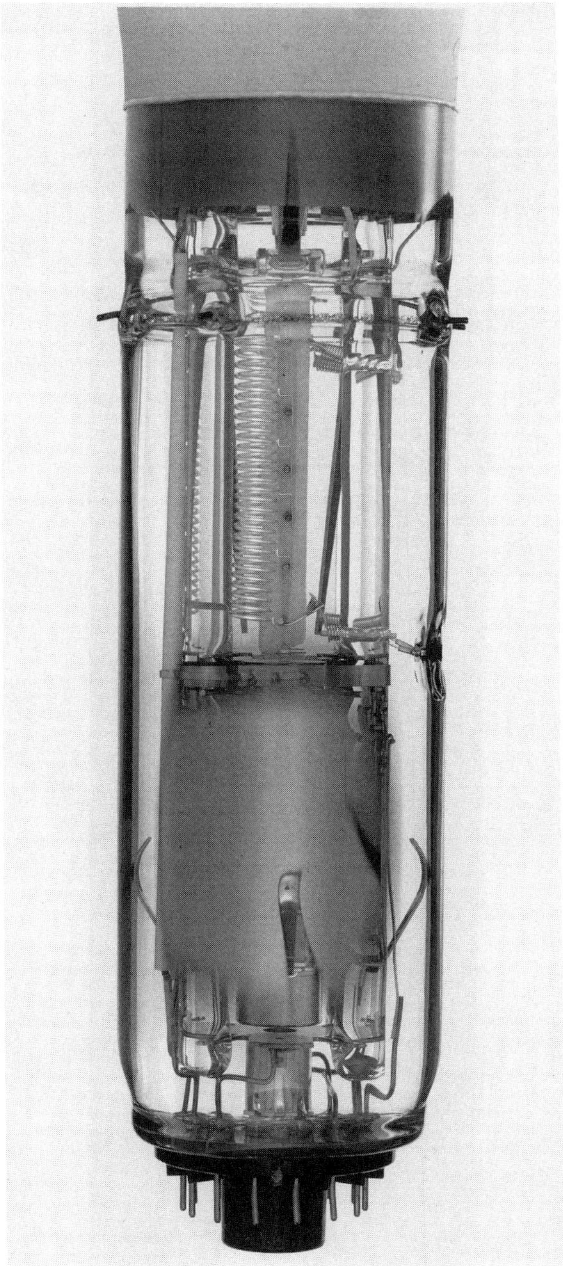

Figure 1.46 Getter "flash" area is silvery area near base of this illustration. (Courtesy of Tektronix, Inc.)

move residual gas from the tube by trapping it with the chemically active barium. The getter continues this clean-up process throughout the tube's life. The appearance of the silvery patch provides a clue as to the tube's condition. Tubes that have operated for many hours or have poor vacuum will have a shrunken getter flash, and tubes that have completely lost their vacuum will have getter coatings that have turned white or completely vanished.

1.9 SCANNING METHODS

Scanning refers to the movement of the spot formed by the electron beam in a repetitive pattern in one or both axes on the CRT screen to form a useful image.[1-6] Many types of scanning have been used over the years, but raster scanning has emerged as the predominant method used today. Information that was previously displayed with the other scanning methods may now be converted digitally to the raster scan format in many applications, including the digital storage oscilloscope.

X–Y scanning–The earliest useful cathode-ray tubes were scanned by applying related signals to the horizontal or "X" axis and the vertical or "Y" axis. The relationship between the frequencies, phases and amplitudes of the two signals can be viewed in the *Lissajous* figures displayed on the screen. The heights and widths of the figures are indicative of their relative amplitudes. A third axis, the "Z axis," also is often used. The Z axis is utilized to turn the electron beam on or off, or to control the beam intensity to apply blanking or video signals to the display. X–Y scanning is the basis for most other scanning methods in that X and Y signals, at least one of which usually is a sawtooth or "ramp" waveform, are applied to the deflection plates or deflection yoke to direct the electron beam to various areas of the screen in some predetermined sequence.

Y–T scanning–Also known as time domain displays, Y–T scanning is the basis for the oscilloscope (also commonly known as an oscillograph until the 1950s) display. Most oscilloscopes use electrostatically deflected CRTs due to the advantage of their ability to respond to much higher frequencies than can be performed electromagnetically. The electrical signal to be measured is applied to the vertical deflection plates (Y axis). A voltage that is linearly increasing with time (T) and known as a sawtooth or ramp waveform (Figure 1.47) is applied to the horizontal deflection plates to provide a time-base. By synchronizing the time-base to the vertical axis signal to be viewed, the amplitude and polarity, with respect to time, may be observed.

Since the oscilloscope uses a CRT which is electrostatically deflected, a very wide range of frequencies may be observed. The inertialess beam is limited in its ability to be deflected rapidly only by the transit time of the electrons through the vertical deflection plates and the deflection plate capacitance. The practical frequency limit may extend from DC to one gigahertz or more, and the time-base may be anywhere between seconds and picoseconds (1×10^{-12} second) per calibration mark on the CRT graticule (calibrated scale on the screen).

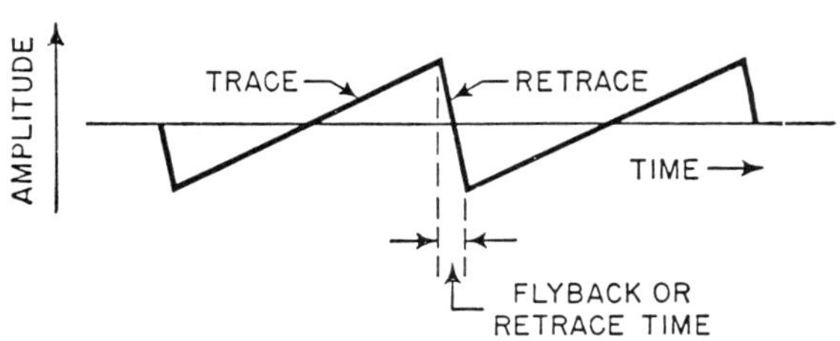

Figure 1.47 Sawtooth waveform used to produce linear scanning on cathode-ray tubes.[4]

Oscilloscopes usually are capable of displaying the X–Y relationships of two electrical signals. An additional input to the horizontal deflection amplifier is provided to allow comparison of two signals as described in the previous X–Y Scanning section.

The earliest radar systems employed an oscilloscope-type display where time represented distance to the target as shown in Figure 1.48. This presentation of radar information was known as an "A" scope indicator.

Y–F scanning–Amplitude versus frequency or Y–F scanning is the basis for the spectrum analyzer.[76] Essentially, the spectrum analyzer is a tuned amplifier/detector combination which is tuned to a narrow frequency band (F). The tuned frequency is rapidly scanned in synchronism with a horizontal time base to display the amplitude of the signals (vertical) versus the frequency (hor-

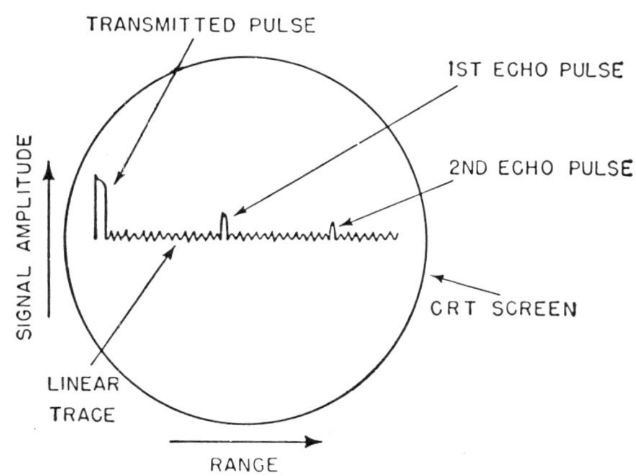

Figure 1.48 A-scope radar indicator.[4]

izontal) on the CRT screen as shown in Figure 1.49. Either audio or radio frequencies may be displayed depending on the instrument frequency range design. These are often referred to as frequency domain measurements.

Radial deflection–Radial deflection consists of the pattern of a circle created by a pair of sine waves phase-shifted 90 degrees from each other and applied to the X and Y axis of a cathode-ray tube.[2,4,5,77] The circumference of the circle is somewhat greater than the length of a straight line that could be displayed on the same size tube. This allows a small CRT to be used to measure timing to a higher degree of accuracy than would ordinarily be possible with a conventional Y–T display as used in oscilloscopes. The pulse, whose timing is to be measured, is superimposed on the circular pattern either elec-

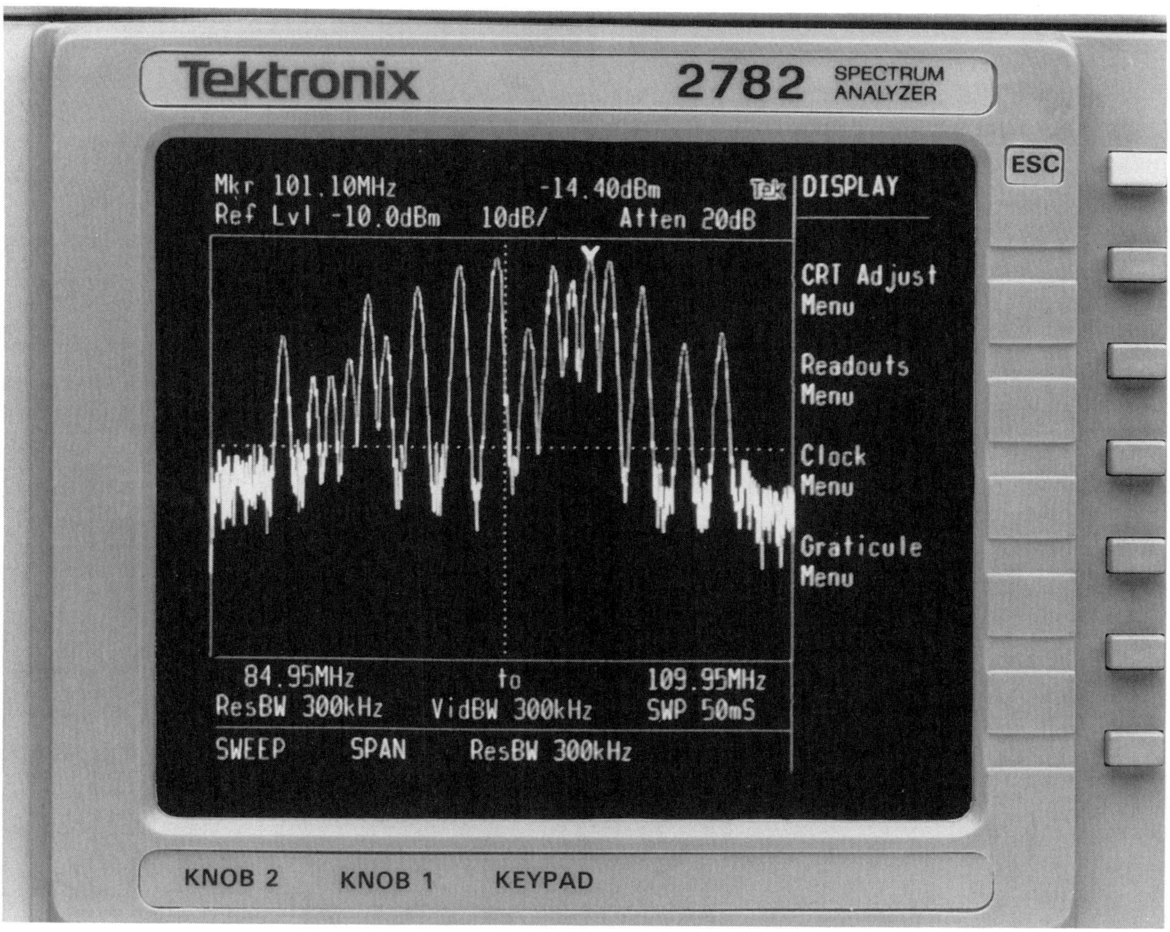

Figure 1.49 Spectrum analyzer display. (Courtesy of Tektronix, Inc.)

tronically in the deflection amplifiers or through a long wire radial deflection electrode sealed into the face of the CRT and extending almost to the deflection plates. The time delay of the pulse from the start of the circular sweep, usually at the top of the circle at 0 degrees, may be measured by observing the angle around the circle at which the pulse is displayed as shown in Figure 1.50. The time for the beam to make one complete circle is known and the delay time computed from the distance around the circle at which the pulse appears.

The radial scan display was developed during World War II as a means to determine distance to a target, since the time required for a radar or sonar pulse to reach the target and return could not otherwise be accurately displayed on a small indicator CRT. In radar terminology, a display using radial deflection is referred to as a "J" scope. Long-range navigation displays (loran) also used radial deflection displays.

Spiral sweep–Basically, spiral sweeps are similar to radial deflection except that as the phase-shifted sine waves trace their circular path around the screen, their amplitude is steadily decreased by a superimposed sawtooth waveform.[5,78,79] The result is a greatly lengthened trace on which timing may be measured (see Figure 1.51). A 10-turn spiral takes 10 times as long to complete as each circular trace, therefore it presents a much longer time span over which a pulse's time delay may be measured.

PPI displays–Radar indicators have traditionally utilized Plan Position Indicator (PPI) displays to show both azimuth angle and distance in a single indicator. The PPI Display uses a linear sweep (sawtooth waveform) to rapidly sweep an electromagnetically deflected CRT from the screen center to the edge. At the same time the deflection yoke is mechanically rotated around the CRT neck in synchronization with the azimuth (horizontal) rotation of the radar antenna by electromechanical coupling. The start of the linear sweep in the center of the screen is synchronized with the transmitted pulse from the radar transmitter. The received return pulse, or echo, is delayed in time propor-

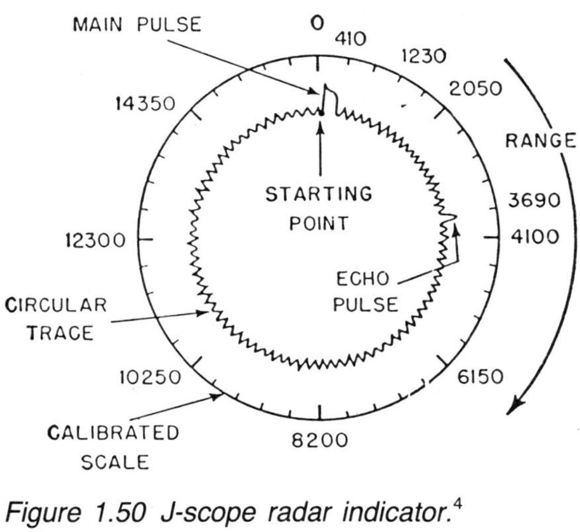

Figure 1.50 J-scope radar indicator.[4]

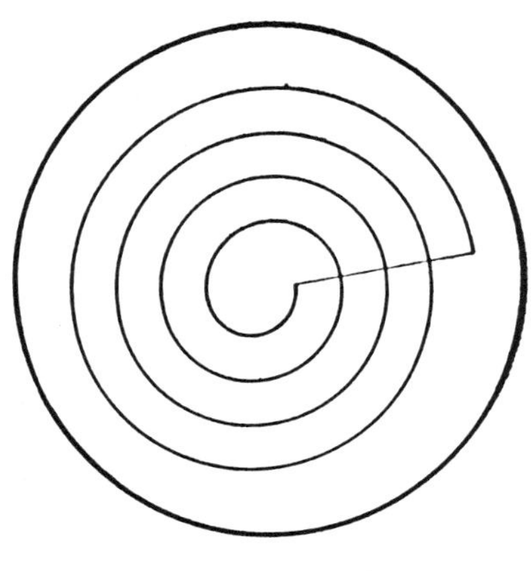

Figure 1.51 Spiral scanning.[4]

tionally to the distance to the target. Therefore the return pulse occurs at some time while the linear sweep is scanning the beam between the screen center and edge. The returned pulse is amplified and applied to the CRT grid to intensify (Z axis modulate) the electron beam to display a bright spot on the CRT screen. The location of the spot indicates the target distance by the distance from the screen center, and the azimuth by the angle about the screen center as indicated by a bearing scale around the screen's periphery (Figure 1.52).

Other radar scanning methods–A number of other scanning methods have been used for radar displays. These are known by letter designations from "A" to "P" as shown in Figure 1.53.[80] These will not be described due to their more specialized nature. The MIT Radiation Laboratory Series books on radar systems published in 1948 is a classic source of further information on the subject of the radar systems developed during World War II.

Vector writing–Earlier computer displays were often vector- or stroke-written, i.e., they produced alphanumeric characters and graphics by deflecting the CRT electron beam in a series of strokes whose position, length and direction were determined sequentially by a vector generator under computer control. Both electrostatically and electromagnetically deflected CRTs were used, depending on addressing speed requirements. Vector writing is essentially X–Y addressing as previously described.[81,82]

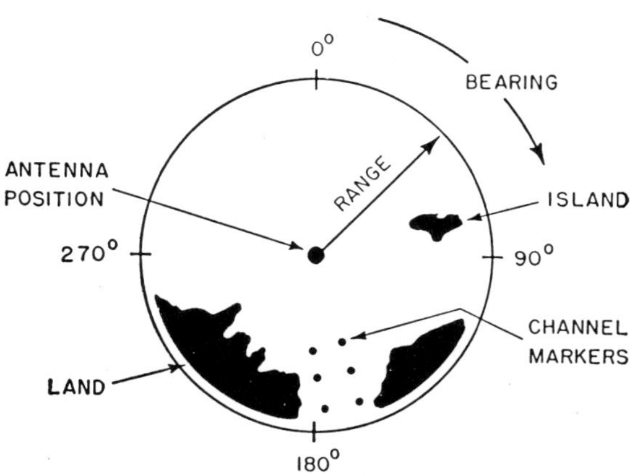

Figure 1.52 Plan Position Indicator (PPI) for radar displays.[4]

Raster scanning–Raster scanning is the most common method in use today. It is used wherever images or pictures must be displayed. Applications include television, graphics displays, computer displays and, recently, digital storage oscilloscopes. Raster scan CRTs are almost universally of the magnetic-deflection type. These have the advantage of relatively low cost and short length for a given screen size.

Raster scanning consists of two sawtooth current waveforms, one at a relatively low frequency and the other at a much higher frequency, applied to both the vertical and horizontal deflection coils of the tube. The result is deflection of the electron beam in a fast left to right motion with a rapid retrace back to the left screen edge. Simultaneously a slower scan is scanning vertically beginning at the top of the screen, progressing to the bottom with a rapid retrace to the top. The result is a series of horizontal lines displayed from the top to the bottom of the screen. The electron beam is turned off (blanked) during both the vertical and horizontal retrace times to prevent the retrace lines from being visible on the screen. The picture information from the video amplifier is applied to the grid or cathode of the CRT to control the intensity of the beam (Z axis modulation) at the lo-

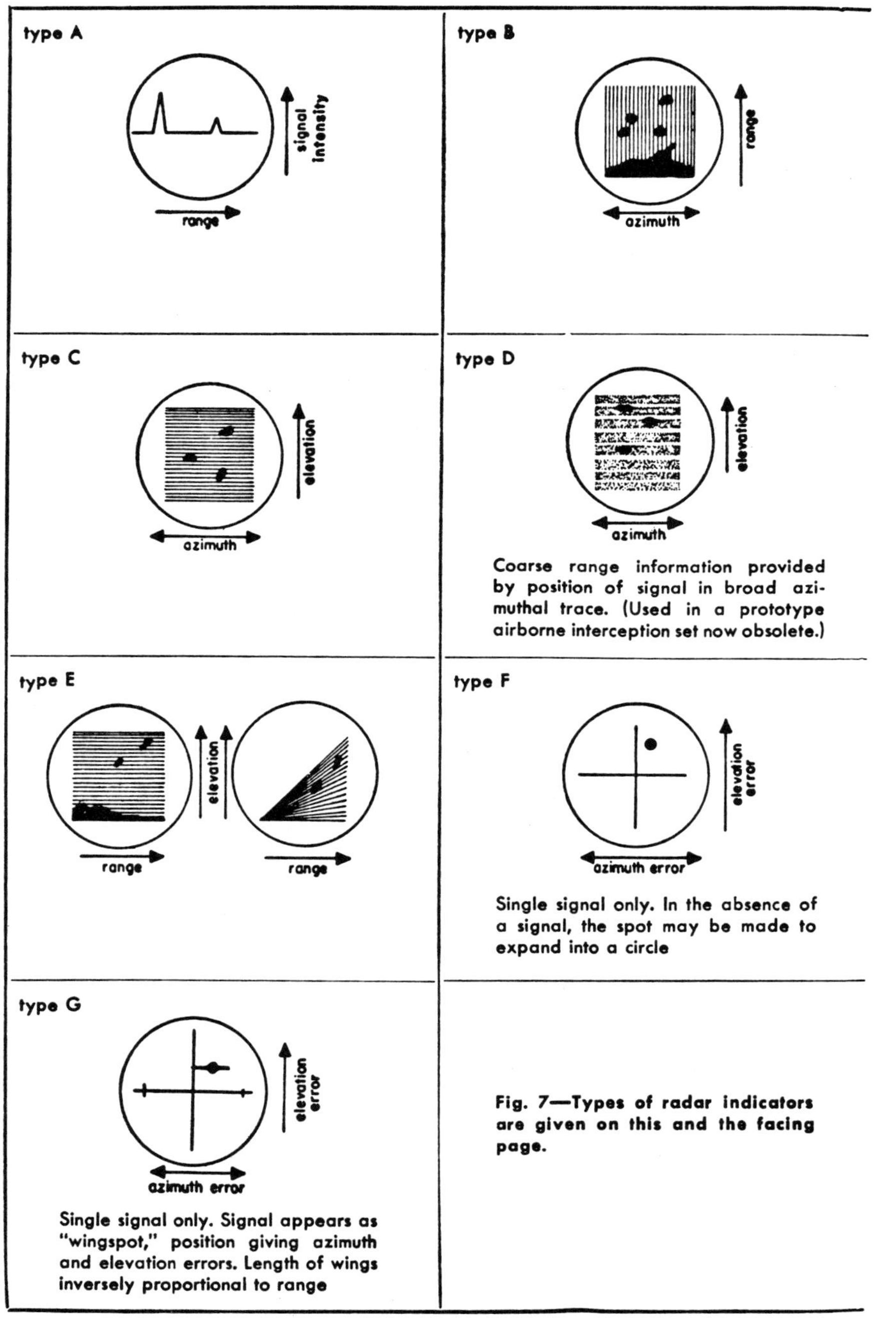

Figure 1.53 Radar indicators.[80]

type H

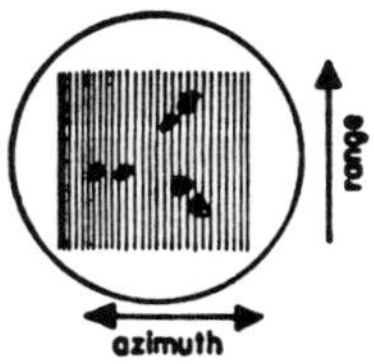

Signal appears as two dots. Left dot gives range and azimuth of target. Relative position of right dot gives rough indication of elevation

type I

Antenna scan is conical. Signal is a circle, the radius proportional to range. Brightest part indicates direction from axis of cone to target

type J

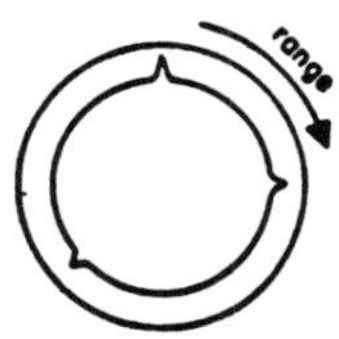

Same as type A, except time base is circular, and signals appear as radial pips

type K

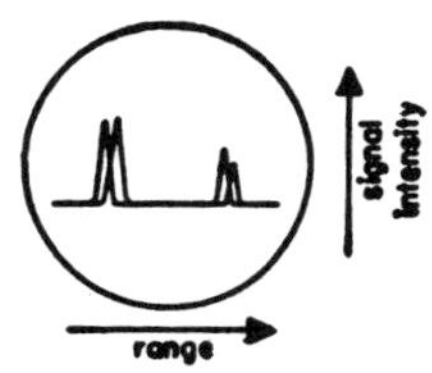

Type A with lobe-switching antenna. Spread voltage splits signals from two lobes. When pips are of equal size, antenna is on target

type L

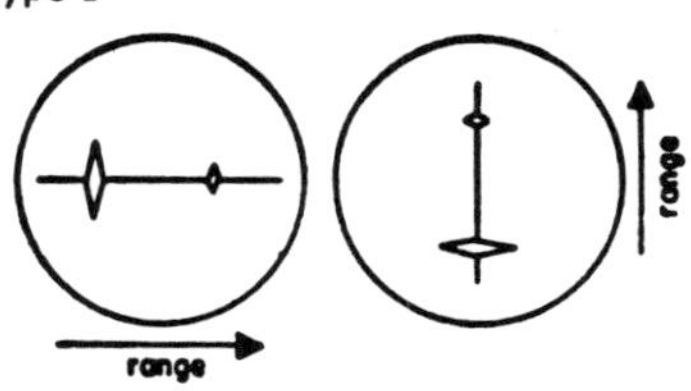

Same as type K, but signals from two lobes are placed back to back

type M

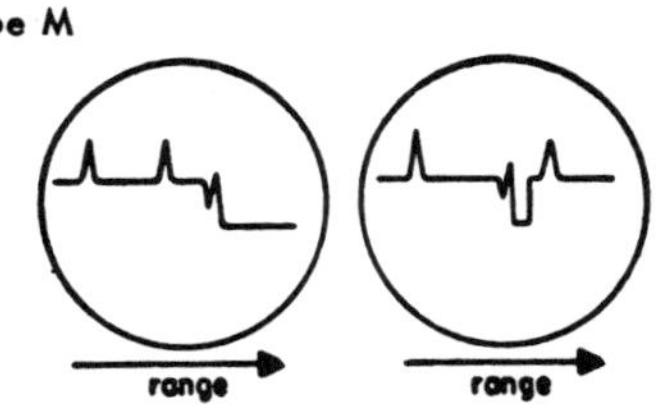

Type A with range step or range notch. When pip is aligned with step or notch, range can be read from dial or counter

type N

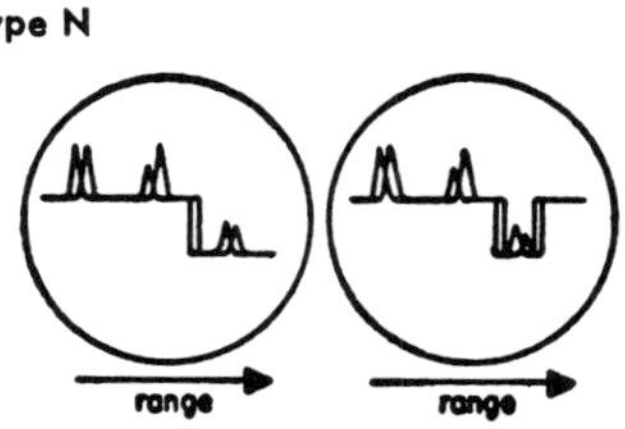

A combination of type K and type M

type P (PPI)

Range is measured radially from center

Figure 1.53 Continued.

cation on the screen that is being struck at any given instant. In the case of television monitors or receivers, the scanning of the CRT is synchronous with that of the camera scanning to recreate an image.

The vertical sawtooth frequency usually is in the range of 30 to 120 hertz and in most cases is 60 hertz or more to reduce flicker. The horizontal frequency is commonly 300 to 3,000 times that of the vertical and determines the number of horizontal scanning lines. Higher horizontal scanning frequencies are used for higher resolution displays to provide greater picture detail in the vertical axis. Higher horizontal frequencies require better deflection circuits and deflection yokes to prevent non-linear scanning. Additionally, they require video amplifiers of much higher bandwidth to provide commensurately greater picture detail in the horizontal axis.

In the 1920s early television systems in the United States used only 30 to 50 lines for the mechanically scanned systems. Image shapes could be recognized but little detail could be reproduced. Electronic scanning with a CRT display allowed an increase to 120 lines in 1931, 343 lines in 1934, 441 lines in 1937 and finally the present 525 lines in 1941[83] as developed by the National Television Standards Committee (NTSC) for the United States. NTSC scanning[83–89] is done at 30 hertz for the vertical and 15,750 hertz for the horizontal. Dividing the horizontal frequency by the vertical frequency determines the number of scanning lines (15,750 Hz/30 Hz = 525 lines). The video bandwidth used is four megahertz to provide horizontal resolution roughly equivalent to the vertical resolution.

To avoid flicker at the 30 hertz vertical scanning frequency used to keep the horizontal frequency and video bandwidth within reason, interlaced scanning is used. In interlaced scanning, the even numbered lines are scanned in the first field while the odd numbered lines are scanned in the second frame (Figure 1.54). The two fields, each consisting of 262.5 lines, combine to produce a complete frame of 525 lines due to the persistence of vision. Each field is refreshed at a rate of 60 times per second to display one complete picture 30 times per second. This is known as 2:1 interlace. Up to 4:1 interlace was proposed for television by Du Mont in 1938,[90–92] but 2:1 became standard for the NTSC system.

The emerging high definition television (HDTV) systems will have double the resolution of current television systems and an aspect ratio (picture width to height) of 16:9 instead of the standard 4:3 ratio now used. New color CRTs with corresponding screen shapes will be required as well as higher resolution electron guns and screen structures.

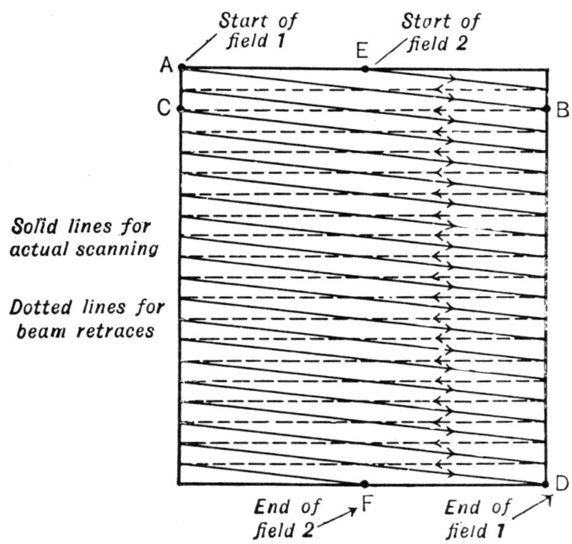

Figure 1.54 Interlaced raster scanning.[88]

The present limit for full-color high resolution displays for computer and graphics displays is about 1280 × 1024 pixels (picture elements) due to limitations of color CRT gun and screen resolution.

High-resolution black and white (variously known as gray scale or monochrome) displays of 2048 × 1536 pixels and greater are finding applications in the fields of radiology, medical imaging and photogrammetry, mapping, photoreconnaissance, page layout for electronic publishing and phototypesetting. High-resolution computer displays often use vertical scanning frequencies of 70 to 80 hertz to avoid flicker. Horizontal scanning at 125 to 250 kilohertz represents the current state-of-the-art in deflection yokes and amplifiers. The number of scanning lines produced is in the range of 2,000 to 3,000. State-of-the-art video amplifiers have bandwidths of between 200 and 500 megahertz, which are capable of displaying 2,000 to 4,000 pixels horizontally. Naturally, displays having such extreme performance are considerably higher in cost and require a precision cathode-ray tube having a spot size of 2,000 to 4,000 times smaller than the screen width.[93]

Ultra-high resolution displays of 4000 × 3000 pixels are limited to text and line drawing uses where the electron beam is either on or off in digital fashion. At present it is impractical to construct linear video amplifiers capable of producing various intensity levels (gray scale) at one gigahertz or more as required to support such extreme resolution. A further limitation is the 30 to 60 volt swing required to drive the CRT grid from full black to full white.

1.10 CRT AND PHOSPHOR NOMENCLATURE

Within the United States, receiving tubes, special purpose tubes, cathode-ray tubes and phosphors have been registered to promote standardization since the 1930s under the auspices of the Radio Manufacturers Association (RMA) which was founded in 1924. This trade association later became the Radio-Television Manufacturers Association (RTMA) in 1950, the Radio-Electronics-Television Manufacturers Association (RETMA) in 1953 and finally the Electronic Industries Association (EIA) in 1956.[94] The engineering council charged with administering the registration system went through its own share of alphabet soup starting at JETEC (Joint Electron Tube Engineering Council), JEDEC (Joint Electron Device Engineering Council) to the current TEPAC (Tube Engineering Panel Advisory Council).

The goal of tube registrations was to ensure common specifications and interchangeability between manufacturers. In the early days of radio, one manufacturer's tubes might perform satisfactorily in a given circuit while another's might not, even though both had the same type number.

From about 1934 to 1940 the earliest cathode-ray tube registrations consisted of three and four digit numbers. The RCA 902 through 914 series, 1802 through 1813 series and the Du Mont 2501 through 2533 series were examples of the straight numbering system. Four-digit numbers also were used for some RCA CRTs, such as the type 4490 registered in the 1960s, although the reason for deviating from conventional practice is unclear. An "A" following the numbers indicated an improved version that could replace the original, e.g., type 902-A. From about 1940 until 1982, the tube type numbers were in the form of 5BP1 and 12AP4, where the first number designates the screen size in inches, five and 12 inches, respectively. The first letter is a sequentially assigned identifier to discriminate one tube from another of the same size, for example 5AP1, 5BP1 and 5CP1. This letter was assigned beginning with "A" and progressing through "Z" after which a second letter was added such as 5ABP1. The "P" followed by a one- or two-digit number indicated the phosphor screen type.[32] Eventually these reached P57 before being superseded by the new Worldwide Type Designation System (WTDS) in 1982.[33] As with the earlier numerically designated tubes, a suffix with "A," "B," etc., designated an improved design which would replace the earlier version.

One additional letter was added to the tube designation in 1966 as a result of a regulation imposed by the Federal Trade Commission.[95] Prior to 1966, CRT screen sizes, such as 21 inches, were an overall diagonal measurement of the glass faceplate. Not all of this dimension represented the viewable screen dimension which was more like 19 or 20 inches. It wasn't a major problem when comparing screen sizes because all manufacturers measured screen sizes in the same way, but it did result in the FTC-imposed requirement to accurately state screen size and whether that size is based on horizontal, diagonal or other dimension. Screen diagonal has remained the usual method of specifying screen size. To differentiate the new tubes clearly from the old, which continued to be manufactured as replacements, the letter "V" for viewable was added following the screen diagonal in inches. An example of the designations used from 1966 to 1982 is the Clinton 15VAUP4. Screen dimensions are often expressed in the form "25V inches." Likewise, glass bulbs, faceplates and funnels are identified as "25V" and "25V90," respectively, with 90 indicating the deflection angle in the latter case.

Since the WTDS nomenclature was adopted in 1982, phosphors are assigned a two-letter code. The first letter designates the color and the second letter is merely a sequential identifier. For example, P1 is now known as "GJ," P22 is "XX" and P31 is "GH." (See Appendix 3 for a list of EIA/WTDS phosphor designation equivalents. This book uses the earlier "P" designations because of its historical perspective.

Picture and display tubes also are assigned designations under the WTDS.[96] These are in the form "A63AAA00XX01" with the following coding:

- A indicates a television picture tube, M indicates a monitor tube.
- 63 is the viewable screen diagonal in centimeters.
- AAA is the CRT family progressing AAA, AAB, AAC, etc.
- 00 indicates the first tube within the AAA family, 01 is a minor variation such as mounting ears or filter.
- XX designates the phosphor type with XX being the designation for P22 tricolor phosphor.
- 01 (optional) indicates an integral deflection yoke supplied with tube.

The WTDS is administered on a worldwide basis by the EIA JT-31 committee in Washington, D.C., through an agreement with the European Pro-Electron and the Electronic Industries Association of Japan (EIAJ) standards organizations.

The EIA also has developed date codes that are usually stamped on the base or label to identify the date of manufacture. These are in the form 5601 or 56-01 with the first two digits indicating the year and the second two digits indicating the week.

Several committees are active in developing CRT measurement standards and defining terms used by the industry.[97,98] The Society for Information Display (SID) membership is participating in joint standards activity with EIA and maintain liaison with other display standards formulating organizations.

During World War II, and for some years thereafter, the government assigned "JAN" designations to indicate compliance with military standards, most notably MIL-E-1. Designations were in the form "JAN CRC 2AP1" with the center letter group indicating a commercial manufacturer. In this example "CRC" denotes a commercially manufactured tube by RCA. A list of JAN codes appears in Appendix 4.

1.11 CRT SAFETY

Several hazards are associated with cathode-ray handling and operation. Primarily these include implosion, high voltages and X-radiation. The degree of hazard varies from one tube type to another, although some risk is associated with any CRT.

Since there is a near-perfect vacuum within the CRT envelope, atmospheric pressure of 14.7 pounds per square inch is present on all surfaces of the tube. This can amount to over two tons of pressure on the faceplate of a 25-inch television picture tube having a 315-square-inch faceplate. To counteract atmospheric pressure, tube surfaces usually are curved to gain strength by placing the glass in the compression mode where its strength is greatest.

Implosion is the opposite of explosion, that is, the highly evacuated glass envelope may first rapidly collapse inward following an impact. From there on it resembles an explosion as pieces of glass are thrown violently in all directions. Some very dramatic high-speed motion pictures have been made showing the effects of CRT implosion. Manufacturers go to great lengths to minimize any hazards, including hydrostatic testing (applying at least three atmospheres of pressure on sample CRTs in a pressure vessel filled with water), impact testing to Underwriters Laboratories specifications, computer stress analysis simulation and integral implosion protection (e.g., laminated safety glass or steel tension bands placing the faceplate in strong compression).

High voltage is present near any operating CRT. Voltages ranging between 500 and 80,000 volts are used to accelerate the electron beam and will be found at the tube connections as well as in the associated electronics. Circuit failure may cause the high voltage to appear in normally low-voltage portions of the equipment. Potentially lethal current levels exist, especially if contacted by moist hands, etc.

X-radiation is generated by any CRT in normal operation, however, it is usually totally absorbed by the envelope and is not a potential problem unless the manufacturer's maximum acceleration voltage is exceeded or shielding recommendations are not heeded. Since the television X-ray scare of the late 1960s, many precautions have been taken to carefully control glass thickness, X-ray absorption coefficient, operating voltage and equipment failure modes which could cause an unexpected increase in accelerating voltage. Annual reporting by manufacturers to meet federal guidelines is required for many CRT types and the equipment utilizing them.

Sylvania Electric used printed labels on

each CRT produced which contained good advice in a concise form. The following is a verbatim copy of that notice for the benefit of newcomers to the world of CRTs:

WARNING

The walls of all cathode-ray tubes are under great force due to the high vacuum they contain. Breakage of these tubes may cause injury from high velocity flying glass fragments. KEEP TUBE ENCLOSED in its shipping container or in a cabinet having protective barriers adequate to contain flying glass in case of breakage. When tube must be exposed WEAR PROTECTIVE CLOTHING, including goggles or face shield and gloves. DO NOT STRIKE OR SCRATCH TUBE. Use only moderate pressure when inserting into or removing from its socket. Remove tube carefully from container. OPERATING PRECAUTION–X-RAY RADIATION SHIELDING MAY BE NECESSARY to protect against possible danger of personal injury from prolonged exposure at close range if tube is operated at higher than manufacturer's maximum-rated voltage or 16 kilovolts, whichever is less. Operation above maximum rated voltage will cause damage to tube and associated equipment. The purchaser of this tube assumes complete responsibility for notifying any person to whom it is resold of the contents of this warning notice.

2

A Laboratory Curiosity

2.1 Background

One hundred years ago when the cathode-ray tube was just a laboratory curiosity, it would have been impossible to envision a "crystal ball" having the tremendous capabilities of today's devices and such a profound impact on everyone's lives. Like that mythical crystal ball, the cathode-ray tube has not only given us the ability to partially see into the future through displays of trendlines and extrapolations, but it has allowed us to shape our future through the advancements made using it.

The cathode-ray tube is used in business and industry at almost every desk as the window for computers or computer terminals to view the contents of data banks, edit text and design items ranging from skyscrapers to advertising layouts. It is used for safety and security in radar displays for civilian air traffic control and military radar displays. Its longest and most visible exposure to the public has been as an entertainment display in television and most recently in video games.

The actual anniversary of the cathode-ray tube is difficult to place accurately since it has undergone a long series of evolutionary changes. Significant contributions may be traced as far back as 1603, although most of the development leading to a device having the basic form of today's cathode-ray tube occurred in the second half of the nineteenth century. Generally, Karl Ferdinand Braun is credited with the invention of the modern cathode-ray tube in early 1897. Braun's CRT contained all of the basic functions of today's tubes including the electron source, focusing, deflection, acceleration, phosphor screen and a sealed off mechanical structure. The term *cathode rays* predates Braun's tube by about 20 years. Figure 2.1 illustrates the evolutionary steps leading up to the Braun tube.

The years up to about 1920 constituted a period of experimentation with few practical uses for the CRT. Despite this lack of practicality, these experiments were the foundation of an understanding of the electron's and atom's nature, as well as for the discovery of X-rays. Many of the great names in physics are associated with this early research and several Nobel prizes were awarded in the process. This chapter briefly describes in chronological order the contributions of each of the experimenters who contributed to the

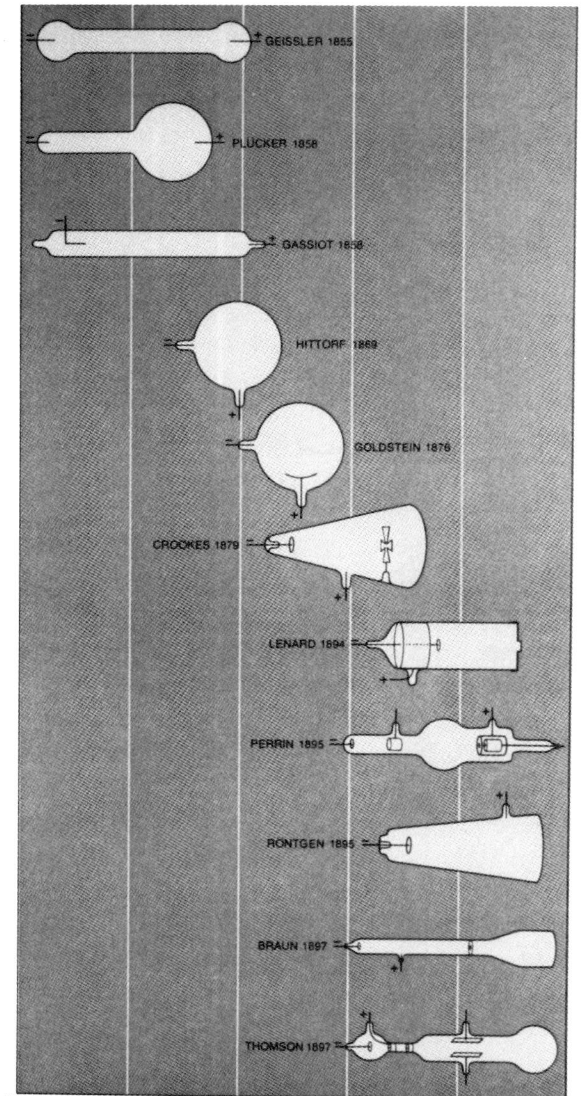

Figure 2.1 Early evolution of the cathode-ray tube.[5] *(From* Ferdinand Braun and the Cathode Ray Tube *by George Shiers. Copyright © 1974 by SCIENTIFIC AMERICAN, Inc. All rights reserved.)*

development of the cathode-ray tube. These developments ultimately led to the modern cathode-ray tube which has become an integral part of everyday life.

2.2 VINCENZIO CASCARIOLO

The first manmade phosphor was made by an Italian shoemaker and alchemist, Vincenzio Cascariolo, in 1603 as a result of trying to create gems or gold from lesser materials. This is considered as perhaps the single most important discovery in the history of inorganic luminescence. Cascariolo pulverized barite, which was found locally in Bologna, and heated it in contact with coal. The resultant material was found to absorb sunlight and cast off a purple-blue glow in darkness. The discovery attracted the attention of philosophers, and especially the Italian mathematician, astronomer and physicist, Galileo Galilei (1564–1642), who was one of the first to learn of it. Galileo later gave samples of the material, lapis solaris or "sun stone" (also referred to as the Bolognian stone, Bononian phosphorus and other variations), to Gulio Cesare Lagalla of the Collegio Romano who reported on it in the book, *De Phaenomenis in Orbe Lunae,* published in 1612. Although other naturally occurring phosphorescent materials and chemical processes had been observed for thousands of years, this material was the first step in synthesizing phosphors. Many years later these phosphors would be required in large quantities for cathode-ray tube screens and fluorescent lamps. The term *phosphorus* was later confined to the element phosphorus. Phosphorus was discovered in 1669 and produced light through chemical reaction. The shortened word *phosphor* is used to identify luminescent materials.[1–3]

2.3 JEAN PICARD

The first reported observation of electroluminescence (light generated by an electric field) in an evacuated glass tube was by Jean Picard (1620–1682), a French astronomer and priest, in 1675. He found that a mercury barometer carried into a dark room exhibited a greenish glow above the mercury column when it was descending but not when it was stationary or ascending.[2,4]

2.4 JOHANN BERNOULLI

In 1700, Johann Bernoulli (1667–1748) conducted further experiments into the *mercurial phosphor* phenomenon observed by Jean Picard in 1675. Bernoulli, a Swiss mathematician, duplicated the effect by shaking a clean glass vial containing mercury.[1,2,4]

2.5 CHRISTIAN F. LUDOLFF

Christian Friedrich Ludolf (1707–1763) was a German physician who, in 1745, was able to ascertain that the luminescence observed by Picard and Bernoulli in glass tubes of mercury was due to electrical charges generated by the mercury's motion. Using fine silk threads suspended in the vicinity of the mercury he detected electrical charges by movement of the threads.[2]

2.6 FRANCIS HAUKSBEE

The first experiments related to the cathode-ray tube were conducted in 1705[5] or 1706[2,6] by Francis Hauksbee (died 1713), a self-educated English scientist. His experiments consisted of rotating an evacuated glass globe at high speed and generating electrical discharges through friction when rubbed by the hand.[1,4,7,8] A faint purple glow was reported to be visible from the globe's interior. It is unclear whether the glow was due to fluorescence of the glass under electron bombardment or (more likely) to ionization of residual gases as a result of an imperfect vacuum.

Hauksbee also conducted studies of the previously described mercurial phosphor, as well as a number of other materials. These experiments included electrical effects created by friction and triboluminescence (luminescence generated by fracturing or other physical stresses).

2.7 WILLIAM WATSON

William Watson (1715–1787) discovered in 1751 that a voltage applied across an evacuated glass tube having two electrodes resulted in a bright light being emitted.[5] This was due to discharge through the remaining gases within the tube and is more closely related to the Geissler tube and its descendant, the neon tube.

2.8 JOHN CANTON

In 1768 an English physicist, John Canton (1718–1772), found that phosphors could be made to emit light when excited by electrical discharges. Early work in this area appeared to be concentrated primarily on development of an electric light.[2–3]

2.9 HEINRICH GEISSLER

For approximately 100 years, few additional contributions to the development of the cathode-ray tube were made due to limitations of vacuum pumps of the period. Heinrich Geissler (1815–1879), a German glassblower, developed an improved vacuum pump in 1855 and later fabricated the sealed off glass tubes used by Julius Plücker to study electrical discharges through gases at low pressure.[2,5,9–11] Plücker himself referred to these tubes as *Geissler tubes* (Figure 2.2). They are still widely known by that name. Geissler discharge tubes in modified form became widely used for neon signs around 1930 and further evolved into the fluorescent lamp in 1938 and, still later, the high-intensity discharge lamps widely used for highway and other outdoor lighting. Many other gases eventually were used, including mercury, argon, xenon, sodium and helium.

2.10 JULIUS PLÜCKER

A German mathematician and physicist, Julius Plücker (1801–1868) of the University of Bonn, found in 1859 that at pressures below about 1/100,000 atmosphere the glass walls of a vacuum discharge tube began to fluoresce rather than display visible gas discharge between the cathode and anode as observed at higher pressures.[12] He observed that the rays emitted by the cathode could be deflected by a magnetic field in the vicinity of the device as opposed to the path of the discharge through the gas which could not be deflected with a magnetic field.[2,5,13,14] This was shown by movement of the fluorescing green spot where the electron stream struck the glass. Since most glasses fluoresce to some degree when bombarded with electrons, this

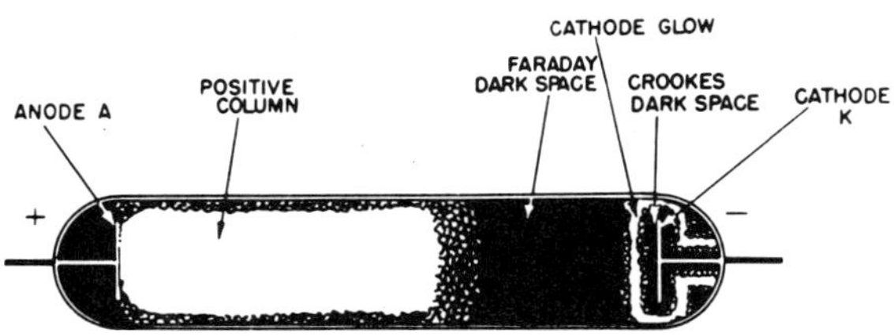

Figure 2.2 Geissler gas discharge tube.[16]

was a convenient way to observe the location of the invisible electrons. Thomson later reported that Plücker was the first to leave a record of having observed cathode rays and attributed his success to the vacuum pump (made by Geissler) which was capable of the highest vacuum up to that time.[13]

2.11 MICHAEL FARADAY

Michael Faraday (1791–1867), an English physicist and chemist, introduced the terms *cathode, anode* and *electrode,* which continue to be used today to describe the elements of a cathode-ray tube. Originally he used them for the process of electrolysis, but they were well-suited to vacuum tubes as well. His experimental work on gas discharge tubes occurred from about 1858 to 1865.[8,14,15] Faraday observed a dark gap near the negative electrode in a low pressure gas discharge. This became known as the "Faraday dark space."[7,16]

2.12 JOHN P. GASSIOT

John Peter Gassiot (1797–1878), often worked with Faraday and sometimes used tubes constructed by Geissler. Around 1858 he conducted experiments in England which showed that the cathode was the source of *negative rays* (cathode rays) that caused the residual gases and glass walls of the tube to glow where struck.[5]

2.13 JOHANN W. HITTORF

A professor of physics and chemistry at the University of Munster, Germany, Johann Wilhelm Hittorf (1824–1914), used lower pressures than Plücker in his investigations of the glow in gas discharge tubes. In 1869 he published a paper in the *Annalen der Physik and Chemie* that described the characteristics of the glow as the pressure was decreased. Of greater significance to the cathode-ray tube were his observations of the cathode being the source of rays that traveled in straight lines from the cathode and which cast a shadow of either conducting or insulating objects placed in the rays' path. He also reported in the same paper that the rays would strike the bend in a tube made from a piece of right angle tubing rather than following the tubing.[5,11–13,17–20]

2.14 Eugen Goldstein

Eugen Goldstein (1850–1930), a physicist with the astronomical observatory at the University of Berlin, conducted investigations of gas discharge tubes in a number of experiments during the 1870s.[5] He demonstrated chemical changes in coated plates under bombardment by the invisible rays. His studies of cathodes, the source of the rays, included materials as well as showing that the rays were emitted perpendicularly from the cathode.[10,13] Goldstein made use of that fact and brought the rays to a focus using a concave cathode.

In 1876 he originated the term *kathodenstrahlen* (cathode rays), which remains the term used to describe these rays.[3,11,12] That same year, Goldstein discovered that the cathode rays could be deflected by an electrostatic field[2] such as is formed by a metal electrode having a voltage applied in the rays' vicinity.

In 1886 Goldstein discovered *canal rays,* which were found to be emitted in the direction opposite to that of the cathode rays.[17,18] This discovery was made using a cathode with holes in the surface so that rays in the opposite direction could be observed. These rays were composed of positive ions created by ionization of residual gases contained within the discharge tube.

2.15 Sir William Crookes

Sir William Crookes (1832–1919) was an English chemist and physicist who was very active in both fields. Using a variety of gas discharge and vacuum tubes, he conducted studies on the previously known properties of cathode rays, as well as their heating effects and mechanical force.[2,5,10–12,14,16,18,21,22]

Upon reducing the gas pressure to the point at which light from the gas discharge became invisible, he observed fluorescence of the tube's glass walls. Crookes used a number of small and large electrodes to determine the effects of both the positive electrode (anode) and the negative electrode (cathode) on generation of the fluorescence and of intervening objects used to cast shadows of the cathode-ray beam. He referred to the cathode rays as *molecular rays* or *projected molecules* and correctly believed them to be negatively charged particles from the cathode.[17]

One experimental tube used by Crookes measured the effect of magnetic fields on the molecular rays. The deflection of the rays was determined to be proportional to the magnetic strength. Another tube incorporated a radiometer consisting of mica vanes suspended on a pivot in front of the beam. Tilting the tube moved the radiometer in and out of the beam. Crookes discovered that it rotated as expected when partially bombarded and that no rotation occurred when located out of the beam or with the beam impinging equally on both sides. Heating effects of the rays also were observed.[17]

The tube most closely associated with Crookes, commonly referred to as the Crookes tube, is illustrated in Figure 2.3. In this de-

Figure 2.3 Crookes tube replica.

vice, a hinged Maltese cross is used to demonstrate the properties of cathode rays by casting a shadow in the fluorescence at the tube's large end.*

2.16 HEINRICH R. HERTZ

Heinrich Rudolf Hertz (1857–1894), a German physicist at Karlsruhe Polytechnic and later at the University of Bonn, was best known for his contributions to electromagnetic wave theory. He also performed some experiments involving cathode rays. One of Hertz's discoveries in 1892 was that the rays can penetrate thin metal sheets.[5,10] This discovery was later further developed by Philipp Lenard in a device known as the Lenard tube.

Hertz tried unsuccessfully to deflect cathode rays electrostatically by use of a pair of parallel metal plates through which the cathode rays passed. Residual gas ions apparently neutralized the charge on the plates, thus pre-

*The experiments and many of Crookes' conclusions are published in "On the Illumination of Lines of Electrical Pressure, and the Trajectory of Molecules." In *Philosophical Transactions,* Part 1, p. 135, 1879.

venting him from observing electrostatic deflection, a significant property of cathode rays. Sir Joseph J. Thomson later successfully repeated the experiment under higher vacuum conditions.

2.17 PHILIPP E. A. LENARD

Philipp Eduard Anton Lenard (1862–1947) was a German physicist who received the 1905 Nobel prize for his work with cathode rays. In 1894, he added to the discovery by Heinrich Hertz, to whom he was an assistant, that cathode rays can penetrate thin sheets of metal foil. Lenard constructed a tube having a thin aluminum window opposite the cathode. The cathode rays penetrated the window and were detected at distances of up to eight centimeters by using a movable phosphor screen. Lenard discovered that the rays were absorbed by intervening materials according to their density and that higher velocity rays were absorbed to a lesser degree.[5,23,24]

Lenard also studied phosphorescent materials, such as calcium sulfide, and found their phosphorescence to be a result of metallic impurities within the crystals.

2.18 ALBERT HESS

Albert Hess used a special Lenard tube in 1894 to study and map magnetic fields.[5,10] This project constituted one of the earliest actual applications for cathode rays.

2.19 JEAN B. PERRIN

Another Nobel prize winner, although not for his work with cathode rays, Jean B. Perrin (1870–1942) was a physical chemist at the University of Paris. In 1895 Perrin proved that cathode rays were a stream of charged particles moving at high velocity, as believed by Crookes and Thomson and in opposition to the theory of Goldstein, Hertz and Lenard that they were vibrations of the "ether," or simply shorter wavelengths of light. Perrin used a tube (Figure 2.4) with a cathode at one end and a Faraday cylinder at the other. The Faraday cylinder was surrounded by a larger cylinder as a shield against external electrical disturbances. The Faraday cylinder would charge negatively when cathode rays from the cathode entered it, thus inferring negatively charged particles. When the cathode rays were deflected away from the cylinder with a magnet, no negative charge was measured.[5,10,12,14,16–18,25,26] This experiment set the groundwork for the discovery of the electron by Thomson two years later.

2.20 WILHELM C. RÖNTGEN

Wilhelm Conrad Röntgen (1845–1923) was a Dutch-born German physicist who was awarded the Nobel prize for physics in 1901 for his discovery of X-rays. Röntgen first began experimenting with cathode rays in June of 1894 at the University of Wurzburg (where he was a physics professor). By October of 1895 he had devoted all of his effort in this direction. His work was more the application of cathode rays than actual development of the cathode-ray tube, however, his discovery of X-rays was one of the most important spinoffs to develop from the cathode-ray tube. It

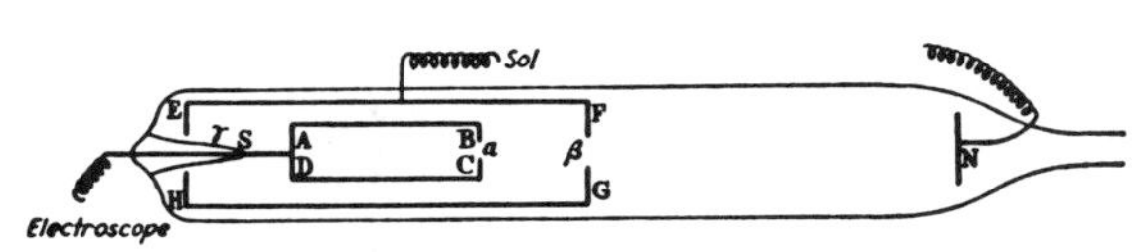

Figure 2.4 Perrin tube with Faraday cylinder used to establish that electrons are negatively charged.

took nearly another 50 years for the cathode-ray tube to evolve to a similar level of importance.

In 1895 Röntgen found that a high-vacuum Hittorf or Crookes tube (Figure 2.5) completely covered by black cardboard would cause bright illumination of a barium platinocyanide-coated paper screen in the vicinity of the operating tube. He named the rays, which were penetrating the black paper and most other objects placed between the tube and screen, X-rays because of their unknown origin. It was clear that they behaved differently than the cathode rays themselves. They traveled considerably farther than cathode rays from a Lenard tube and they were not influenced by a strong magnetic field. Nor did they have many of the properties of light.

Röntgen could not refract, reflect, focus or polarize the rays with any of the means usually applied to ultraviolet, visible or infrared light. Photographic film in black wrappers was

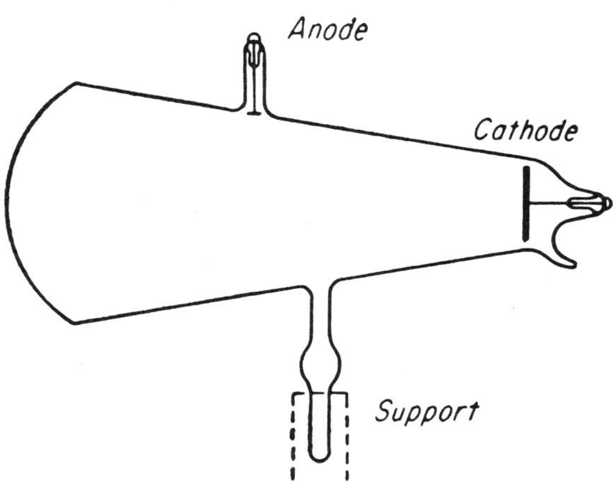

Figure 2.5 Tube used for the discovery of X-rays by Röntgen. (From Applied X-Rays *by G. L. Clark © 1955 by McGraw-Hill Book Company.)*

discovered to be exposed by the X-rays and Röntgen made radiographs of a number of objects, including his hand. He observed that the rays originated not from the tube's cathode, but from the fluorescing spot on the glass where the cathode rays impinged, and that the X-rays were generated by bombardment of other materials besides glass. The X-rays, unlike the cathode-ray beam, diverged in all directions from the fluorescing spot on the glass.[5,17,27–29]

The discovery of X-rays resulted in one of a number of important branches in the development of the cathode-ray tube. The next few years witnessed intense development of tubes for X-rays and their eventual application in the medical field as a diagnostic tool. At present it is still opening up new applications for the cathode-ray tube for displaying digitally stored radiographs without the need for film. The images may be viewed instantly at many locations, including the physician's office, while saving expensive storage space in hospitals that was previously needed to archive film. This emerging market for the CRT comes almost 100 years after Röntgen's discovery of the X-ray.

2.21 SIR JOSEPH J. THOMSON

An English physicist at the University of Cambridge, Sir Joseph John Thomson (1856–1940), was another recipient of the Nobel prize in physics in 1906 for his work on the conduction of electricity through gases. In 1897 he carried Perrin's determination that cathode rays were negatively charged particles a major step farther. The experiment to deflect cathode rays with an electrostatic field applied to a pair of metal plates through which the cathode-ray beam passed, originally tried unsuccessfully by Hertz, was repeated under a much higher vacuum. Thomson obtained deflection of the cathode-ray beam with voltages as small as two volts and observed that the degree of deflection was proportional to the voltage applied to the deflection plates.

An additional series of experiments by Thomson compared the beam's deflection with electrostatic and magnetic fields and measured the charge deposited by the beam, as well as its heating effect on a thermocouple. These trials enabled him to calculate the ratio of charge to mass of the particles in the cathode-ray beam[5,10,11,13,14,16–18,20,22,30–33] (Figure 2.6). The results showed the mass of the electrons to be less than one thousandth of that of a hydrogen atom which until that time had been the smallest known particle.[12] These results pointed to the existence of a subatomic particle that might be a building block for all atoms. Thomson called these particles *corpuscles* and it was a number of years before the term *electron* came into general use, although that term had been coined in 1881 by G. Johnstone Stoney.[12,16,33] R. A. Millikan (1868–1953) conducted many experiments about 1909 to calculate the precise charge of the electron.[144,146]

2.22 KARL F. BRAUN

Another Nobel prize for physics recipient was Karl Ferdinand Braun (1850–1918) whose name is most closely associated with the cathode-ray tube. His Nobel prize was awarded jointly with Guglielmo Marconi in 1909 for their work in wireless telegraphy. Braun held a chair of physics and directorship at the University of Strasbourg at the time of his experiments with the cathode-ray tube in 1896.

Braun was the first to combine all of the parts of the modern cathode-ray tube into one device. Central power generation stations were coming into use during the late nineteenth century and the newer alternating current systems posed problems with measuring voltage, current and phase. The inertialess beam of electrons offered significant advantages over the mechanical oscillographs using mirrors and light beams then in use. These oscillographs were restricted in frequency to 50 or 60 hertz due to their mechanical limitations. Higher frequencies were being investigated for power generation and Braun saw the possibilities for using the cathode-ray tube as a measurement tool.[2,5,10–12,16,18,22,20,34–38]

The Braun tube (Figures 2.7, 2.8) was constructed by Franz Müller of Bonn who was a successor of Heinrich Geissler. As shown in Figure 2.7 the Braun tube incorporated a cathode (K) and anode (A) as had previously been used in the Crookes tube. An aluminum diaphragm (C) was inserted in the tube's neck to limit the electron beam diameter so that a

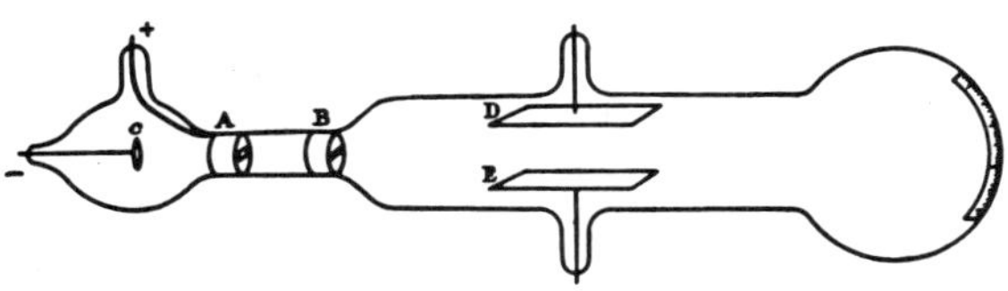

Figure 2.6 Tube used by Thomson in 1897 to determine the electron's mass.

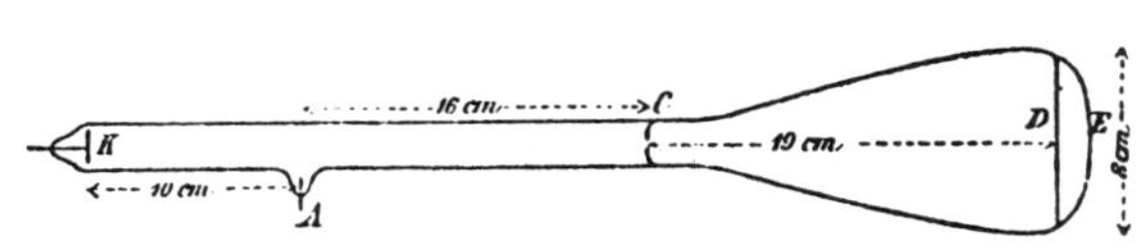

Figure 2.7 Braun tube used for the first oscillograph in 1897.[39] (K) cathode, (A) anode, (C) aluminum diaphragm, (D) screen.

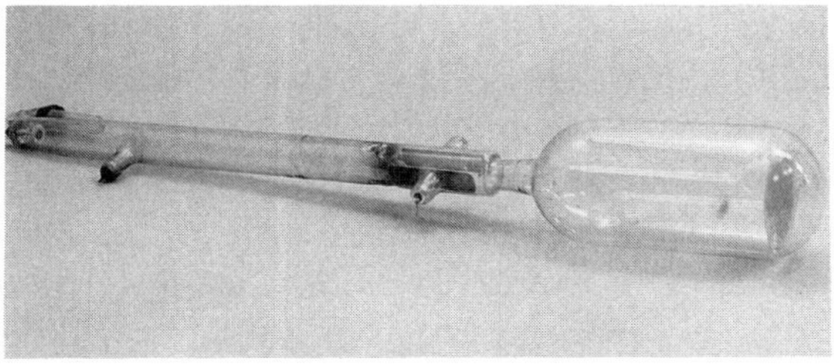

Figure 2.8 Braun tube replica. (Courtesy of National Museum of American History, Smithsonian Institution.)

clearly defined spot would be formed at the screen. The screen (D) consisted of a transparent sheet of mica coated with a phosphor on the side toward the cathode. This was the first use of a phosphor screen within the tube, replacing dependence on the weaker glass fluorescence to make the cathode rays visible.

Deflection of the beam in one axis was accomplished with placement of an electromagnetic coil next to the tube's neck. The voltage to be measured was applied to the coil and resulted in a green line measuring approximately one inch in length on the screen. A rotating mirror in front of the screen, as used with the mechanical oscillograph, provided scanning in the other axis to allow the waveshape to be observed. A rotating magnet mounted below the tube in place of the rotating mirror in front of the screen also was used. The rotation speed was adjustable to allow synchronization with the frequency of the voltage applied to the deflecting coil in the other axis. Later experiments used a second deflecting coil mounted at right angles to the first. This allowed a second voltage to be applied for comparison to the first voltage by displaying Lissajous figures.

From the above description it is evident that all six elements of the modern cathode-ray tube as we know it were in place in the Braun tube. They included: (1) The source of cathode rays (cathode), (2) a means to provide a focused spot (diaphragm), (3) a means of accelerating the beam of cathode-rays (anode), (4) a deflection system to direct the beam to various locations on the screen, (5) a phosphor screen to make the cathode-ray beam visible and (6) a sealed off evacuated envelope.

Braun published a paper in 1897 describing the investigations of electrical waveforms, but only a brief description of his tube is included.[39] Yet, he deserves credit for the modern cathode-ray tube, as well as for the first cathode-ray oscilloscope, one of the CRT's many important applications.

2.23 JONATHAN ZENNECK

An assistant to Karl Braun, Jonathan Zenneck (1871–1959), continued work on the cathode-ray tube as an indicator. By 1889 he had developed an improved tube using an additional aperture in the neck to improve focusing and he photographically recorded waveforms from the screen. Zenneck developed an electromechanical time-base circuit which allowed waveforms to be displayed on the screen without the rotating mirror. This completed the present-day cathode-ray oscillograph concept. He became dedicated to finding as many applications as possible for the oscillograph in later years.[5,20,34,35,40]

An improved Zenneck CRT of more compact dimensions (Figure 2.9) also was produced for J. T. MacGregor-Morris in 1902 by A. C. Cossor, Ltd. (a major name in cathode-ray tubes and oscillographs in England for many years).[41]

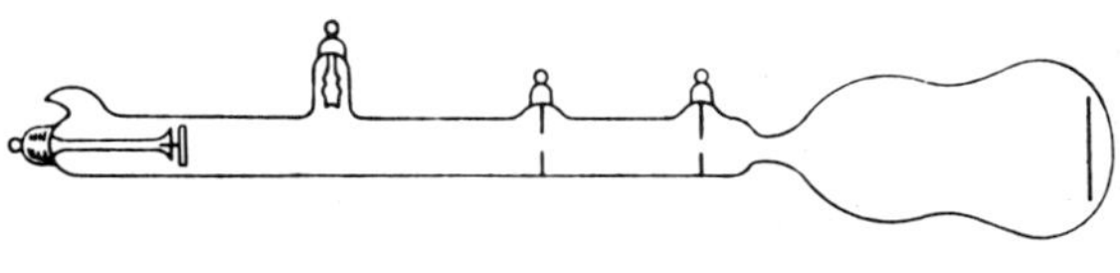

Figure 2.9 Cathode-ray tube made by A. C. Cossor, Ltd. for J. T. MacGregor-Morris in 1902.[41]

Dr. Zenneck later was the station engineer at WSL, a transatlantic wireless station at Sayville, Long Island, New York, where he was reportedly convicted of violation of the U.S. neutrality act for transmissions to German U Boats in 1915.[147]

2.24 HARRIS J. RYAN

Professor Harris J. Ryan also reported on the use of the cathode-ray oscillograph using orthogonally mounted magnetic deflection coils for alternating current studies at an American Institute of Electrical Engineers Convention in July 1903. He indicated that he had begun work in 1900 using tubes constructed by Mr. Miller-Uhri of Braunschweig, Germany, who had attempted to make six-inch diameter Braun tubes, but finally succeeded in making a pair of five-inch tubes. Ryan also noted that Miller-Uhri was now prepared to export similar tubes at a price of $20 each.[11,24,42]

2.25 ARTUR R. B. WEHNELT

One of the greatest advances that made the modern cathode-ray tube possible was the development of the oxide-coated hot-cathode in 1903 to 1904 by Artur Rudolph Berthold Wehnelt, a Brazilian-born German physicist. The oxide-cathode was an outgrowth of the "Edison effect" noted by Thomas Alva Edison in 1883. In that process a current was measured in an electrode in close proximity to a hot filament in an incandescent lamp. The current's cause was the emission of electrons from the filament. Wehnelt discovered that a coating of alkaline earth oxides on the filament greatly increased the emission of electrons.[11] He constructed cathode-ray tubes with oxide-cathodes as shown in Figure 2.10. These had the advantage of having strong emission with anode voltages of only a few hundred volts compared to earlier tubes which required up to 100,000 volts to sustain emission. Cathode life was a problem since the imperfect vacuum attainable at that time resulted in bombardment of the cathode by positive gas ions. This limited the operating voltage of the tubes to about 1,000 volts.[10,14,16,37,43]

In 1904–1905 Wehnelt also invented the Wehnelt cylinder, which is today called the control grid. This creation was in the form of a cylinder surrounding the cathode and having an aperture in the end through which the electron beam could pass through. Applying a negative voltage to the cylinder reduced the number of electrons in the beam by repelling them back toward the cathode.[16]

2.26 DIECKMANN/GLAGE/ROSING/CAMPBELL-SWINTON

During the period of 1906 to 1911, several individuals proposed the use of the Braun tube

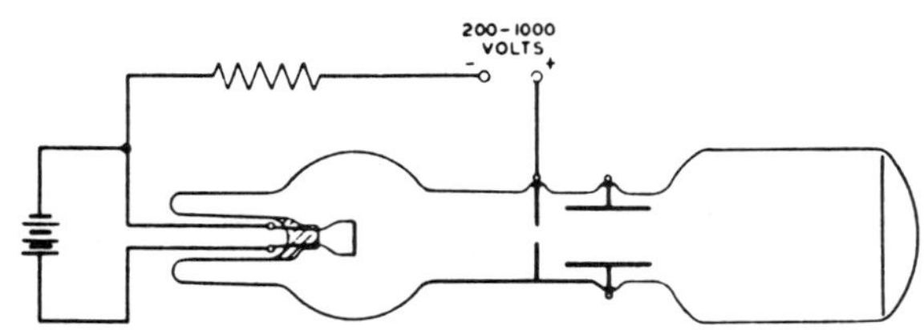

Figure 2.10 Wehnelt cathode-ray tube with oxide-coated cathode.[43] *(Courtesy of Howard W. Sams, Inc.)*

to display television images in all electronic systems.[5,45] Max Dieckmann and Gustav Glage, who also were assistants to Professor Braun, patented the first television system using the Braun tube in 1906. A short time later, in 1907, this achievement was followed by similar systems developed by a Russian, Boris Rosing, and an Englishman, Alan Campbell-Swinton.[10,11,46–49] In Germany Dieckmann had an actual operating system at that time, although it was quite crude by today's standards. Only silhouette images could be displayed with no halftones. Braun was not enthusiastic about use of his tubes for television since it was not a scientifically respected subject at that time.[35]

For the next few years, development effort was concentrated on the mechanical scanning system for television by pioneers such as C. F. Jenkins and J. L. Baird.[11,45] This work was based on the *Nipkow disc* invented by Paul Nipkow of Berlin and described in a German patent application of January 6, 1884.[50] It was not until about 1929 that the all-electronic system began to gain a toehold against the mechanical scanning systems of the period.

2.27 A. DUFOUR

A. Dufour is widely associated with the high-speed oscillograph through his development in 1913 (in France) of a high-voltage, continuously pumped cathode-ray tube. The Dufour CRT, shown in Figure 2.11, operated at up to 60,000 volts and was used to record oscillograms directly onto photographic plates inserted in the photographic chamber. A gas discharge cathode was employed and construction was primarily metal. Twelve minutes were required to pump the tube to a sufficient vacuum for operation. Good exposures

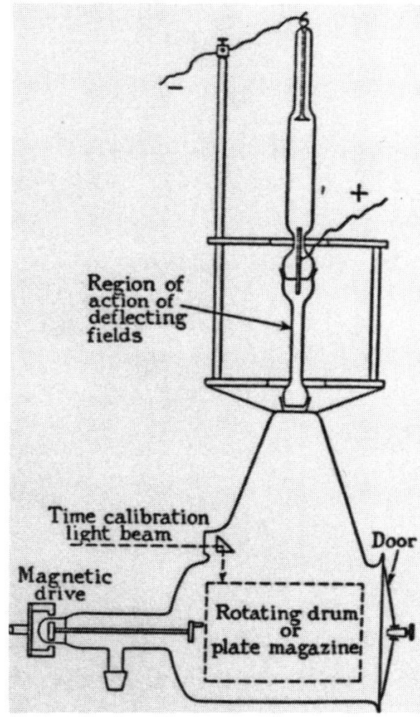

Figure 2.11 Dufour oscillograph with continually pumped vacuum system for recording high-speed transient waveforms directly on photographic plates.

were obtained at scanning velocities up to 100 miles per second. Its size, operating complexity and cost limited the Dufour CRT to research applications.[12,37,51–53]

Later refinements of the high-speed oscillograph were reported by W. Rogowski and F. Malsch in 1927 or 1928, H. Norinder in 1928, E. S. Lee of General Electric in 1928, R. H. George in 1929, Finch of Cambridge Instrument Co. and M. Knoll in 1930, F. P. Burch and R. V. Whelpton of Metropolitan-Vickers Electric Co. in 1932, and J. L. Miller and J. E. L. Robinson of Feranti, Ltd. in 1934. Most were related to the discharge tube used to produce the electron beam although Rogowski and George adapted the Wehnelt hot-cathode.[12,54]

2.28 JOHN B. JOHNSON/H. J. VAN DER BIJL

The first commercial CRT in the United States was the Western Electric type 224-A developed in 1921 by H. J. van der Bijl (1887–1948) and John B. Johnson (1887–1970). It is referred to in literature of the period as a "cathode-ray oscillograph" rather than tube.[10,11,14,16,55–65] The use of a small quantity of gas within the tube to provide focusing of the electron beam had been suggested by Dr. van der Bijl in a 1920 U. S. patent application.[11,66] The 224-A used argon or mercury vapor, although a number of other gases were found to produce the same effect. When the electron beam strikes the gas atoms on its way to the screen, ions are generated. Since the region of the tube where they are formed is surrounded with a positive potential, the negative ions are pulled away from the beam and the positive ions are concentrated within the beam's core. The fields created by this ionization cause the electrons (which are negative) to be converged to the point by the time they reach the screen. The filament voltage was adjusted to control the number of electrons emitted (beam current) to optimize the gas focusing effect.

To prevent damage to the Wehnelt cathode by positive ion bombardment, the cathode and anode were enclosed in a glass cylinder to limit the ions' mean free path. Even then, tube life was only on the order of 200 hours. Two sets of deflection plates were used to obtain electrostatic deflection, although a magnetic deflection coil also could be used around the neck for magnetic deflection. The result was a compact self-contained tube suitable for portable instrument use (Figure 2.12). Crisp, bright traces on the screen could be obtained with operating voltages of 300 to 400 volts. Previous tubes required either very high operating voltages or beam-limiting apertures to obtain focusing.

One side effect of gas focusing was a distortion on the screen known as *origin distortion* or *gas cross*. This distortion manifested itself as a slowing and consequently an apparent brightening of the displayed pattern as the beam crossed the centerline of the vertical and horizontal axes. It was caused by ionization of the gas in the deflection plate region.[14,16,22]

Later, in 1924, L. T. Jones and H. G. Tasker demonstrated the use of electrostatic focusing in a magnetically deflected CRT and indicated that it had been earlier suggested by Wood and von Baeyer.[67] Other names associated with the application of electrostatic focusing include Lilienfeld (Figure 2.13) in 1915,[10] Rogowski and Grosser in 1926,[68] R. H. George in 1927 to 1928[54] and F. Holweck

Figure 2.12 Johnson/van der Bejl cathode-ray tube. (Courtesy of National Museum of American History, Smithsonian Institution.)

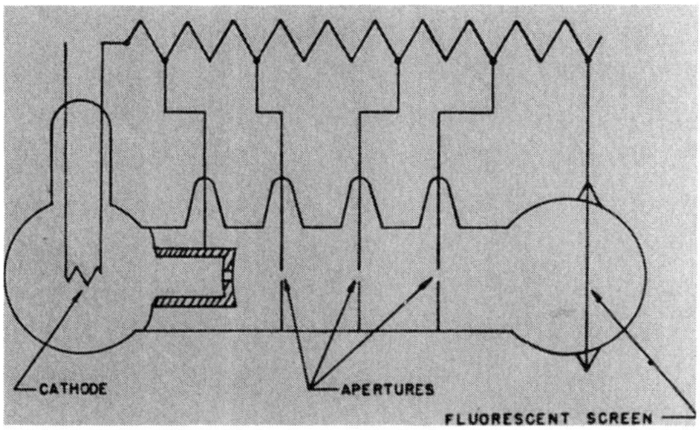

Figure 2.13 Lilienfeld cathode-ray tube with electrostatic focusing.[10]

and Chevallier in 1928.[11] Gas-focusing as developed by van der Bijl remained in general use for electrostatically deflected CRTs until the early 1930s.

Dr. Johnson was active for many years in the field of electron physics with about 21 papers and 30 patents to his credit. His studies and understanding of electron-generated noise at Bell Telephone Laboratories led to the term Johnson Noise being universally applied to the phenomena.[69] Johnson was again involved with cathode-ray tubes in 1948 with the development of an unusual tube with a cylindrical screen around the periphery of what normally is the screen. The tube was mechanically rotated to provide a continuous display of voice waveforms for studies of speech characteristics.[70] After retiring from Bell Laboratories in 1952, he joined the Thomas A. Edison Research Laboratory as head of the physics department.

Dr. van der Bijl is best known for his book, *The Thermionic Vacuum Tube and Its Applications,* which was published in 1920 and was the first significant book on the subject. His gas focusing concept was apparently his last contribution to the art of electron physics. He returned to his native South Africa in 1920 where he set up the first iron mine and steelworks. He was later responsible for establishing a grid system encompassing the country's independent electricity generating companies.

2.29 KENJIRO TAKAYANAGI

Kenjiro Takayanagi, of the Hamamatsu Higher Technical School in Japan, experimented with the use of a gas-focused oscillograph CRT made by Tokyo Denki to display 4 × 5 cm television images from a Nipkow scanning disc camera in 1927.[11,71] He later demonstrated television pictures with 15 and 30 centimeter diameter CRTs in 1930. Takayanagi was at the forefront of Japanese television efforts during the 1930s and held patents on both cathode-ray tubes and television camera tubes.

2.30 VLADIMIR K. ZWORYKIN

Russian-born Vladimir Kosma Zworykin (1889–1982) was a student of Boris Rosing[47] at the Institute of Technology in St. Petersburg. Zworykin was strongly influenced by the early work in television to which Rosing exposed him. He and Rosing constructed a cathode-ray tube and used it in a television demonstration with a selenium detector for image pickup. In 1919 when he arrived in the United States, he began looking for a position in television research. Eventually he was employed by Westinghouse Research Laboratories and filed for patents in 1923 and 1925 covering an all-electronic television system.[48,72,73]

Zworykin demonstrated an all-electronic television system while still at Westinghouse in 1929. This system employed a cathode-ray tube which he called a *kinescope* as the heart of the receiver.[10,11,18,45,74,75] The Zworykin kinescope was of the high-vacuum type with electrostatic focusing (Figure 2.14). After 1929, all-electronic television rapidly gained acceptance over the previous cumbersome mechanical scanning systems. Zworykin joined RCA in 1929 where he did considerable research in the development of television and invented the device most closely associated with his name, the iconoscope (which made the television camera practical). During this time, he made numerous improvements to the kinescope[76–78] and developed a projection television CRT capable of producing 18 × 24-inch pictures.[79,80]

2.31 PHILO T. FARNSWORTH

A strong rival of Zworykin, Philo T. Farnsworth (1902–1971) was also best known for the invention of a television camera tube, in this case, the image dissector. Farnsworth was an American inventor who began television development in 1926 in the Crocker-Bishop Laboratory in San Francisco.[11,81,82] In 1927, Farnsworth demonstrated all-electronic television systems based on his own patents[83] which were challenged, and in some cases upheld, by Zworykin. His principal contribution to the cathode-ray tube art was the introduction in 1930 of magnetic focusing for his tube known as the *Oscillite*.[84,85] Although the Oscillite (Figure 2.15) was unsuccessful due to very dim images when compared to the Zworykin kinescope,[11] magnetic focus coils became widely used in the early postwar years of television.

Most of Crocker-Bishop's San Francisco laboratory was moved to Philadelphia in 1931. While working there, Farnsworth licensed Philco to manufacture his television receivers for the public.[81] Little came of this association. In 1939 a manufacturing plant was established in Fort Wayne, Indiana, under the auspices of Farnsworth Television and Radio Corporation. This plant was soon involved in the war effort instead of television manufacturing and finally was acquired by Interna-

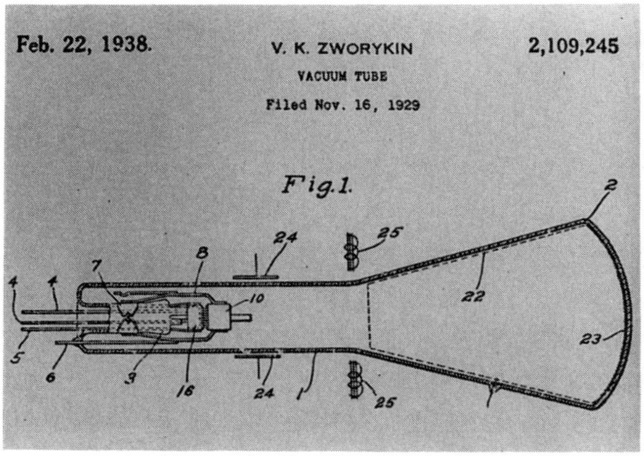

Figure 2.14 Zworykin kinescope in U.S., patent filed in 1929.[78]

Figure 2.15 Farnsworth "Oscillite," circa 1930. (Courtesy of National Museum of American History, Smithsonian Institution.)

tional Telephone and Telegraph in the late 1950s.

2.32 MANFRED VON ARDENNE

Manfred von Ardenne (1907–) was a prolific researcher and author in the field of cathode-ray tubes in pre-World War II Germany. Numerous references to developments at his company, Leybold and von Ardenne, appear in technical publications of the 1930s. His book, *Cathode-Ray Tubes,* originally published in 1933 in German and updated and translated into English in 1939 by G. S. McGregor, is an excellent review of the state-of-the-art in prewar cathode-ray tubes.[86] A companion book titled *Television Reception* originally published in 1934 and translated into English in 1936 by O. S. Puckle contains further information on cathode-ray tubes as they apply to television reception.[87]

Von Ardenne demonstrated an all-electronic television system in Berlin about six months before Zworykin's. Upon meeting five years later, the two inventors found that they had very similar systems and that their conclusions were identical. Subsequent inventions by both often were independently made within a few weeks of each other.[88] Von Ardenne investigated the use of early gas-focused cathode-ray tubes for television in 1929[89,148] and was the supplier of the gas-focused 478-A for the General Radio 535-A Electron Oscillograph introduced in 1931.[11,90,91] Figures 2.16 and 2.17 show details of the construction of an early electrostatically deflected five-inch cathode-ray tube by von Ardenne.

High-vacuum cathode-ray tubes were developed by von Ardenne in 1928 to 1929. An example of a five-inch diameter von Ardenne

Figure 2.16 Von Ardenne electrostatic deflection cathode-ray tube, circa 1932. (Courtesy of National Museum of American History, Smithsonian Institution.)

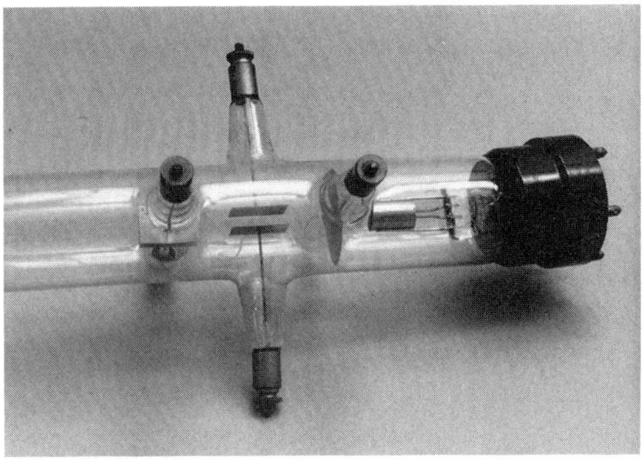

Figure 2.17 Electron gun of von Ardenne design, circa 1932. (Courtesy of National Museum of American History, Smithsonian Institution.)

CRT from 1932 is displayed in the Transportation Museum at Nuremburg, Germany. Other significant CRT developments by von Ardenne included use of a CRT for flying-spot scanning in 1930,[11] a projection oscillograph CRT sometime before 1936,[12] a dual beam oscillograph tube in 1936,[92] a flat face CRT in 1937[93] and a high frequency micro-oscillograph in 1939.[149]

A strong believer in multidisciplinary study, von Ardenne moved from electron-physics to nuclear research in the late 1930s. The Russians, during their occupation of Germany in 1945, recognized his value and encouraged him to relocate to the USSR where they set him up as director of the Sukhumi Institute. Von Ardenne was awarded the Lenin Prize in 1953 for his contributions to development of the atomic bomb for the USSR. He returned to East Germany in 1955 where he engaged in many types of research of which medicine and cancer research became the most important.[88]

2.33 ALLEN B. DU MONT

Dr. Allen Balcom Du Mont (1901–1965) was a U.S. pioneer in the development and application of cathode-ray tubes. His earliest work following an electrical engineering degree from Renssalaer Polytechnic Institute in 1924 was with the Westinghouse receiving tube department. In 1928 he became chief engineer for the De Forest Radio Company in Passaic, New Jersey, where he developed manufacturing capability for radio receiving tubes. After becoming vice-president, he devoted most of his efforts toward television development with the company acquired from C. F. Jenkins.[11,94–96,103]

In 1930 he invented a tube that was described as an electronic motor.[97–102] The tube contained a turbinelike anode suspended from a bearing. Pressure from the electrons from the cathode mounted in the center caused the anode to rotate as they struck the vanes. Needless to say, this was not one of the period's more profound inventions.

Du Mont left De Forest Radio in 1931 to start his own company for the development of the cathode-ray tube in the basement of his home in Upper Montclair, New Jersey. There, he began what grew into Allen B. Du Mont Laboratories which, until the early 1950s, was devoted exclusively to the development and manufacture of cathode-ray tubes and associated equipment. Du Mont manufactured an extensive line of cathode-ray oscillographs (oscilloscopes) beginning in late 1931[11,103] and continuing until the early 1960s when Du Mont Labs was acquired by Fairchild Camera and Instrument. Early Du Mont CRTs were of the gas-focused type, similar to the Johnson tube, although they were rapidly replaced by high-vacuum types embodying many improvements in the mid-1930s.[104–107]

Du Mont's experience at De Forest Radio convinced him of the advantages of the cathode-ray tube for television as opposed to the mechanical scanning system used at De Forest's experimental television station (W2XCD) in Passaic in 1930. As early as 1932, he was investigating the application of CRTs to television.[108] In 1938, Du Mont introduced the largest screen television receiver in the United States for its time using a 14-inch CRT.[95] It was based on a similar tube manufactured by A. C. Cossor, Ltd. which he had seen on a trip abroad in 1937.[109] The Du Mont tube shown in Figure 2.18 was all the more unusual in that it was an electrostatic deflection type with its inherently greater length. This was soon followed by a 20-inch diameter electrostatically deflected CRT of gigantic proportions measuring almost 28 inches in length.[110]

Du Mont conducted considerable research during the early television years[111,112] with several patents resulting.[113,114] Du Mont Laboratories also manufactured television re-

Figure 2.18 Dr. Du Mont holding 14-inch diameter electrostatic deflection cathode-ray tube as used in early television receivers circa 1939. (Courtesy of Jacob Ruiter.)

ceivers, television studio equipment and transmitters. They established and operated television stations in Passaic, New Jersey (experimental W2XWV in 1939), New York (WABD in 1944), Washington (WTTG in 1946) and Pittsburgh (WDTV in 1949). The ill-fated Du Mont television network operated under Du Mont control from 1947 to 1955 when it was spun-off to later become the successful independent, Metro-Media Broadcasting.[115,116]

Other work by Du Mont included the *Cathautograph* in 1932,[117–119] a device developed to transmit handwriting on an electronic tablet to a remote cathode-ray having a long persistence screen at a remote location and the "Magic Eye" tuning indicator.[120,121] The rights to the Magic Eye were later sold to RCA to finance a new Du Mont plant. Additional work included a simple cathode-ray tube in a receiving tube envelope (Figure 2.19), a combination CRT for both television pickup and viewing,[122] the elimination of pattern distortion in electrostatic CRTs[123] and some early color CRT development.[124] More information on the history of Allen B. Du Mont Laboratories may be found in a 1989–1990 series of articles in *Radio Age* magazine.[125–129,145,150]

2.34 JOHN L. BAIRD

A well-known name in British television development, John Logie Baird (1888–1946), concentrated on mechanical scanning systems during the 1920s and early 1930s. The Baird Television Company, Limited was established in 1927. Early innovations included mechanically scanned stereoscopic and color television pictures.[130]

By 1935, Baird too had turned to all-electronic television using the cathode-ray tube and, under license from Farnsworth, the image dissector camera tube. At that time, he demonstrated high-resolution television of 700 lines while the competing Marconi-EMI sys-

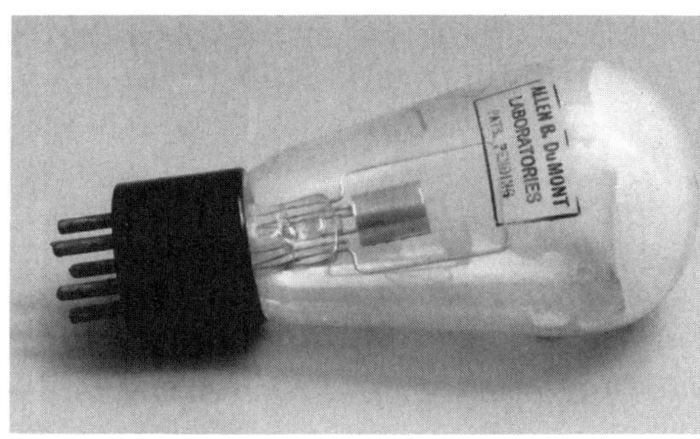

Figure 2.19 Du Mont Type 22 cathode-ray tube circa 1933. (Allen B. Du Mont Collection, National Museum of American History, Smithsonian Institution.)

tem was limited to 405 lines. During the late 1930s and early 1940s, Baird demonstrated several color and stereoscopic television systems using cathode-ray tubes.[130–141]

Baird was producing large-screen, magnetically deflected television picture tubes trade-named *Cathovisors* (Figure 2.20) and receivers of 22-inch diameter in 1937 and reportedly 26-inch diameter in 1939.[130]

Figure 2.20 Twelve-inch diameter electromagnetic deflection Cathovisor television picture tube manufactured in 1939 by Baird Television, Ltd. (Courtesy of C. E. "Sonny" Clutter.)

The Baird Television Company was absorbed into what is today Rank-Brimar Limited through many acquisitions during the period of 1933 to 1986. Baird had somewhat of a reputation for being a promoter, at least partly due to poor choice of associates.[130] Baird's contributions to the CRT's development were more in the area of its application although he did demonstrate several color CRTs and systems.

2.35 THE END OF THE BEGINNING

This completes the development of the cathode-ray tube through 1940. Practical CRTs were then commercially available and were being employed in limited quantities for oscillographic and television applications. The stage was set for the explosive growth and refinement that was to come as the CRT was drafted for military electronics and radar with the onset of World War II. This rapid growth phase continued after the war as television, aerospace systems, color television and, finally, computer display markets emerged in rapid succession. There are no signs of this growth abating. Total cathode-ray tube production figures dramatically illustrate its progress. In 1939, approximately 50,000 CRTs were produced; by the end of 1944, more than 2,000,000 per year were being produced.[142] As impressive as those figures may be, they are dwarfed by recent figures which show over 100,000,000 CRTs manufactured worldwide in 1987.[143]

Until the 1930s, cathode-ray tube advancements were largely made by individuals, often in an academic environment. After this time, major advances began to emerge from the commercial cathode-ray tube manufacturers such as RCA, Du Mont, Cossor, Telefunken, Sylvania and Tektronix, as the commercial and military applications developed. Still later in the 1960s, advances from Japan became widespread as the consumer color television market expanded. Chapters 3 through 9 will discuss the developments from 1940 to the present by application category with approximate chronological order maintained within each chapter.

3

Radar Indicator Cathode-Ray Tubes

3.1 Background

Heinrich Hertz first observed the reflection of radio waves from solid objects in 1887 during his famous experiments with electromagnetic waves.[1,2] In 1904, a German engineer, Christian Hulsmeyer, was granted patents in several countries for an obstacle avoidance system for ships using the radio-echo principle.[2–4] Guglielmo Marconi addressed the Institute of Radio Engineers at a dinner on June 20, 1922, where he was awarded the IRE Medal of Honor. He briefly discussed his observations of radio reflections by metallic objects many miles away and suggested that ships could be detected in heavy fog or bad weather by such means.[2,5,6]

The pulsed method that was essential to the measurement of distances of reflecting objects was first applied in 1925 to measurement of the ionosphere's height by Gregory Breit and Merle Tuve of the Carnegie Institution in Washington.[2] During the next 10 years, the principle was applied independently to the detection of ships and aircraft in the United States, Germany, France, Russia and Great Britain. Due to government requests to delay patent applications until after World War II for reasons of national security, the actual inventor of radar remains a hotly debated subject. The parallel efforts in several countries and the problem of defining exactly what constitutes a radar system have not made it any easier to determine the true inventor.

The first reflections from aircraft were obtained by the Naval Research Laboratory (NRL) in 1934. Demonstrations at sea were conducted by the NRL in 1937 on the destroyer U.S.S. Leary. In 1939 the U.S. Navy successfully tested the prototype model XAF radar set developed by the NRL on board the battleship U.S.S. New York. This led to the first production units of the CXAM radar built by RCA in 1940.[2,4,7,8]

The U.S. Army began tests of pulsed radar in 1936 which resulted in the development of the SCR-268 radar in 1938 for coast artillery. The SCR-268 was used to aim searchlights since, at that stage of development, the resolution was inadequate to directly control the guns.[9] In 1937, the Army requested development of a long-range radar for aircraft detection which led to the prototype SCR-270 demonstrated in 1939. Production contracts were issued in 1940.[2,8] The SCR-270 was the radar that detected the attack on Pearl Harbor on December 7, 1941, although human error led to discounting the observation's significance. Two models were lumped together under the SCR-270 designation: the SCR-270 mobile version and the otherwise identical SCR-271 fixed-station version. In 1944, the SCR-584 mobile microwave radar provided the resolution required for direct automatic tracking of antiaircraft batteries.[8,10–15]

Meanwhile, after a later initial start, the British were developing radar with a greater sense of urgency due to events transpiring in Europe.[8] In 1934–1935, Sir Robert Watson-Watt of the National Physical Laboratory submitted a plan for the detection of aircraft using the pulsed method. An experimental radar station was first established in 1935 on a small island on England's east coast. By 1938 a chain of five stations to protect the mouth of the Thames River was operational. These stations, known as the British Chain Home (CH), were used throughout the war and played a key part in England's defense.[2,4,10,16,55] Without radar, the Battle of Britain might not have been won.[56]

The outbreak of World War II and the Japanese surprise attack on Pearl Harbor spurred the rapid development of ground-based, shipborne and airborne radar by the major powers. By December 7, 1941, 79 radar sets were installed on American warships. Where only a handful of systems existed in 1940, $2.7 *billion* worth of radar systems had been delivered to the U.S. military by June 1945.[2,4] The Massachusetts Institute of Technology Radiation Laboratory was established in November 1940 as a civilian research organization to pursue research and development of radar for the war effort. During its five years of existence many of the most important wartime American radar developments were made there under the directorship of Lee A. DuBridge.[2,4] The Radar Section of the Naval Research Laboratory in Washington, D.C.; the Evans Signal Laboratory of the Army Signal Corps at Fort Monmouth, N.J.; and the Aircraft Radio Laboratory at Wright Field, Ohio, were also important contributors to the wartime radar development effort.[17] Canadian radar development was concentrated at the radio branch of the National Research Council and Research Enterprises, Ltd.[54] (An excellent, in-depth account of the development and use of radar is presented in Henry Guerlac's two-volume text, *Radar in World War II*.[18] An overview of radar systems over the last 50 years is presented in *Radar Applications,* edited by Merrill Skolnik for the Institute of Electrical and Electronic Engineers.[8])

From the very beginning, the cathode-ray tube was an essential component of all radar systems. Radar spurred the introduction of high-speed, mass-production methods to cathode-ray tube manufacturing. In 1939, approximately 50,000 CRTs were produced per year. By late 1944, more than 2,000,000 CRTs per year were being produced.[19] This production capacity at war's end and the knowledge gained certainly aided the rapid introduction of television in the following years.

3.2 OSCILLOSCOPE INDICATORS

The standard form of radar indicator circa 1940 was the cathode-ray oscilloscope, or deflection modulated display, which received the designation of *A-scope*.[20] A-scope indicators were used with most of the early radar systems, such as the SCR-268, SCR-270 and

SCR-271, and continued in later years as an essential piece of test equipment for maintenance and calibration purposes. Since distance to a target is measured by the time of the radar pulse's travel to the target and its return, the oscilloscope was useful to measure the delay interval between the transmitted pulse and the received echo as shown in Figure 1.48. Each microsecond of delay for the round-trip represents a distance of 164 yards.

Figure 3.1 RCA 5BP4/1802-P4 electrostatic deflection radar indicator CRT manufactured in 1943. (From the author's collection.)

Electrostatically deflected cathode-ray tubes such as the RCA 900 and 1800 series that were introduced beginning in 1934 were used for early radar systems. Few commercial CRTs were available at the time except for a few oscilloscope tubes and some limited production television picture tubes. In 1939, the newer RMA CRT type designations started to be used. The type 1802-P1 became the 5BP1, the 906 became the 3AP1 and the 1805-P1 became the 5AP1. RCA double-branded (e.g., 3EP1/1806-P1) some CRTs up to about 1944. The U.S. Army Signal Corps started assigning their own nomenclature to vacuum tubes during World War I. These were "VT" numbers and a few radar CRTs received nomenclature under this system during the early World War II years. (The VT-111 was one example.) It was the 5BP4/1802-P4 (Figure 3.1) which was used for the indicator in the SCR-268 and SCR-270 radar systems.[21] The Joint Army-Navy nomenclature system later standardized tube type numbers for both services using the RMA numbering system in 1943 (see Appendix 2).

Several new tubes were introduced during World War II which were used for A-scope indicators and oscilloscopes for radar. They included the RCA types 2AP1, 3BP1, 3EP1, 3FP7, 5CP1 and 5HP1; Du Mont 3GP1, 5JP1 (Figure 3.2) and 5LP1; and Sylvania 5GP1 and 5NP1. Post-deflection acceleration was used in the types 3FP7, 5CP1 (Figure 3.3), 5JP1 and 5LP1 for greater brightness, a requirement that became increasingly important as CRTs were used in airborne applications. The 3BP1, 3FP7 and 5CP1 used the new 14-pin diheptal base to prevent arcing or flashover at high altitude for airborne systems, while the 5HP1 and some others used a yellowish color, mica-filled phenolic base which was nonhygroscopic and therefore better suited to shipboard equipment. Late in World War II, improved electron guns with

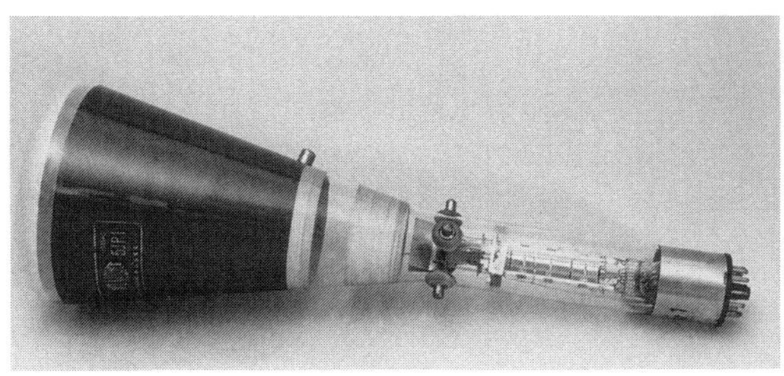

Figure 3.2 Du Mont 5JP1 manufactured in 1944 and used in radar range scopes. (From the author's collection.)

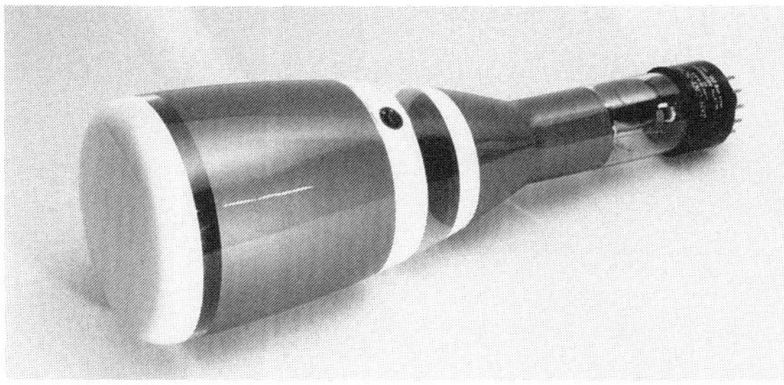

Figure 3.3 North American Philips (Norelco) 5CP1 circa 1944. (From the author's collection.)

zero-focus current and improved sharpness were developed by RCA[22] and most popular tube types were changed over to the new guns. These are identifiable by their "A" suffix, e.g., type 5BP1-A. Large quantities of most of the aforementioned tubes were manufactured from 1942 to 1945 on U.S. Navy and U.S. Army Signal Corps contracts issued to RCA, Du Mont, Sylvania, General Electric, National Union and Norelco (North American Philips). (See Chapter 4 for additional information on these tubes.)

The A-scope was limited to measuring distance to the target; elevation and azimuth could be obtained only by moving the antenna for the strongest echo, and the number of targets could not be determined unless they were widely separated. This led to the use of multiple indicators as well as new types of information presentations using various combinations of distance, elevation and azimuth formats on the same CRT. (See Figure 1.53 in Chapter 1 for many of the radar display formats developed during World War II.[23])

3.3 RADIAL-DEFLECTION INDICATORS

Radial-deflection CRTs, which were used in polar coordinate indicators known as J-scopes and which were another example of deflection-modulated displays, were demonstrated in 1936 by Leybold and von Ardenne in Germany.[24,25] The special CRT used a truncated-cone deflection electrode with an additional electrode in its center just following the deflection system as in Figure 3.4. The beam, which was scanned circularly, passed through the circular gap between the additional electrodes and a voltage applied between these elements would cause an outward deflection of the beam. Similar tubes were manufactured by A. C. Cossor, Ltd. in Great Britain during World War II for coast artillery radar indicators and later in the United States by Electronic Tube Corp.[26] The British also made use of a conventional 12-inch electrostatic deflection CRT (type ACR-12) with a one-inch diameter metal disc fastened to the tube's

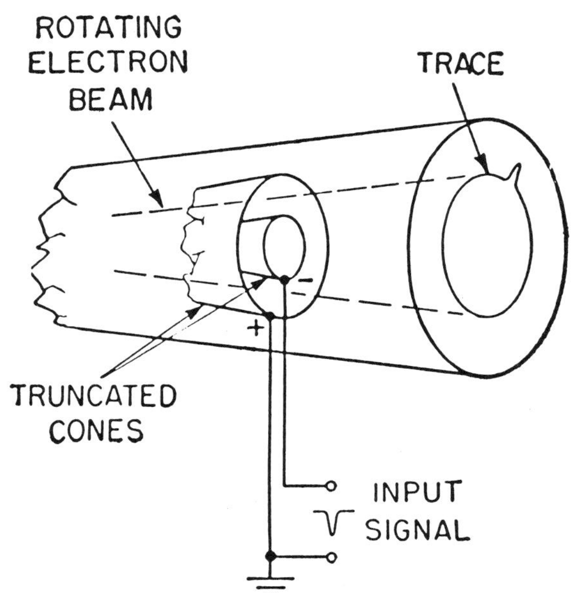

Figure 3.4 Early radial deflection cathode-ray tube construction circa 1936 to 1940.[28]

exterior in the center of the face to accomplish the same result. About 1,000 volts were required for suitable deflection.[27]

Radial-deflection tubes presented a distinct advantage over the conventional A-scope display since the circular scan resulted in an increased-length time-base without an increase in tube size. The beam was scanned in a large diameter circle around the screen with a return echo being indicated by an outward pulse on the circle at an angle corresponding to the target distance (see Figure 1.50). The same type of radial scan indicator also could be calibrated in azimuth angle. Other J-scope applications during World War II included radio altimeters and radar range calibrators.

Radial-deflection tubes in the United States consisted of a standard electrostatic deflection CRT with an additional deflection electrode rod sealed through the face's center (Figure 3.5).[26,28] The first tube introduced in 1941 was the three-inch RCA 3CP1/1808-P1 which was identical to the 3EP1/1806-P1 oscilloscope CRT except for the radial-deflection electrode. Usually the 3CP1 was supplied with a cemented calibration scale on the tube face to facilitate measurement of azimuth angle or distance. A similar tube was the 3DP1 which was the same as a 3BP1 except for the center electrode. Both the 3BP1 and 3DP1 had large 14-pin bases which allowed them to be used in airborne radar and navigation indicators because of their wider spacing between pins. This overcame the problem of high-voltage flashover or arcing between pins at high altitude. 3DP1s were manufactured by RCA, Du Mont and General Electric. Figures 3.6 and 3.7 illustrate the 3CP1 and 3DP1 radial-deflection CRTs. The 3DP1 was superceded in 1954 by the

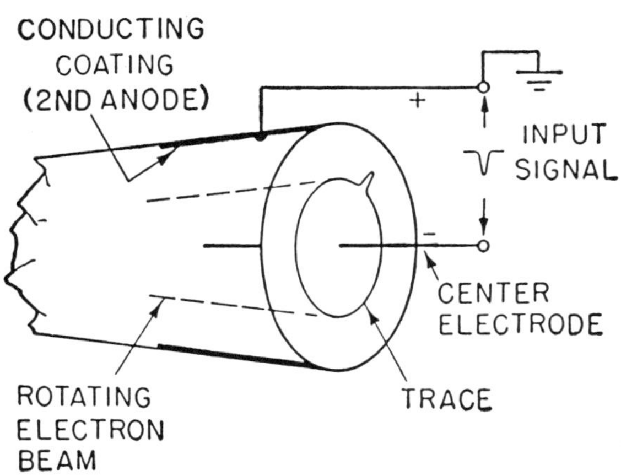

Figure 3.5 Radial deflection cathode-ray tube construction used during World War II.[28]

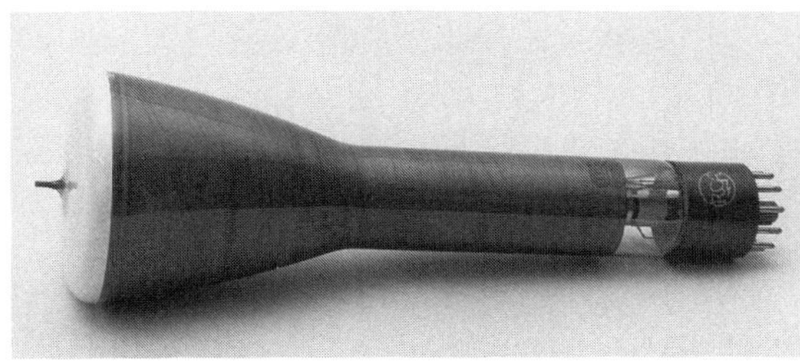

Figure 3.6 RCA 3CP1/1808-P1 radial deflection cathode-ray tube manufactured in 1943. (From the author's collection.)

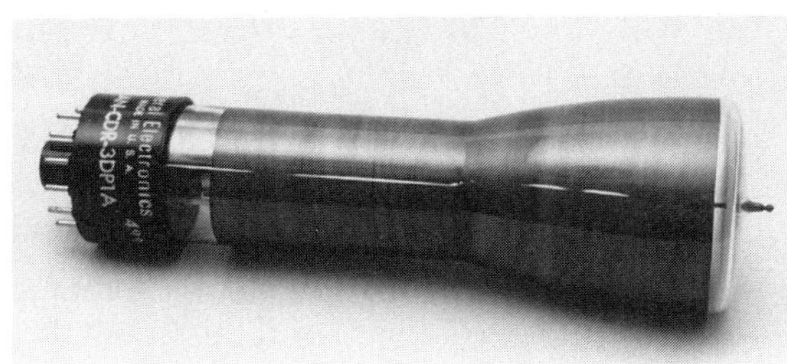

Figure 3.7 General Electronics 3DP1-A radial deflection CRT manufactured in 1954. (From the author's collection.)

3DP1-A as improved electron guns were developed.[22]

3.4 LARGE-SCREEN ELECTROSTATIC DEFLECTION CRTS

Large-screen electrostatic-deflection CRTs suitable for various radar displays were developed from oscilloscope origins beginning in the late 1930s. Later, electrostatic deflection allowed vector-generated alphanumerics to be displayed along with the conventional PPI information to identify target and navigation aids. Vector-generated character writing (also called stroke writing) requires the beam to be moved at higher speeds randomly to different screen locations, a difficult task for a magnetically deflected tube and deflection yoke to perform quickly.

Great Britain produced several 12-inch diameter, electrostatic-deflection CRTs beginning with the pre-war VCR-84 and VCR-85 types used in the Chain Home radar system. These were similar to the same period's early Du Mont 14-inch television picture tubes. Later in the war, the improved VCR-131, VCR-511 and VCR-519 tubes appeared. Improved long-persistence screens and better focus were featured in these tubes.[27]

North American 12-inch tubes included the 12FP7 (Figure 3.8) and 12HP7 manufactured by Research Enterprises Ltd. of Toronto, Canada in 1943 and 1944, respectively, and the 12GP7 by General Electric in 1944. The 12FP7 and 12GP7 were post-deflection acceleration types and used 14-pin di-heptal bases and medium caps for the post-accelerator. The 12HP7 used an 11-pin magnal base. These tubes were not widely used.

Several additional round, large-screen, electrostatically deflected CRTs for military radar applications were produced beginning

Figure 3.8 Research Enterprises, Ltd. 12FP7 large-screen, electrostatic deflection cathode-ray tube circa 1945. (From the author's collection.)

in about 1958 with the 12ACP19 by Du Mont. Others of 10-, 12-, 16- and 19-inch diameter and the generous overall length (up to 28 inches) necessitated by electrostatic deflection were registered in the period up until 1965. All were available predominantly with the common long-persistence radar phosphors such as P2, P7, P14, P19 and P25. From one to three guns were used. Also registered with EIA in the period of 1958 to 1965 were Du Mont's 10ALP-, 12AHP-, 12AMP-, 12AQP-, 12ATP-, 12AUP- and 12AVP-; Sylvania's, 12AXP-; and GE's 12AKP-, 12ANP-, 12ARP-, 12ASP- and 16AMP-.

The only known 19-inch diameter electrostatic deflection CRTs were the M-1125 manufactured by Electronic Tube Corporation and the General Electric Z-4917 and Z-4928 in the 1960s. All except the 10ALP- utilized post-deflection acceleration to prevent the deflection amplifier voltage swing requirements for full-screen deflection from getting out of reason. The 10ALP- and the companion 7AGP- were the most unorthodox of the group. They used a 22-pin collar midway along the neck, instead of a conventional base at the neck's end. The collar permitted all connections to be made through a single socket while minimizing deflection plate capacitance and decreasing overall tube length. The others used the 10-pin duodecal, 14-pin diheptal or 25-pin pentaquintal bases. A number of other similar tubes were developed with house numbers. Figure 3.9 illustrates a representative large-screen electrostatically deflected CRT.

Finally, the largest electrostatic deflection CRT appears to be the General Electric rectangular Z-4909 with a 20-1/2 × 16-1/2-inch screen cataloged in 1969.

3.5 PPI DISPLAYS

The plan position indicator (PPI) was developed as a means of presenting target range, azimuth and, to a limited extent, size on a single polar coordinate indicator. PPI displays are the ones that the general public associates with radar (see Figure 1.52). Range is indicated by distance outward from the screen's center and azimuth angle is shown by a scale calibrated in degrees around the screen's periphery. Targets are intensity-modulated on the CRT to produce a bright spot or "blip" indicating its location in a maplike presentation.[28] Up to the present time, PPI indicators have been the mainstay of most radar systems, especially in the civilian air traffic control system. Now they are being upgraded by the FAA to computer-based, raster scan displays.[29]

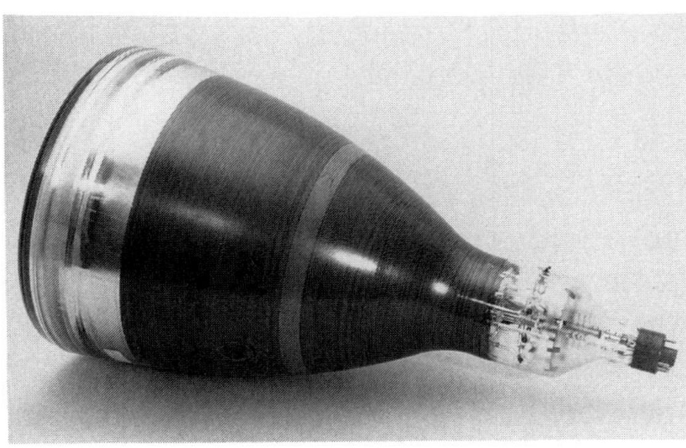

Figure 3.9 Thomas Electronics 12E24P19 electrostatic deflection CRT manufactured in 1980. (From the author's collection.)

The first PPI indicator was installed in 1941 on an aging destroyer, the U.S.S. Semmes, as part of an experimental sea-search radar system. The test's successful results led to widespread use of the PPI indicator during World War II.

Electromagnetically deflected cathode-ray tubes are well-suited for PPI displays. Wartime displays used mechanical rotation of the deflection yoke about the tube neck in synchronism with the rotation of the radar antenna in azimuth. By the early 1960s, mechanical yoke rotation had been replaced by sine/cosine vector generators to produce the rotating scan line.[29] Radar CRTs are generally similar to television picture tubes except for the long persistence phosphor and for airborne applications they may be operated at lower accelerating voltages to minimize external high-voltage breakdown problems.

The only commercially available electromagnetic deflection CRTs available in 1940 in the United States were the 7AP4, 9AP4 and 12AP4 television picture tubes which were ill-suited to radar requirements. The primary limitation was that the phosphor had too short persistence to be used in a display that required many seconds to "paint" a complete scan of the antenna. This problem was the focus of attention for many years as attempts were made to hold the displayed information long enough to fully interpret it, especially in bright ambient light applications such as air traffic control towers and airborne radar.

By early 1942, RCA had registered the first of a family of CRTs designed expressly for radar PPI indicators. These included the widely used 5FP7 (Figure 3.10), 3HP7 (Figure 3.11), 7BP7 (Figure 3.12) and 12DP7 (Figure 3.13). All were magnetically focused and used the common "octal" eight-pin radio tube base. The first three used small ball contacts for the high-voltage anode connection while the 12DP7 used a medium cap similar to radio receiving tubes.

The major innovation of these four tubes was the P7 cascade phosphor screen initially developed in Great Britain in 1938 and 1939,[27] and later refined in the U.S. by RCA and General Electric.[18] The P7 screen provided image retention between successive sweeps of the antenna, especially in low ambient light environments. At the onset of the war, only

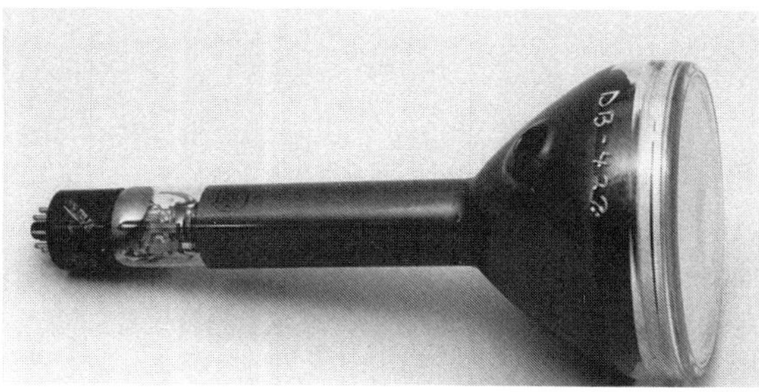

Figure 3.10 National Union 5FP7 magnetic deflection radar CRT circa 1944. (From the author's collection.)

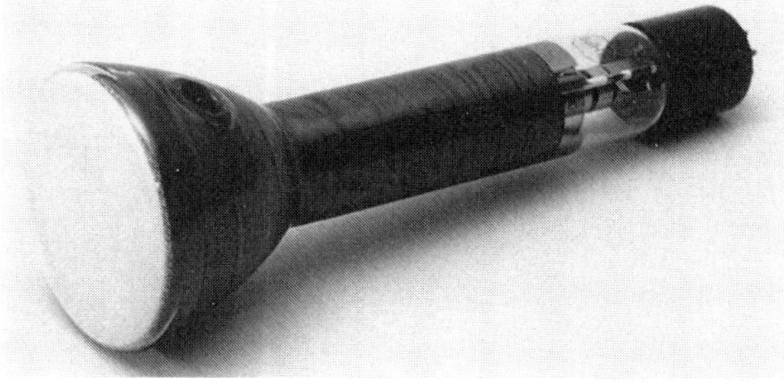

Figure 3.11 RCA 3HP7 magnetic deflection radar CRT circa 1944. (From the author's collection.)

Figure 3.12 General Electric 7BP7 magnetic deflection radar CRT circa 1944. (From the author's collection.)

Figure 3.13 RCA 12DP7-A magnetic deflection radar CRT manufactured in 1949. (From the author's collection.)

six standard phosphor screens were registered with RMA as P1 through P6. Of these, only P2 (zinc sulfide, copper-activated) had reasonably long persistence. That phosphor was imported into the U.S. from Germany in five-pound lots for theatrical special effects and limited CRT uses. Obviously, the demands of radar combined with the loss of P2 supply from Germany stimulated phosphor synthesis in the U.S. and Great Britain.[31] By the end of the war's hostilities, over 200,000 pounds of phosphor had been produced by the allies.

Cascade screens using phosphors with dual activators were described by A. Wackenhut in a German patent in 1936, but were largely unsuccessful due to low efficiency.[31] The P7 cascade screen uses the same fundamental concept, but consists of two separate phosphors for higher efficiency. P7 screens use a long-persistence yellow-green phosphor in contact with the glass faceplate and a short-persistence, near-ultraviolet emitting phosphor deposited on the back of the long-persistence layer. The electron beam excites the near-ultraviolet emission, which in turn stimulates the long-persistence layer. The long-persistence phosphor exhibits longer persistence under near-ultraviolet excitation than under direct electron beam excitation, hence longer persistence is obtained by the two-step excitation process.[20,26,28,30] The similar British P9 and long-persistence single layer white P8 screens were registered with RMA during the war by General Electric Co., Ltd. for Canadian use.[31]

As with the electrostatic A-scope tubes, contracts to supply large quantities of the new radar CRTs were issued to RCA, Du Mont, Sylvania, General Electric, National Union, Norelco and Research Enterprises, Ltd. In 1943 and 1944, several nine-inch tubes were registered by General Electric and Research Enterprises, Ltd. of Toronto, Ontario. They included the 9GP7, 9HP7, 9LP7 and 9MP7, none of which was widely used. "A" suffix versions of the earlier 5FP7, 7BP7 and 12DP7 using improved electron guns for better focus became available in the mid-1940s.

Following World War II, a series of improved radar CRTs were announced using the new television round glass envelopes, 12-pin duodecal bases and cavity anode connectors. These included the GE 10KP7 in 1947 (Figure 3.14), the Sylvania 12SP7 in 1949, the RCA 7MP7 in 1950 and the Philco/Lansdale 7SP7 and 10QP7 in 1951. Also announced in 1951 by RCA was the metal-cone 16ADP7 (Figure 3.15). All of these radar CRTs were similar except for screen size and the use of the improved focus electron guns.

The next stage in the evolution of the PPI radar CRT occurred in the mid-1950s with a series of glass CRTs having low-voltage electrostatic focus guns similar to those being introduced within the television industry at that time. The tubes included the Du Mont 5AHP7, 7ABP7 (Figure 3.16), 10WP7 and 12ABP7 and the Westinghouse 16AKP7. Also drawing upon the recent television developments, these tubes were available with aluminized screens for improved brightness. Until about 1951, radar cathode-ray tubes were little more than standard television picture tube designs with P7 phosphor, smaller spot size and tested to more stringent MIL standards for military use.

After 1951, radar tubes began to take on more of an identity of their own, although they still resembled the then-obsolete round screen monochrome television picture tube in many respects. A unique radar indicator from

Figure 3.14 RCA 10KP7 magnetic deflection radar CRT manufactured in 1954. (From the author's collection.)

1952 was the Philco 4CP7 (Figure 3.17). This four-inch, flatface, round screen tube was designed for operation at up to 20 kilovolts for high brightness and had an integral molded glass corona shield around the anode button as was relatively common in European CRTs. Although applications were not reported in any literature the author has found, this tube appears to have been particularly well-suited to airborne heads-up displays, radar displays

Figure 3.15 Raytheon 16ADP7 metal cone radar CRT. (From the author's collection.)

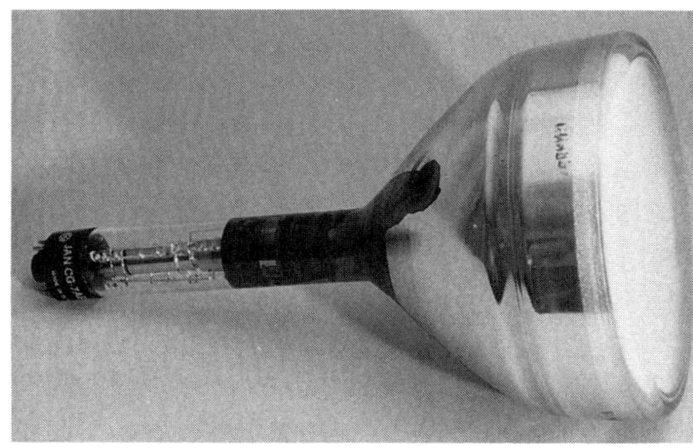

Figure 3.16 General Electric 7ABP7-A magnetic deflection radar CRT manufactured in 1962. (From the author's collection.)

for use in bright ambient light, or for projection radar displays.

Meanwhile, a wider range of registered phosphors had been developed and several of these were made available in this latest series of tubes. P14 and the very long-persistence, but inefficient and fragile, P12, P19 and P25 phosphors were among the choices suitable

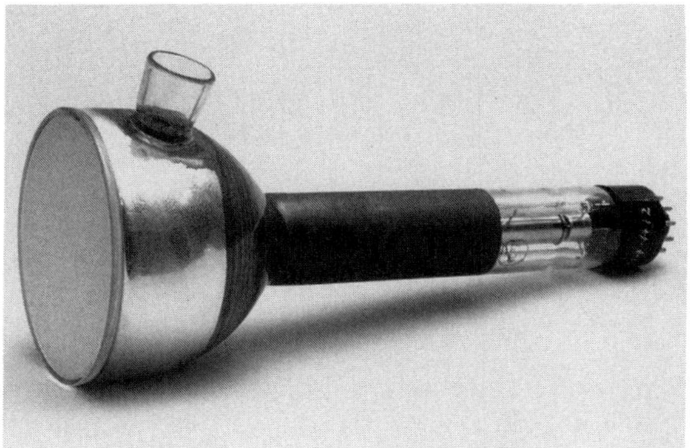

Figure 3.17 Philco 4CP7 radar cathode-ray tube manufactured in 1953. (Courtesy of U.S. Army LABCOM, E T and D Laboratory, Ft. Monmouth, N.J.)

Figure 3.18 Du Mont 22-inch metal cone CRT for radar consoles. (From the collection of James Richardson.)

for radar indicator CRTs. From the 1950s onward, most industrial and military CRTs could be supplied in almost any of the standard registered phosphors on special order.

Another spin-off from monochrome television picture tube technology was the 12-inch Rauland 12AFP- and 12AGP-; the 16-inch RCA 16ADP7; the 22-inch Rauland 22CP- and 22DP-; and Du Mont 22GP-series of metal cone indicator tubes of the 1951 to 1956 period (Figure 3.18). The Rauland tubes were available with P7, P14, P19 and, later, P25 phosphors having long persistence. While using the standard television metal cone envelope, special near-flat glass faceplates made the tubes useful for air traffic control radar consoles. All but the 12AGP-, 22DP- and 22GP- were magnetically focused. "A" suffix versions of the 12-inch Rauland tubes were produced with aluminized screens for brighter displays, while the 22GP- was aluminized in the standard version.

Round screen, electrostatic-focus PPI display CRTs similar to the 16AKP7 and 22DP7 continue in use today as part of the various systems installed in the 1970s, including the ASDE (Airport Surface Detection Equipment) and ARTS (Automated Radar Tracking System). The current tubes are not EIA-registered types and instead go by manufacturer's "house numbers" such as the 16-inch Raytheon CK-1352 and the 22-inch Thomas Electronics 22M30. The primary differences between these and the earlier tubes are the permanently attached high voltage anode leads and contrast enhancement filters laminated to the glass faceplates. Terrain, minimum altitudes and airway overlays are superimposed on these displays by use of a flying-spot scanner cathode-ray tube (see Chapter 9) and a series of selectable transparencies. The overlay video output is combined with the radar video signals for simultaneous viewing. At the same time, alphanumeric characters are added to the display by the computer during the radar receiver's inactive time to display aircraft identification, altitude and controller responsibility.[29]

3.6 SECTOR-SCAN DISPLAYS

Sector-scan indicators are similar to PPI displays except that only a pie-shaped segment of the 360-degree scan is displayed. Most CRTs used in this mode are identical to tubes used for PPI indicators. Phosphor persistence requirements are eased somewhat since it requires less time to complete the partial scan than a full sweep. Airborne weather radar is an example of sector scanning. Some tubes with offset necks, such as the five-inch Du Mont 5EAP7 and 5EBP7 and the house-numbered Sylvania SC-3180 and SC-3179, were produced specifically for sector scanning in the early 1960s.

During the period of 1944 to 1960, a variety of specialized CRTs were developed to overcome the problems of image retention during the time between successive scans of the antenna, and to make radar displays easier to interpret rapidly. Some of these CRTs' descriptions follow.

3.7 SKIATRON

The one remaining electromagnetically deflected radar CRT to be produced in the United States during World War II was the RCA 4AP10 Skiatron dark-trace storage tube registered by RCA in 1944 (Figure 3.19). A similar tube, the VCR-520, was produced in Great Britain around the same time.[27] Instead of the usual phosphor screen, the 4AP10 used a screen of potassium chloride which is a material known as a *scotophor* or *cathodochromic*.

Figure 3.19 General Electric 4AP10 Skiatron dark trace storage CRT circa 1944. (From the author's collection.)

The scotophor exhibits a white screen color until struck by the electron beam. The bombarded area turns to a magenta color under electron excitation and remains so indefinitely until infrared energy is applied to erase the image.[30,31] The tube was designed with a spherical faceplate for use in the optical system of projection radar displays. Drawbacks included relatively long infrared erasure times (greater than one second), low contrast and low sensitivity.

National Union developed the type NU-2112 after the war. The NU-2112 featured potassium chloride deposited on a thin mica sheet suspended behind the glass faceplate with an erase filament behind the mica sheet. The lower heat capacity of the suspended screen permitted faster heating, hence the time for erasure was reduced to five to eight seconds. Du Mont used a similar approach in which the mica sheet was replaced with a fine metal mesh. This was constructed using their multiband accelerator and electrostatic deflection oscilloscope CRT. It was designated type 5RP10. This design suffered from self-erasure due to ambient temperature and flaking of the potassium chloride under handling and vibration. The Philco Corporation experimented with a more conventionally screened tube where the scotophor was deposited on the glass faceplate, which had an electrically conducting, transparent, thin film layer between the glass and potassium chloride. A very high current derived from a motor-generator, which could supply a burst of power through the conductive layer, provided erasure. The low-mass scotophor layer was heated rapidly while the much higher mass of the faceplate remained relatively unaffected. A. Rosenthal and Dollman of Freed Corporation succeeded in adding material to the potassium chloride or other scotophors to permit natural ambient heat to cause the dark-trace image to slowly fade away. All of these tubes were developed prior to 1956.[32] In 1957, C. Lorenz AG of Germany reported development of a rectangular screen dark-trace storage tube with the suspended mica screen technique combined with an electrically conductive transparent coating on the mica.[33]

Thomas Electronics developed 10-inch diameter Skiatrons (10M55P10) in the early 1960s for U.S. Air Force weather radar displays transmitted over telephone lines in a slow-scan format. These tubes used an approach similar to that of National Union with a suspended mica substrate screen and heating element. A special green circline fluorescent backlight was used to enhance contrast of the background to the magenta stored image by illuminating the scotophor through the funnel's rear. These tubes were also tested for displaying medical X-rays in doctors' offices via telephone lines, but results were disappointing. Contrast and resolution were not high enough for the critical requirements of radiologists. Other problems associated with the tube were cost and availability of 10-inch mica sheets of optical quality.[34]

The Skiatron was one more attempt to solve the problem of retaining a radar presentation long enough to extract the desired information from it.

3.8 GRAPHECHON

RCA developed the Graphechon in 1949 as a means of storing an image for a short period of time using scan-conversion techniques.[26,35] Applications included PPI radar and oscillograms. The Graphechon basically consisted of a cathode-ray tube and an iconoscope in one envelope (as shown in Figure 3.20). The CRT electron gun was used to "write" slow patterns on the charge-sensitive, high-capacitance storage target while the "read" electron gun scanned the target at television scan rates. The video signal was derived from the target electrode connection and a continuous image could be displayed on a television monitor. Since the Graphechon is not a directly viewed CRT, it is mentioned only as an intermediate stage to the display of radar information on a conventional television picture tube.

3.9 CHARACTRON

The Charactron was originally developed in 1952 by Consolidated Vultee Aircraft Company (Convair) as a computer readout CRT.[36] It was later adapted to military and air traffic control radar systems by the Stromberg-Carlson Division of General Dynamics, the successor to Convair. The Charactron had the ability to present alphanumeric data such as flight number, speed and altitude superimposed on the radar map.

Figure 3.21 illustrates the Charactron's functioning. Two extra pairs of deflection plates are incorporated within the electron gun

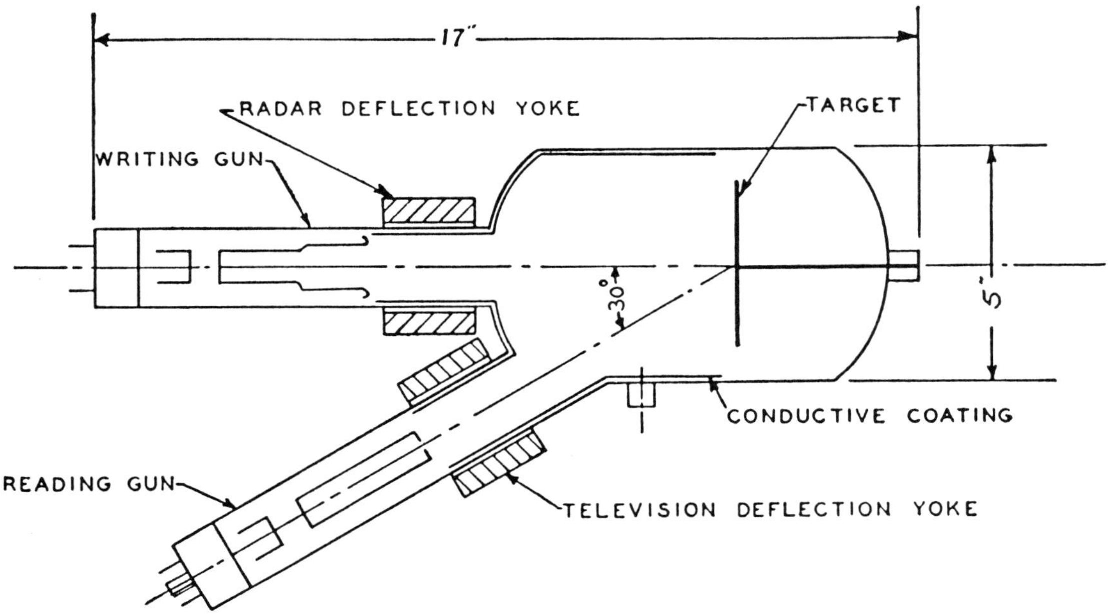

Figure 3.20 RCA Graphechon with magnetic deflection and single-sided target.[35] (Courtesy of RCA/ Thomson.)

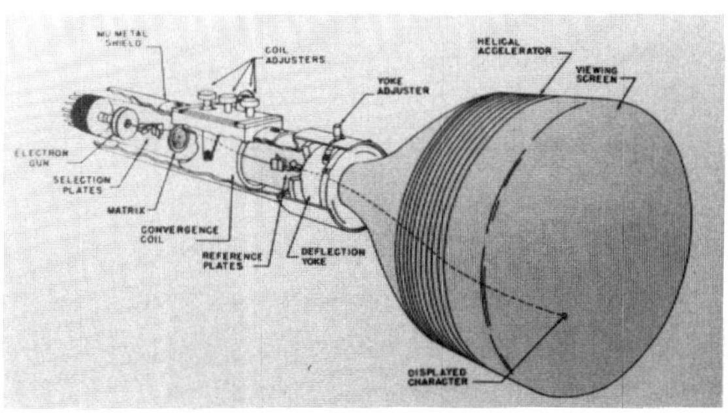

Figure 3.21 Charactron cathode-ray tube.[26]

on either side of a metal stencil or *character matrix*. The beam is offset an appropriate amount by the first set of plates to pass through the proper character and is returned to the CRT center axis by the second set of plates before conventional magnetic deflection is applied to position the character at the desired screen location.[26,36]

3.10 DIRECT-VIEW STORAGE TUBE

RCA developed some of the first storage tubes applicable to radar with the first being described in 1953 and 1954.[37,57] Recognizing the need for large-screen storage displays for air traffic control, a developmental 10-inch screen storage tube was described in 1956 along with a projection storage tube.[38]

A number of direct-view storage tubes (DVST) have been produced for radar indicators in screen sizes from three- to 12-inch diameter by Hughes Aircraft, Du Mont, RCA and IT&T. Most have had "house numbers" although a few were registered under the EIA four-digit special purpose tube numbering system. The RCA 6866 (Figure 3.22) and Du

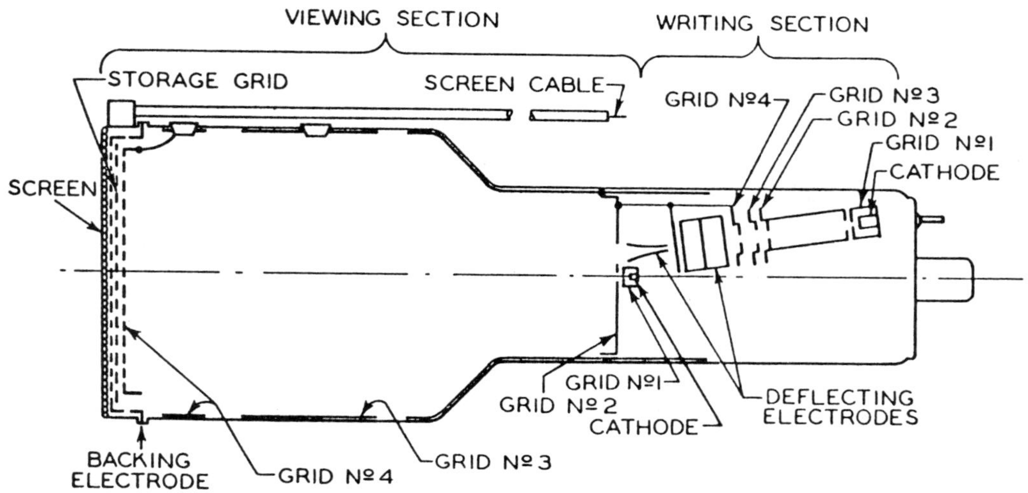

Figure 3.22 Direct-view storage tube.[26]

Mont 7268-B, 7423 and 8931 are examples of standardized types. Storage tubes offered bright images for longer periods than could be accomplished by long-persistence phosphors. The trade-offs were lower resolution, higher cost and more circuit complexity (and adjustments). DVSTs were produced primarily from the mid-1950s to the mid-1970s, although a few are manufactured today for the replacement market. They have been largely displaced by digital computer memory, which has become lower in cost and has greater flexibility for data overlays and the ability to selectively update portions of the display without erasing the entire screen and rewriting the previous information.

3.11 TONOTRON

The Tonotron developed by Hughes Aircraft in about 1956 was a direct-view storage cathode-ray tube capable of storing and displaying bright halftone images. The Tonotron (Figure 3.23) was used primarily for airborne radar applications and will be discussed further in Chapter 8.

3.12 PORTHOLES

Just like the post-war Buick, radar cathode-ray tubes went through a stage of development where portholes were in style. In the case of the CRT, rear ports were used for two distinct purposes. The first was to allow photography of the information on the screen without blocking the operators view of the screen. The second purpose was to optically project information, such as map overlays.

The first ported cathode-ray tube available was the 5VP7 registered in 1948 by the Radio Valve Company of Canada, Ltd. The 5VP7 was similar to the electrostatic 5BP1/5GP1/5HP1 CRT family except for using P7 phosphor and featuring a clear region on the funnel's rear with good quality glass where no conductive dag coating was applied for the anode. A camera could be positioned to view

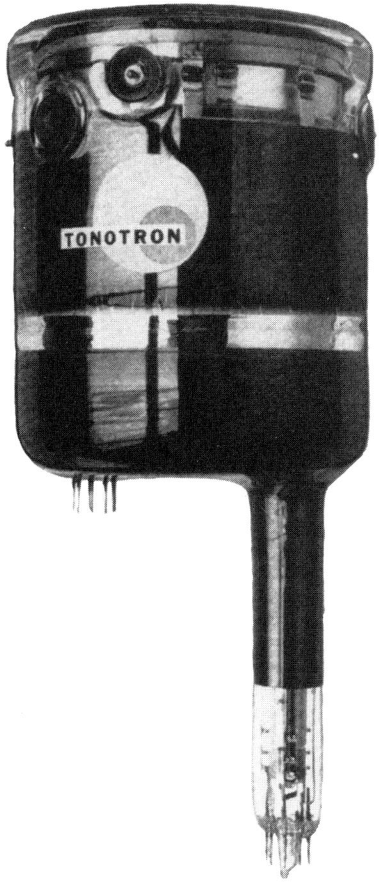

Figure 3.23 Hughes Tonotron storage CRT. (Courtesy of Hughes Aircraft Company.)

the rear of the phosphor screen through this port.

A number of ported tubes became available in the late 1950s, especially those offered by Sylvania (as typified by their SC-3821). Raytheon, Westinghouse and Thomas Electronics also manufactured tubes with optical ports. These tubes had optical glass windows sealed into the rear of the CRT envelope which was shaped specifically for that purpose (Figure 3.24). The port's angle usually allowed the projection of overlays with the optical path perpendicular to the window surfaces. As far as is known, all tubes of this type were not registered with EIA but instead bore house-numbers which were unique to the manufacturer. Since manufacturing quantities were fairly small, interchangeability with other manufacturers' CRTs and second sourcing was relatively unimportant. Ported tubes became less necessary as flying-spot scanned overlays, scan conversion and digital storage techniques became available.

3.13 WAMOSCOPE

The Wamoscope (wave-modulated oscilloscope), developed for the U.S. Navy after World War II by Sylvania and declassified in 1956, combined a traveling-wave amplifier tube into the neck of a radar indicator CRT[39,40,58,59] as in Figure 3.25. The result was comparable to a microwave radar receiver

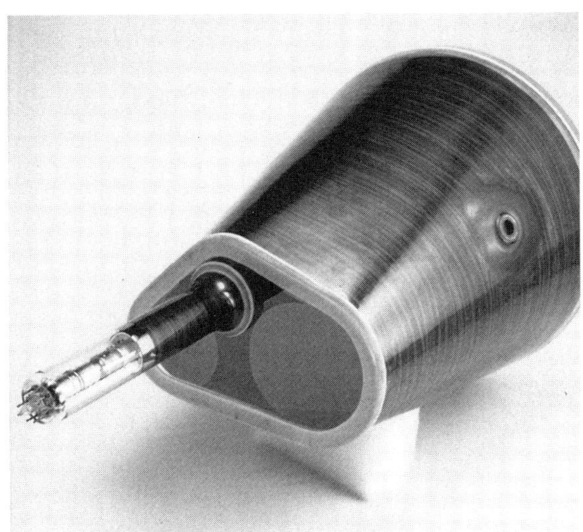

Figure 3.24 Thomas Electronics rear-ported CRT. (Courtesy of Thomas Electronics.)

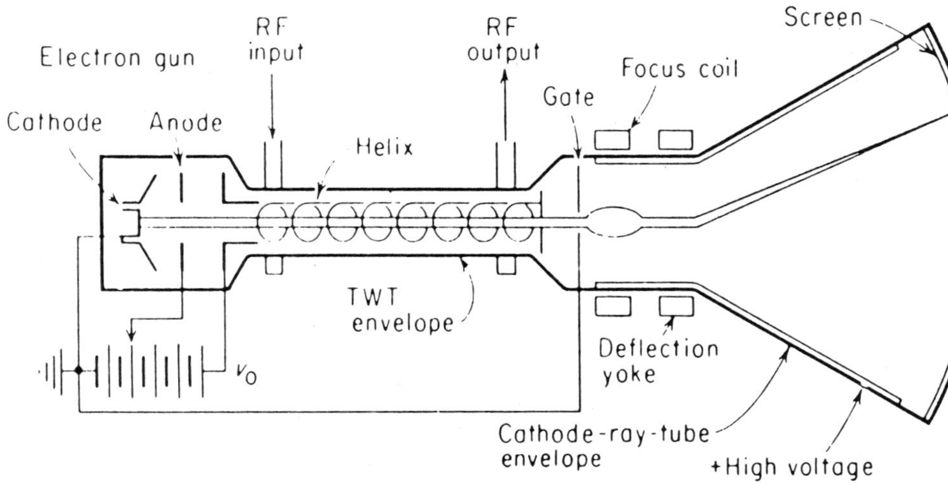

Figure 3.25 Sylvania Wamoscope traveling wave amplifier cathode-ray tube, circa 1956. (Courtesy of Philips Display Components Company.)

within one envelope. The Wamoscope is more of an interesting curiosity than a major innovation since it has not greatly influenced the CRT's evolution. One other distantly related CRT was the Westinghouse 10AMP7 registered in 1960. The 10AMP7 was a 10WP7 with a high transconductance triode amplifier built into the electron gun to reduce the usual 60 volt drive signal to be reduced to 3.5 to 6 volts while maintaining wide Z axis bandwidth.

3.14 PENETRON

L. R. Koller of General Electric described the voltage-sensitive penetration phosphor screen and suggested its use for radar in a patent application filed in 1950.[41] Sylvania developed one of the first commercial Penetron CRTs, the SC-4876, in the late 1960s as a means to display relatively high-resolution color images for improved readability.[42] Other Penetron CRTs have been produced domestically by Du Mont, Thomson CSF and Thomas Electronics, and abroad by Brimar in England.

The Penetron requires only one electron gun with the specially constructed screen. The early screens consisted of two different color phosphor layers separated by a thin dielectric barrier layer (Figure 3.26). At low accelerating voltages, the electron beam was totally absorbed by the phosphor layer closest to the gun, usually red, and only that color was displayed. At a higher accelerating voltage, the beam penetrated the first phosphor layer into the second, usually green, which was brighter due to being on the screen's viewing side. Intermediate accelerating voltages produced a mixture of red and green which is yellow or orange.[43,44]

Later penetration screens were made of "onion-skin" phosphors.[45] These screens consisted of a green phosphor core coated with a barrier layer and finally a surface coating

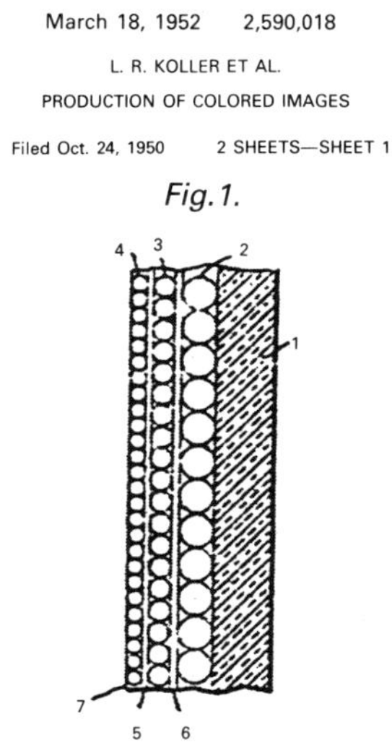

Figure 3.26 Construction of penetration phosphor screen.[41]

of red phosphor. The same principles of operation apply. Screening is simplified since only one settling operation is required and an improved color range is achieved. Mixed phosphors also have been used with somewhat lower brightness.[46]

Other color combinations may be employed, although the range of colors that may be reproduced will be further limited. Similarly, two phosphors having different persistence characteristics may be used.

The accelerating voltage must be changed rapidly in order to produce the proper color at the desired screen location. At the same time, the control grid voltage is compensated to balance the brightness between red and green and the deflection amplifier gains are adjusted to keep the identical pattern size, despite the alteration in beam stiffness with change in acceleration voltage. All of this action must occur in as little as 100 microseconds for higher performance displays.

The Penetron's color gamut is limited by only having two color primaries compared to the three of the shadow mask television picture tube. This is offset by the lack of the shadow mask screen structure, which allows much higher resolution limited only by the beam size and to a lesser degree, the phosphor screen.

Further development work on the penetration color CRT has resulted in evaporated transparent phosphor screens announced by Sperry Flight Systems[47] and a dual-gun, dual-necked CRT described by the Rauland Division of Zenith Radio.[48,49] The latter operated with different accelerating voltages on the two guns so that each produced one color, thus eliminating the need for high voltage switching.

3.15 RASTER-SCAN DISPLAYS

Through two distinct approaches scan-conversion techniques solved the phosphor persistence problem in the late 1960s. In the first method the slow radar sweeps were written on the storage target of a scan conversion tube and read out at normal television scan rates to present a continuous display of the radar information. The scan converter tube consisted of a CRT and camera tube in one envelope. The second method consisted of a five-inch CRT similar to a flying-spot scanner type with the PPI scan display. A television camera with a Vidicon pickup tube having a long lag time viewed the CRT using conventional raster scan, thus converting the PPI display to a television-type display. The CRT requirements were greatly eased since even a conventional black and white television picture tube was suitable in some instances. Good brightness and low cost were features of the CRT indicators. In addition, as many television monitors as desired could be fed simultaneously via coaxial cable, or even used remotely through a transmitting/receiving system. This technique was used in the FAA BRITE (Bright Radar Indicator Tower Equipment) indicators for radar and ground surveillance. Sixteen-inch diameter all-glass CRTs, such as the type CK-1352-P31 manufactured by Raytheon and equivalents by Du Mont and others, are used in the BRITE application.[29]

The 1970s witnessed the change to digital computers for storage of radar information with digital conversion to television scan rates for greater information handling capabilities. All sorts of useful information such as altitude, ground speed, transponder codes, etc., were easily displayed as overlays, and the system could compute collision avoidance information. Standard radar and television-type CRTs were still used at this stage.

Higher resolution displays using faster scan rates than commercial television and color displays were developed during the 1980s. Square CRT faceplates, electron guns having smaller spot size and fine-pitch color shadow mask CRTs all became practical. The developments in computer data display CRTs and monitors were applied to radar systems as well.[50] These developments are expected to become operational in the 1990s. Although the PPI display cathode-ray tubes in use today are relatively unchanged from those of World War II, the next generation of radar indicators will use highly advanced CRT technology for performance gains.[51]

One of the most talked about displays, of which the Sony 20 × 20-inch square color Trinitron CRT is the key element, appears destined for the next generation of air traffic control systems. A full-color display of 2048 × 2048 pixels with a relatively flat face CRT is expected to replace the present 22-inch diameter round screen monochrome displays in ATC consoles.[52,53]

4

Oscilloscope Cathode-Ray Tubes

4.1 Background

In 1940, the oscilloscope was just emerging as a tool for radio and electronics research, design and servicing. The development of television and radar hastened its acceptance as an essential piece of test equipment in the 1940s. Until then, only a few commercially available oscilloscopes were available and their ability to make quantitative measurements was limited by a general lack of calibration for either the vertical amplifiers or the time-base.

Major oscilloscope manufacturers up to 1940 included Du Mont, RCA, General Radio, Clough-Brengle, A. C. Cossor Ltd., General Electric, Hickok, Triumph, Triplett, National and Supreme. Most oscilloscope manufacturers used the RCA one-, two-, three-, five- or nine-inch CRTs except Du Mont who manufactured their own series of similar size tubes used in their expanding line of oscilloscopes. Other manufacturers of oscilloscope CRTs up to 1940 included A. C. Cossor Ltd., Ediswan General Electric, Hygrade Sylvania, National Union, Leybold & von Ardenne, Western Electric and Westinghouse.[1]

The gas-focused CRTs of the 1920s and early 1930s had been displaced by high-vacuum types with their longer life and better performance by 1940.[2–4] Radar provided the primary impetus for the improvement of the oscilloscope and associated CRTs during World War II. In the early post-war years, television became the driving force in CRT developments and the aerospace industry which was just getting off the ground demanded constantly higher performance oscilloscopes. Oscilloscope performance is closely tied to that of the CRT, which went from the comparatively simple device of the 1940s to the most complex types manufactured. Electron guns sprouted a number of additional electrodes to provide adjustment for optimum operation, manufacturing tolerances were trimmed and clean rooms for processing CRT parts and assemblies became the norm. This chapter traces the major developments leading to today's precision oscilloscope cathode-ray tube.

Note that in many early reference books and papers, the term *oscillograph* is used in place of the currently used *oscilloscope*. The term oscilloscope was first proposed by Bedell and Reich in 1927 with the statement,

"As the instrument developed for this purpose is primarily intended for visual observation, we have given it the name '*oscilloscope*'" [author's emphasis].[5] Du Mont Labs continued to use *oscillograph* for their instruments until 1960. Oscillograph has come to be used more correctly to designate a recording instrument that produces a permanent paper graph.

4.2 CIRCA 1940

Gas-focused cathode-ray tubes such as the five-inch Western Electric 224-A developed in 1921, the five-inch General Radio 478-A* and the Du Mont 22, 34, 54 and 94 (the first digit denotes the screen size in inches) introduced in the early 1930s had all been swiftly superseded by the high-vacuum CRT by 1940. The first high-vacuum types for oscillography were the five-inch General Electric FP-53 and Westinghouse Electric RC-593 (both of 1932).[6–8] High-vacuum CRTs were such an improvement that they even provided the basis for a Tom Swift book in 1933, although the capabilities of the high-vacuum CRT may have been exaggerated just a little.[9] Several high-vacuum CRTs followed with the majority being manufactured by RCA and Du Mont in the United States. Some of these tubes will be discussed briefly to illustrate the state of development at the beginning of World War II. Much of the following data is from JEDEC Registration Files[10] and manufacturers' data sheets. It is interesting to note that a number of tubes from the 1940s were copied by the Russians and were still listed as recently as 1975 in the Electronorgtechnica catalog.

*Actually manufactured by Leybold & von Ardenne in Germany.

Western Electric 325 and 326–These tubes were introduced in 1937 with five- and seven-inch screens, respectively. An unusual collar on the neck provided short lead lengths to the deflection plates for good high-frequency characteristics (Figure 4.1).

RCA 902–The two-inch 902 was announced in 1938 (Figure 4.2). Features included operation at as low as 400 volts, P1

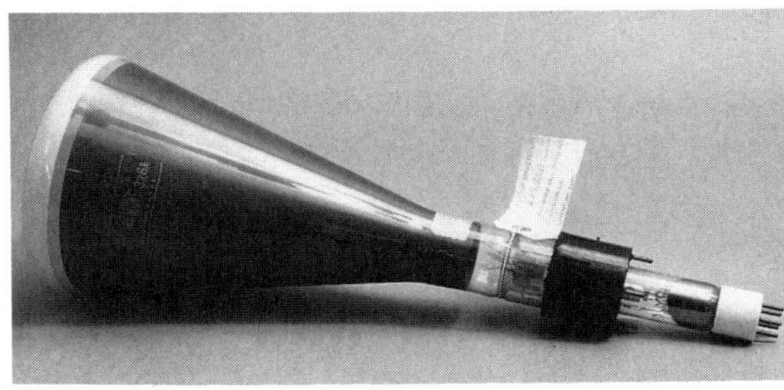

Figure 4.1 Du Mont 326A circa early 1940s identical replacement for the Western Electric 326A. (Allen B. Du Mont Collection, National Museum of American History, Smithsonian Institution.)

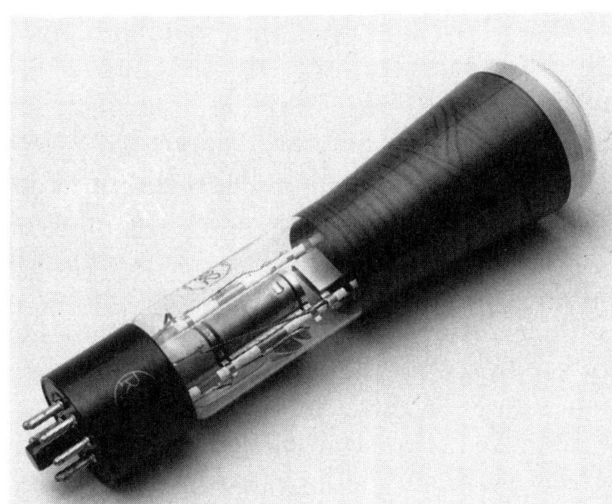

Figure 4.2 RCA 902 two-inch CRT manufactured in 1943. (From the author's collection.)

phosphor and an octal base with one plate of each pair of deflection plates tied to the second anode to minimize the number of base pin connections.[11] The 902 was first used in the RCA Model 151-2 oscilloscope, also introduced in 1938. National Union manufactured 902s for the U.S. Navy during World War II under the designation 902-P1.

RCA 905–The five-inch 905 dates back to circa 1934 and featured deflection plate connections brought out through the neck for good high frequency response, a radio tube five-pin base and P1 phosphor for general purpose use (Figure 4.3). Variations included the type 907 which was identical except for a short-persistence blue phosphor (P5) for photography of oscilloscope waveforms and the type 909 which had a bluish-white long-persistence phosphor. The 905 series of CRTs had no conductive anode coating.

RCA 906–The 906 developed circa 1933[4] was probably the most widely used oscilloscope CRT of the 1930s (Figure 4.4). Its three-inch screen was normally supplied with P1 phosphor for general oscillographic use, but was available with white phosphor for television as the type 906-P4, short-persistence blue phosphor for photography (P5) as the type 908 and a bluish-white long persistence phosphor as the type 910. The base was similar

Figure 4.3 RCA 905 five-inch CRT manufactured in 1944. (From the author's collection.)

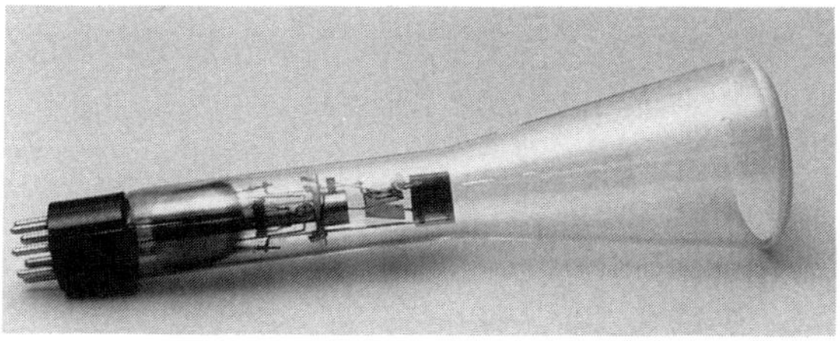

Figure 4.4 RCA 906 three-inch CRT circa 1935. (From the author's collection.)

to the standard radio tube seven-pin receiving tube base and one of each of the pairs of vertical and horizontal deflection plates was tied to the second anode to reduce base pin count. Early 906s had clear glass bulb walls with no conductive anode coating. The 906 was later registered by RCA with the RMA in 1941 as the 3AP1 and 3AP4. Many 3AP1s were manufactured during World War II and RCA continued to supply replacements at least up to 1970. Classic oscilloscopes using the 906 included the RCA models TMV-122-B and 155, United Sound Engineering CR-3, Supreme 555, National, Hickok RFO-1, Triumph Model 800 and the Clough-Brengle Model CRA.[3,12]

RCA 913–The 913 (Figure 4.5) was a one-inch, octal-based tube employing a unique metal envelope and was introduced in 1936.[13] At that time the advertised price was $5 with the amateur experimenter in mind. The 913 was used in the RCA 151 oscilloscope introduced the same year, as well as the National, Triumph Model 77 and the Clough-Brengle Model 105. The metal envelope appeared similar to that of the metal envelope 6L6 beam power amplifier tube used in many radios and audio amplifiers of the period. A glass screen was sealed into the envelope's end in a manner that was a forerunner of that used for the

Figure 4.5 RCA 913 one-inch metal envelope CRT circa 1937. (From the author's collection.)

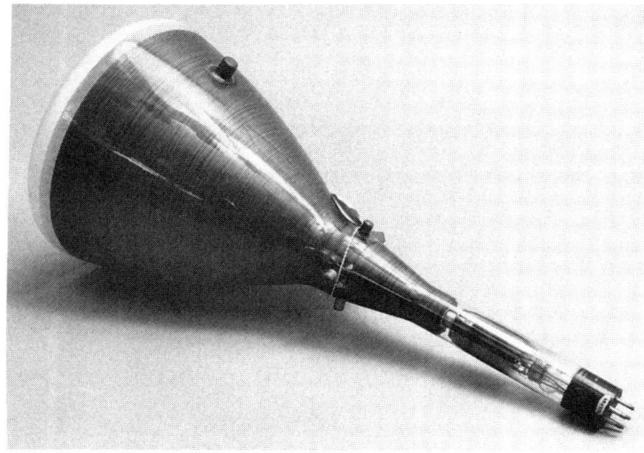

Figure 4.6 RCA 914 nine-inch CRT manufactured in 1943. (Courtesy of National Museum of American History, Smithsonian Institution.)

large-screen, metal-cone television picture tube introduced in 1949. Other features included direct electrical interchangeability with the two-inch RCA type 902, operation at anode voltage as low as 250 volts for high deflection sensitivity (but poor brightness) and shielding against external magnetic fields by the metal envelope. P1 phosphor was the only phosphor available, although the U.S. Navy began RMA registration proceedings for a version with P5 phosphor under the designation 1AP5 during the 1940s. Subsequently they withdrew that request.

RCA 914–The type 914 (Figure 4.6) was a nine-inch CRT having deflection plate connections placed through the neck to reduce capacitance for viewing high-frequency television waveforms. The second anode could be operated at up to 7,000 volts for displaying fast transient waveshapes and a cap-type connector was provided for the high voltage connection near the tube's front. P1 phosphor was supplied and a radio tube six-pin base provided connection to the other gun electrodes. The 914 was used in the RCA Model TMV-136B[14] oscilloscope and Model 304 in 1937, and the Model 327 in the early 1940s.

RCA 1802-P1–The five-inch type 1802-P1 (Figure 4.7) of 1938 was another widely used oscilloscope CRT beginning with the RCA models 158 and 160-B. The 1802-P1 was reregistered with RMA in 1939 as the type 5BP1 and saw wide usage during World War II in oscilloscopes and radar. All connections to the tube were made through a newly designed 11-pin "magnal" base specifically for CRT use. Although P1 phosphor was standard, P4 white phosphor also was available for television applications and was used for early radar indicators. National Union and Hygrade Sylvania also manufactured the 1802.[12]

National Union 1805-P1–The 1805-P1 (Figure 4.8) was a shortened version of the 1802-P1 which was more suitable for television than general oscillographic use. It too

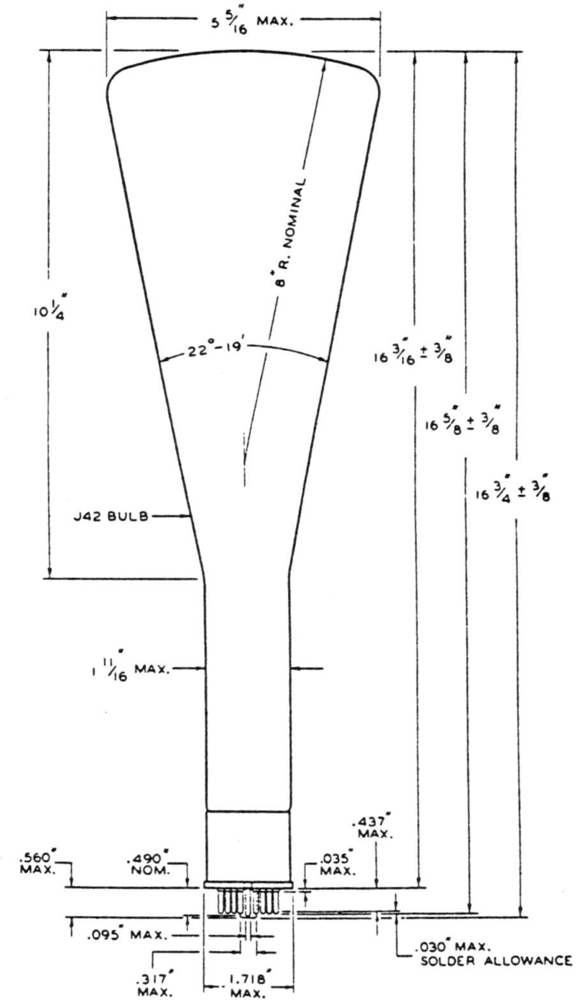

Figure 4.7 RCA 1802-P1 five-inch CRT of 1939. (Courtesy of Electronic Industries Association.)

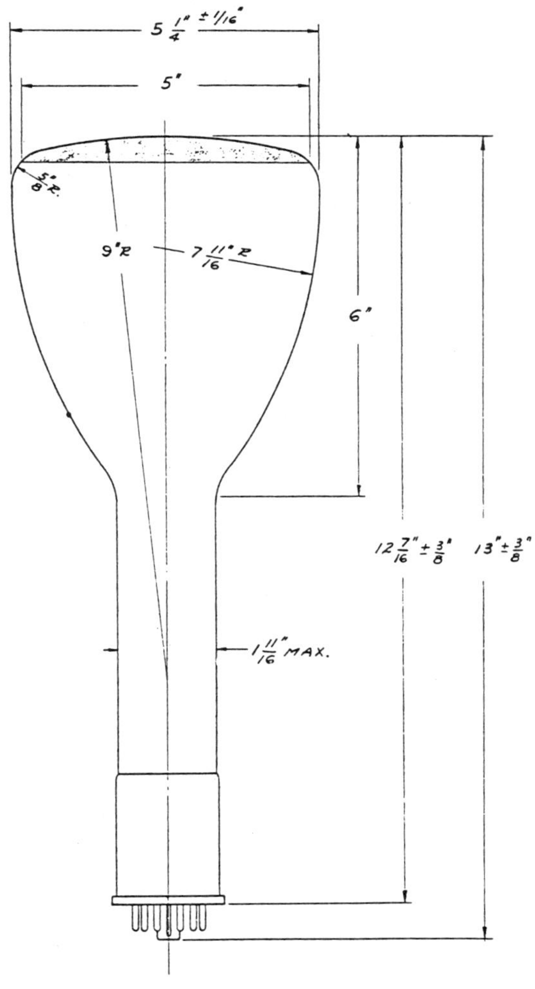

Figure 4.8 National Union 1805-P4 five-inch CRT of 1939. (Courtesy of Electronic Industries Association.)

was available with P4 phosphor as the 1805-P4 which was used in some pre-war television receivers. It was reregistered with RMA in 1939 as the 5AP1 and 5AP4.

National Union 2001–Available data indicates that the 2001 was essentially identical to the RCA 913. The first reference to the 2001 known to the author was in 1937.[1]

National Union 2002–The 2002 appears to be the National Union equivalent to the two-inch RCA 902 and was first listed in an article from 1937.[1,12]

National Union 2003–Equivalent to the RCA 906.[1]

National Union 2005–The 2005 probably was similar to the Du Mont Type 54-XH. The 2005 had P1 phosphor, a 2.5 volt, 2.1 amp heater and was 16-3/4 inches long. It was listed in 1939[12] and what probably is a picture of it is shown in the lower right hand corner of page 5 of John F. Rider's 1935 book, *The Cathode-Ray Tube at Work.*[3] In any event, the photo is erroneously identified as the three-inch type 908.

Du Mont 24-XH–In June 1937, Allen B. Du Mont Laboratories announced the two-inch type 24-XH (Figure 4.9) with a price of $7.50.[15] It is similar to the RCA 902 with the exception of not having a dag-coated envelope. A 6.3 volt heater, P1 phosphor and a large eight-pin octal base were employed.[1]

Du Mont 34-XH–Except for slightly different deflection sensitivity, the 34-XH (Figure 4.10) was similar to the RCA 906.[1] The 34-XH dates back to 1936 and was used in the Du Mont Model 154 oscilloscope. It was a high-vacuum version of the earlier gas-focused type 34. A shorter version using an octal eight-pin base was also marked type 34-XH (Figure 4.11) although this could have been an experimental tube using a reclaimed bulb.

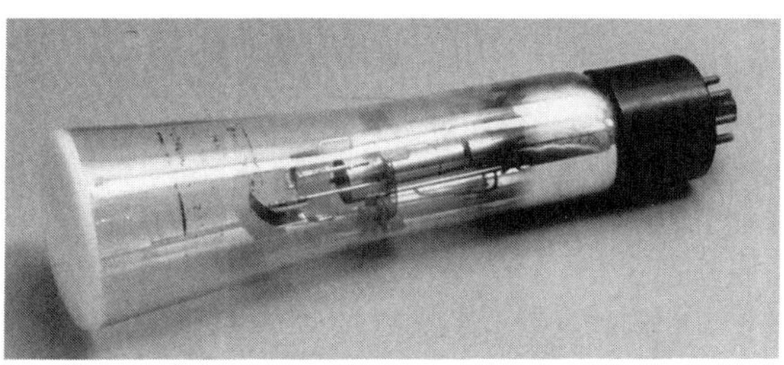

Figure 4.9 Du Mont 24-XH two-inch CRT similar to the RCA 902. (Allen B. Du Mont Collection, National Museum of American History, Smithsonian Institution.)

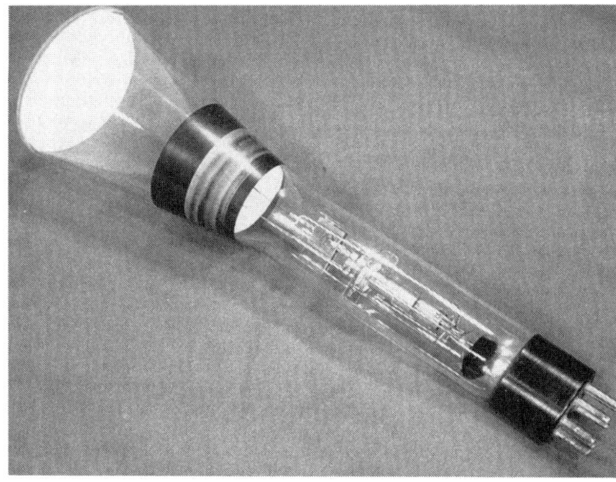

Figure 4.10 Du Mont 34-XH three-inch CRT (circa 1935) similar to RCA 906. (Courtesy of Jake Brain.)

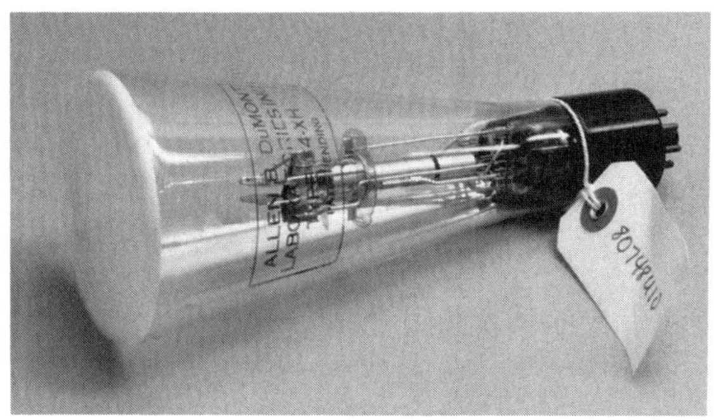

Figure 4.11 Du Mont Type 34-XH, short version. (Allen B. Du Mont Collection, National Museum of American History, Smithsonian Institution.)

Du Mont 54-XH and 54-8H–The five-inch 54-XH (Figure 4.12) used an electron gun and basing similar to the 34-XH.[1] The overall length was 15-1/2 inches. There does not seem to be an equivalent type by another manufacturer, although the National Union 2005 appears to be similar. The 54-XH was available with P1 or P4 phosphor. A version having separate deflection plate connections through the neck was known as the type 54-8H and was available with short-, medium- or long-persistence phosphors. The 54-8H was first used in the Du Mont Model 142 oscilloscope in 1934. The 54-XH was used in the Model 148 in 1935 and the Model 168 in 1937.

Du Mont 94-XH (Figure 4.13) **and 94-8H**–These were the nine-inch versions of the type 54-XH and 54-8H. Both the 94-XH and 94-8H were available with short-, medium- and long-persistence phosphors. The Du Mont Model 158 oscilloscope introduced in 1936 used the 94-8H.

Other Tubes–A series of electrostatically deflected CRTs ranging in size from three to 20 inches in diameter were introduced in early 1942 by Du Mont. All were in the series of 2501 to 2537 with a suffix such as A5. The suffix served to indicate the phosphor type and the screen diameter in inches. The phosphors were: A = P1, B = P2, C = P5 and D = P4. Several of the tubes in this series were soon given standard RMA type numbers.[16]

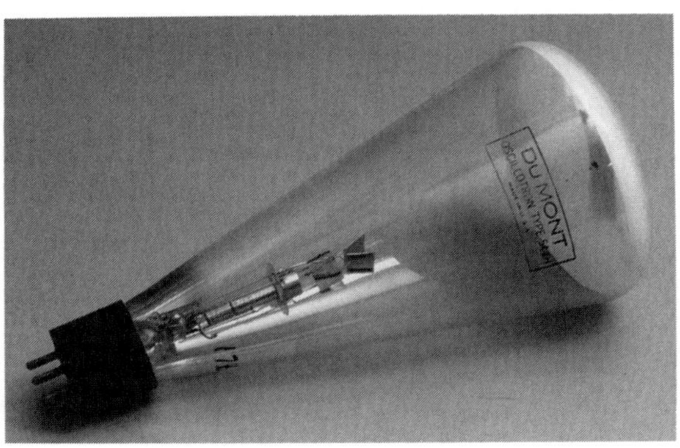

Figure 4.12 Du Mont Type 54-XH CRT, circa 1935. (Allen B. Du Mont Collection, National Museum of American History, Smithsonian Institution.)

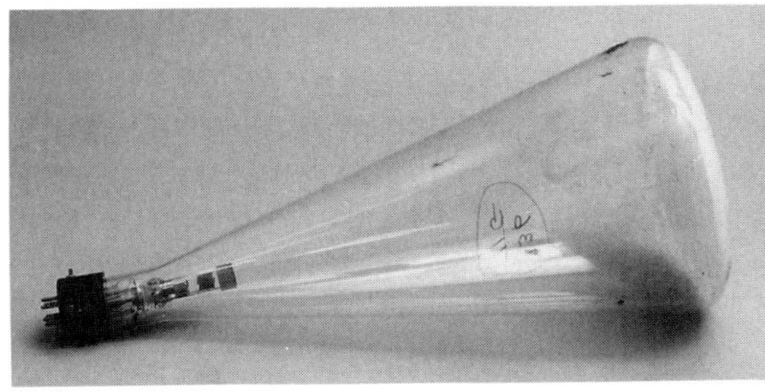

Figure 4.13 Du Mont 94-XH nine-inch CRT. (Allen B. Du Mont Collection, National Museum of American History, Smithsonian Institution.)

The aforementioned tubes are indicative of the state of electrostatic deflection CRTs at the beginning of World War II. We will now examine the subsequent evolutionary stages leading up to today's precision oscilloscope cathode-ray tubes.

4.3 WARTIME DEVELOPMENTS

Several electrostatic deflection CRT types for oscilloscope applications were developed during World War II. Most were produced in large quantities under U.S. Navy and Signal Corps contracts issued to RCA, Du Mont, National Union, North American Philips (Norelco), Sylvania Electric, General Electric and Research Enterprises Ltd. (Canada). The stringent military demands for improved specifications and environmental ruggedness at the same time that production quantities were being greatly increased led to improvements in existing CRT types and the introduction of the following new types. All wartime CRTs had round screens with relatively large curvatures and all-glass blown bulbs.

2AP1–The widely used two-inch diameter 2AP1 (Figure 4.14) developed by RCA in 1943 used the bulb from the older RCA 902 and an 11-pin magnal base. The magnal base offered two advantages: First there was wider spacing between pins which allowed higher anode voltage for improved brightness. Secondly there were separate connections for all deflection plates which permitted push-pull deflection amplifiers to be used for better focus uniformity and reduced pattern distortion over the entire screen. The 2AP1 was used primarily for portable oscilloscopes and indicator applications. Most 2AP1s were manufactured by RCA with smaller quantities supplied by General Electric and North American Philips.

3AP1/906–The RCA type 906 was registered in late 1939 as the 3AP1 (Figure 4.15). The three-inch 3AP1 was used by almost all major oscilloscope manufacturers at one time or another. Wartime tubes were supplied by RCA, Du Mont, National Union, Sylvania, General Electric and probably North American Philips and Ken-Rad. P4, P5 and P11 phosphor versions were available from Du Mont. See the type 906 for further information.

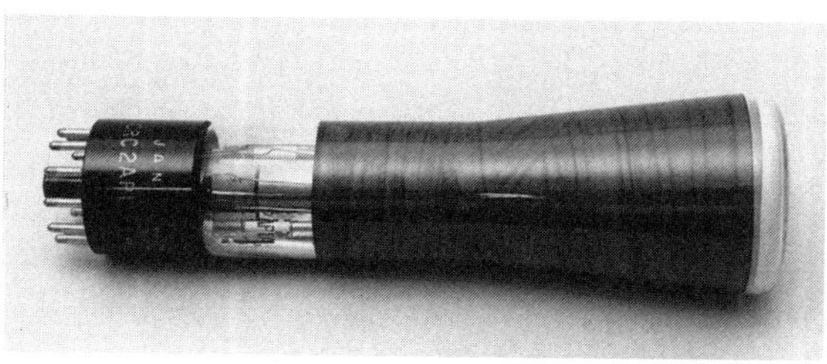

Figure 4.14 RCA 2AP1-A CRT manufactured in 1945. (From the author's collection.)

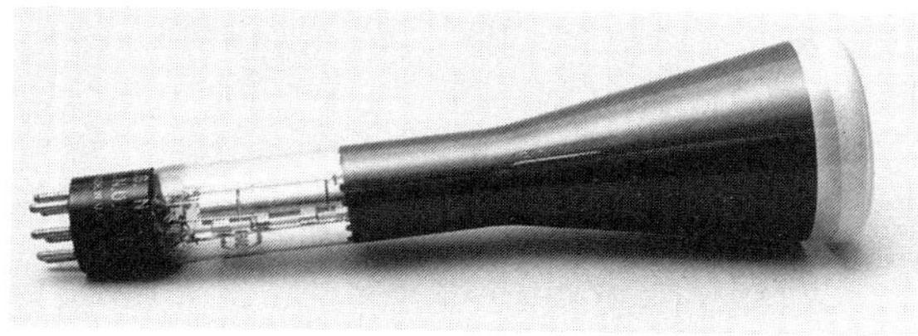

Figure 4.15 National Union 3AP1 manufactured for U.S. Navy in 1943. (From the author's collection.)

3BP1–RCA announced the three-inch 3BP1 (Figure 4.16) in January 1943 as being available for WPB (War Production Board) rated orders. The most significant feature was the large 14-pin diheptal base which was designed to permit operation at higher anode voltages and altitudes without high-voltage breakdown between pins. The 3BP1 was used for a number of oscilloscope, radar and aircraft in-flight engine analyzer applications. Wartime 3BP1s manufactured by RCA, Du

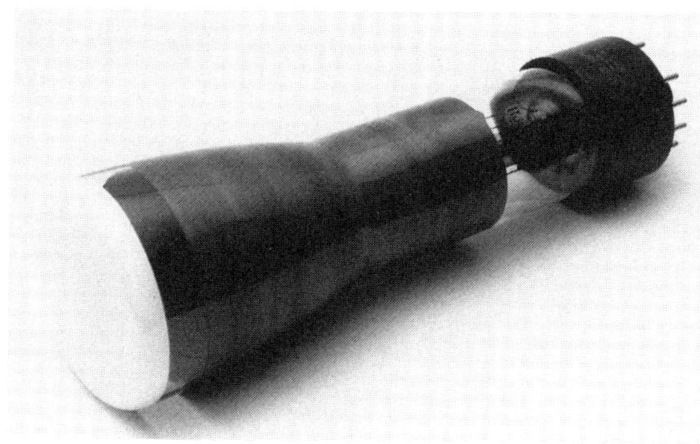

Figure 4.16 Sylvania 3BP1 manufactured in 1945. (From the author's collection.)

Mont, Norelco, National Union and Sylvania still may be found in the electronic surplus market and a 1945 advertisement by General Electric also lists it as part of their product line. The 3BP1A with its improved electron gun was still being manufactured for military replacements as recently as 1988 by Thomson-CSF (now a part of Hughes Aircraft).

3EP1/1806-P1–The 3EP1 (Figure 4.17), which was introduced in late 1942 by RCA, was essentially similar to the 3BP1 except for a smaller neck diameter and an 11-pin magnal base. Additionally, it is identical to the 3CP1/1808-P1 radial deflection CRT described in Chapter 3, except for the omission of the radial deflection electrode in the screen's center. The 3EP1 was never used widely and has only been observed with RCA markings and (in post-war years), Waterman "Rayonic" markings. GE also cataloged the 3EP1/1806-P1 in 1945.

3FP7–Although used primarily as an airborne radar indicator for fighter aircraft, the 3FP7 (Figure 4.18) was essentially an oscilloscope CRT similar to the 3BP1 except for an added post-deflection accelerator electrode consisting of a separate conductive dag band in the bulb's forward end. Connection was made to the post-accelerator through a small-ball anode connector. The post-accelerator design allowed sufficient anode voltage to be applied to make use of the long-persistence of the P7 cascade phosphor screen without too great a decrease in deflection sensitivity. The 3FP7 was made in large quantities by at least RCA, Norelco and National Union.

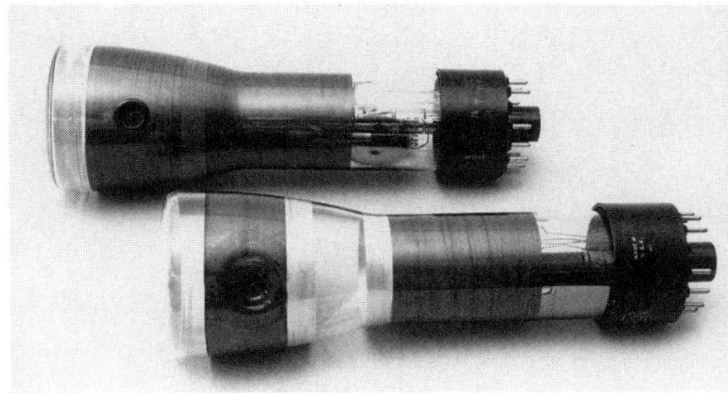

Figure 4.18 RCA 3FP7 CRTs manufactured in 1944 (front) and 1973 (rear). (From the author's collection.)

3GP1–The 3GP1 (Figure 4.19) was introduced in 1942 and Du Mont was the only supplier of the tube. An 11-pin magnal base was used for connection to the tube. The Du Mont 224 oscilloscope, also of 1942, was the first instrument to employ the 3GP1. After World War II, Du Mont used the 3GP1 as a waveform monitor CRT in their line of television studio and transmitting equipment. It

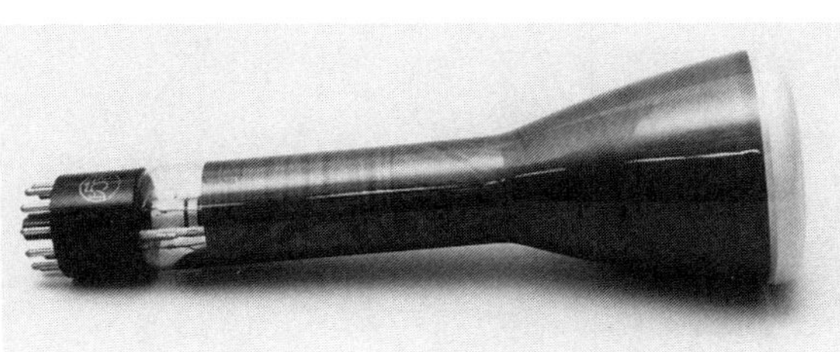

Figure 4.17 RCA 3EP1 CRT/1806-P1, circa 1943. (From the author's collection.)

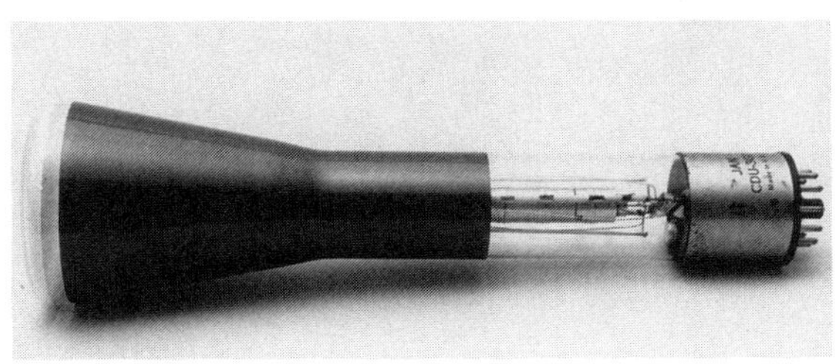

Figure 4.19 Du Mont 3GP1 manufactured in 1944. (From the author's collection.)

was cataloged with P1, P4, P5 and P11 phosphor.

5AP1/1805-P1–The five-inch diameter 5AP1 (Figure 4.20) evolved from the National Union type 1805-P1 in 1939. The 5AP1 was similar to the 5BP1 except for its short length and correspondingly lower deflection sensitivity. Never widely used, the 5AP1 did find use in the Hallicrafters Model S-35 panoramic receiver, "panadaptors" and presumably other applications where its short length facilitated use in rack-mounted equipment. National Union, Sylvania and Radio Valve Co. Ltd. of Canada were the principal producers of the 5AP1. The P4 version, the 5AP4/1805-P4, was used in some pre-war television receivers.

5BP1/1802-P1–Like the 3AP1/906, the 5BP1 (Figure 4.21) was used in many military and civilian applications. RCA registered the 5BP1 with RMA in the winter of 1939 to 1940. This five-inch tube utilized the classic funnel shape, an 11-pin magnal base and P1 phosphor. A P4 version was available as the 5BP4/1802-P4 and was used in some of the earliest radar indicators and pre-war commercial television receivers. The 5BP1 was available on the surplus market after the war in large quantities at very low prices ($2 to $5) and was used by Heathkit, Eico and others as the heart of many inexpensive kit-type oscilloscopes. The 5BP1s were manufactured by RCA, Du Mont, Ken-Rad, GE, Norelco, Sylvania and National Union. Du Mont cataloged the 5BP- with P2, P4, P5 and P11 phosphors.

5CP1–The 5CP1 (Figure 4.22) was one of the most significant cathode-ray tubes to be designed during World War II and its fundamental design was the basis for similar tubes that were used for a period of almost 20 years in higher-performance oscilloscopes, radar range scopes, television waveform monitors,

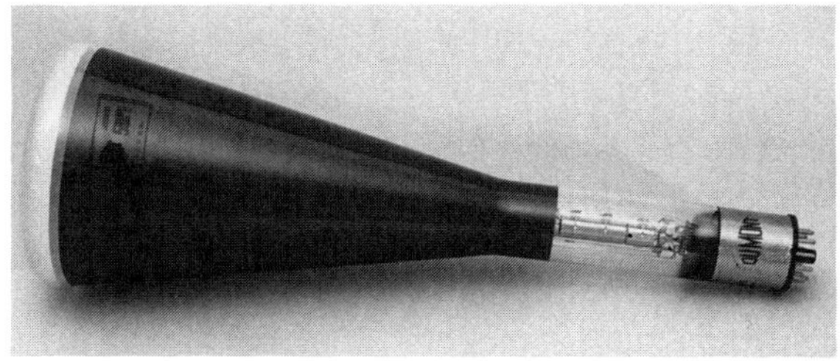

Figure 4.21 Du Mont 5BP1 manufactured in 1943. (From the author's collection.)

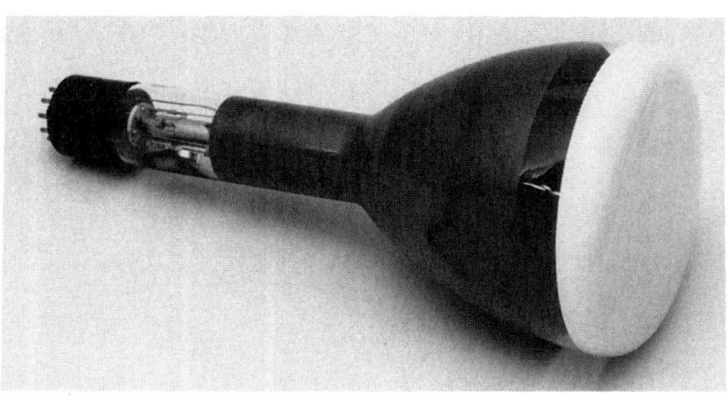

Figure 4.20 Sylvania 5AP1, manufactured in 1945. (From the author's collection.)

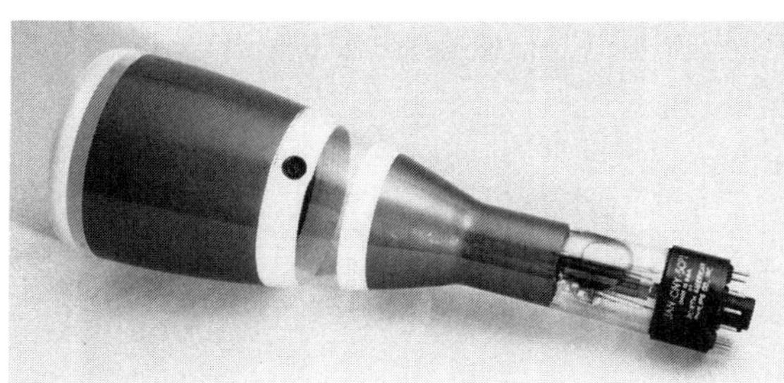

Figure 4.22 North American Philips (Norelco) 5CP1, circa 1944. (From the author's collection.)

spectrum analyzers and pulse-height analyzers. The 5CP1 was a post-deflection acceleration CRT which provided bright traces while retaining good deflection sensitivity. Anode voltages up to 4,000 volts were permissible compared to approximately 2,000 volts for the best tubes without post-deflection accelerators. Further enhancing the higher voltage capabilities of the 5CP1 were a large 14-pin diheptal base and small-ball anode connector about midway along the bulb's flared side. The 5CP1 was registered with RMA in 1942. P1 phosphor was standard, however, P2, P4, P5, P7, P11 and P12 phosphors also were available to fulfill requirements for a variety of colors and persistences. Virtually all major CRT manufacturers have supplied the 5CP-type at one time or another. The wartime Du Mont Model 247 oscilloscope and radar A/R range scopes used the 5CP1. The 5CP1 was available in large quantities on the surplus market after World War II and was used in an improved Model O-8 Heathkit oscilloscope in 1953.

5GP1–The 5GP1 essentially was interchangeable with the 5BP1 except for higher vertical deflection sensitivity and slightly different interelectrode capacitances. Sylvania Electric registered the 5GP1 with RMA in early 1942. As far as is known, only Sylvania manufactured this type.

5HP1–RCA registered the 5HP1 in 1942. It is virtually identical to the type 5BP1 except for a mica-filled (Micanol) wafer 11-pin base which was nonhygroscopic, hence more resistant to electrical breakdown in humid environments, e.g., shipboard applications. Both RCA and Sylvania manufactured the 5HP1. It also was available, at least from RCA, with P4 phosphor as the 5HP4.

5JP1–The 5JP1 (Figure 4.23) was a high-

Figure 4.23 Du Mont 5JP1 manufactured in 1944. (From the author's collection.)

performance CRT that found principal use in oscilloscopes and synchroscopes having wide bandwidths for pulse, radar and television measurements. The wide bandwidth was achieved through neck deflection plate cap connections and their resultant lower capacitance. A post-deflection accelerator further eased deflection amplifier requirements while producing a bright trace for low repetition rate pulses. The 5JP1 used the same five-inch bulb and 11-pin magnal base as the 5BP1. It was registered by Du Mont in early 1942. The 5JP1 was also manufactured by North American Philips and, after the war, by Electronic Tube Corporation of Philadelphia. The 5JP1 was formerly known as the Du Mont type 2511-A5. P2, P4, P5, P7 and P11 phosphors also were available. The wartime Sylvania Model 5 synchroscope and two megacycle (megahertz) bandwidth Du Mont Model 241 oscilloscope employed the 5JP1 CRT.

5LP1–Formerly known as the type 2511-A5, the 5LP1 (Figure 4.24) registered in 1942 by Du Mont was used primarily in the famous Du Mont type 208 oscilloscope also introduced in 1942. The Sylvania Model P4 synchroscope for radar work also used the 5LP1. This tube was a post-deflection accelerator type similar to the 5JP1 except that the

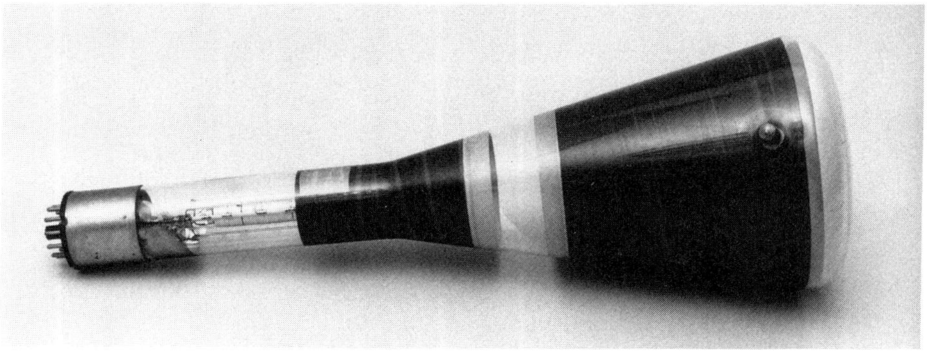

Figure 4.24 Du Mont 5LP5 manufactured in 1945. (From the author's collection.)

deflection plate connections were through the 11-pin magnal base. P2, P4, P5, P7 and P11 phosphors were available. Norelco appears to be the only other supplier of the 5LP1.

5MP1–The rare Du Mont 5MP1 was registered with RMA in 1942. Previously it was identified as the Du Mont type 2505-A5 before becoming the 5MP1. The five-inch diameter 5MP1 was of the classic tapered funnel shape with a seven-pin radio tube base similar to the 3AP1/906. One deflection plate of each pair was internally connected to the second anode to reduce base pin count. P4 and P5 phosphors were available. The only known usage of the 5MP1 was for replacements for the Du Mont Model 148 oscilloscope introduced in 1935 and the Model 168 of 1937.

5NP1–The 5NP1 and its P4 version, the 5NP4, appear nearly identical to the 5BP1 and 5BP4 according to the RMA registration data filed in late 1943 by Sylvania Electric. Further information is required to determine the reason for registering such a similar appearing CRT. The 5NP was never manufactured in significant quantities.

9NP1–Research Enterprises, Ltd. of Toronto registered the nine-inch 9NP1 (Figure 4.25) in 1943. Except for bulb shape and minor electrical differences, the 9NP1 was similar to the RCA type 914. This is another rarely encountered CRT.

12FP7, 12GP7 and 12HP7–These 12-inch diameter, electrostatically deflected CRTs were more likely intended for radar displays based on their registrations with P7 phosphor. They also may have seen oscillographic use, but at present little is known about them. The 12FP7 was registered by Research Enterprises, Ltd. of Toronto (Leaside), Ontario, in 1943 and had post-deflection acceleration with a 14-pin diheptal base. The 12GP7 (Figure 4.26) registered by General Electric in 1944 was very similar. The 12HP7 and sister tube, the 12HP1 (Figure 4.27), which would have been the most likely choice for oscillographic applications, were registered in early 1944 by Research Enterprises, Ltd. The 12HP- used a metal shell 11-pin magnal base and no post-accelerator electrode. Its 5,500 volt maximum rated anode voltage may have been stretching the capabilities of its magnal base. See Chapter 3 for additional information on these tube types.

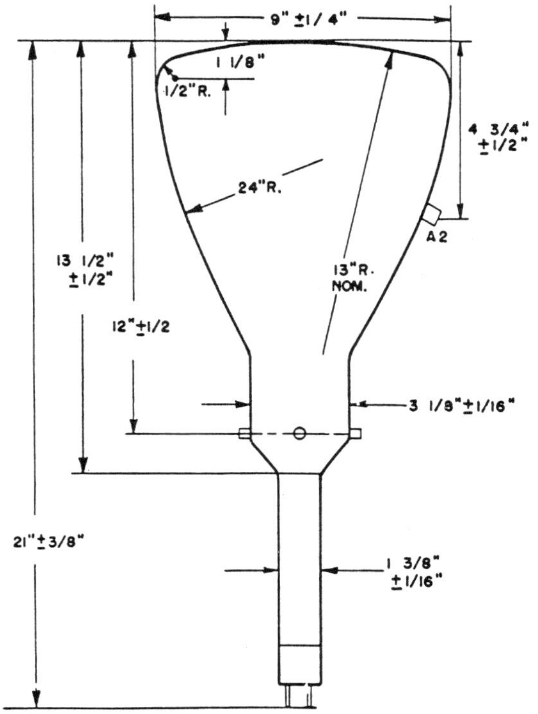

Figure 4.25 Research Enterprises Ltd. 9NP1 of 1943. (Courtesy of Electronic Industries Association.)

14AP1 and 20AP1–These Du Mont CRTs registered in 1942 were incredible tubes for their time. Their balloon shape, 14- and 20-inch screen diameters and greater than two-foot length made them the largest electrostatically deflected CRTs produced commercially in the United States. They both used an English 12-contact peripheral base with their post-accelerator connection made through the base. Both tubes also were available with P2, P4 or P5 phosphor instead of P1. The 14AP1 was formerly the type 2531-A14 and the 20AP1 was the type 2532-A20. Du Mont was the only manufacturer of either tube. The 20AP1 (Figure 4.28) was used in the Du Mont

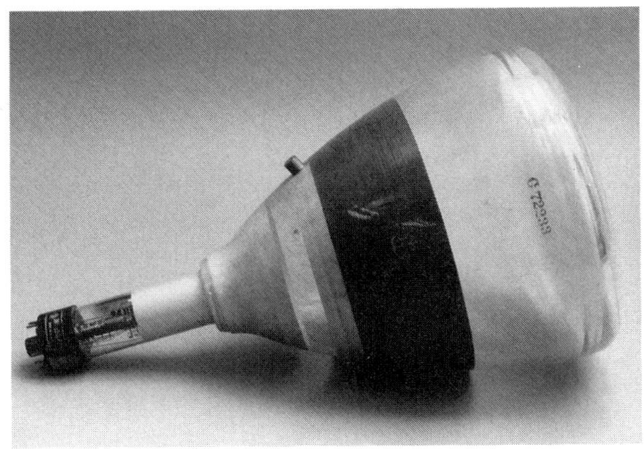

Figure 4.26 General Electric 12GP7 of 1944. (Courtesy of National Museum of American History, Smithsonian Institution.)

Model 233 oscilloscope, an instrument of truly monumental proportions intended for classroom demonstration purposes.

4.4 ZERO-FIRST ANODE CURRENT GUNS

Many of the aforementioned CRTs acquired an "A" suffix in late 1944 and early 1945 with the adoption of an improved electron gun.[17–19] The advanced gun utilized a new Einzel focus electron-optical system that had no interception of beam current in the focus electrode. This permitted freedom of interaction between the focus and brightness controls, in addition to better overall focus. Tubes upgraded to "A" versions included the 902-A, 914-A, 2AP1-A, 3AP1-A, 3BP1-A, 3FP7-A, 3GP1-A, 5BP1-A, 5CP1-A, 5HP1-A, 5JP1-A and 5LP1-A. Of these, the most important was the 5CP1-A, which was the foundation in 1947 of the 511 oscilloscope manufactured by the then recently founded

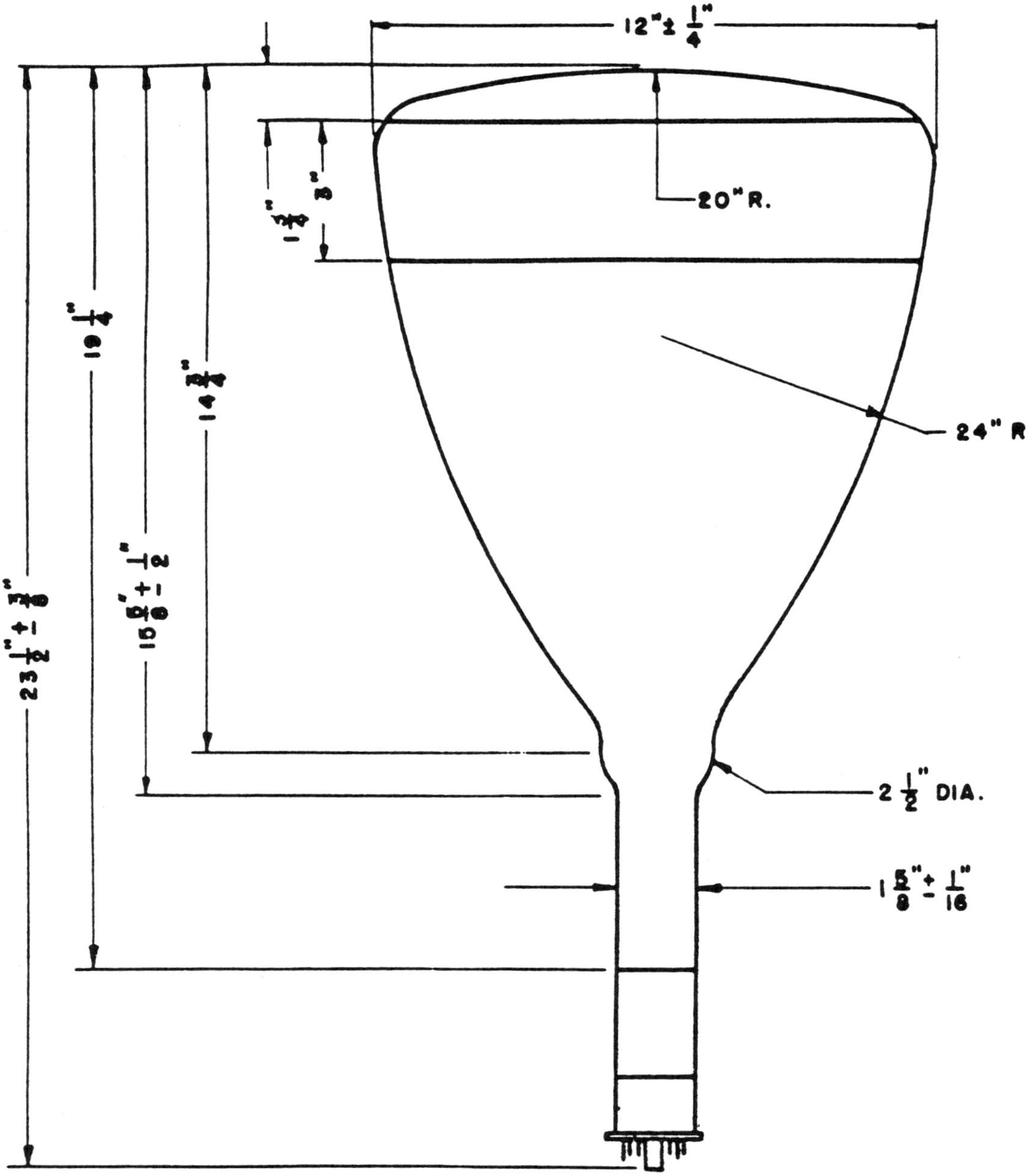

Figure 4.27 Research Enterprises Ltd. 12HP1 of 1944. (Courtesy of Electronic Industries Association.)

Figure 4.28 Du Mont 20AP1, circa 1942. (Allen B. Du Mont Collection, National Museum of American History, Smithsonian Institution.)

Tektronix and the 1949 Model 304 oscilloscope by Du Mont.

4.5 POST-WAR CRTS

Several cathode-ray tubes were introduced shortly after the end of wartime hostilities. The more conventional types will be discussed in this section and following sections will discuss more advanced technological developments.

The three-inch, post-deflection accelerator type, the 3JP1, was introduced in late 1945 by RCA. The 3JP1 used a three-inch screen and a 14 pin diheptal base. It was a replacement for the wartime type 3FP7 with improved spot size and was eventually available with P1, P2 (Figure 4.29), P7, P11 and the very long-persistence P12 phosphor. The latter type, the 3JP12, appears in a Joint Army-Navy list of tubes removed from classified material status in November 1945. The 3JP-series also was manufactured by Du Mont, General Electric and Sylvania. Du Mont later produced the tight-tolerance version, the 3JP-A (beginning in 1954).

The RCA 3KP1 (Figure 4.30) of early 1946 was a general purpose, three-inch oscilloscope CRT that also was applied to low-cost post-war television receivers in the 3KP4 version. The 3KP1 had good deflection sensitivity due to its moderate length. An 11-pin magnal base allowed individual connection

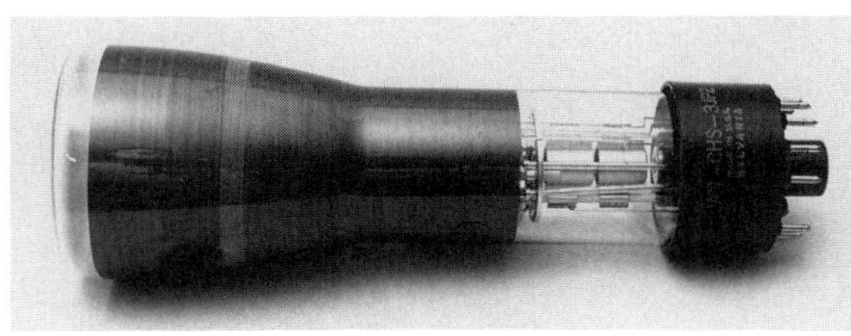

Figure 4.29 Sylvania 3JP2 manufactured in 1950. (From the author's collection.)

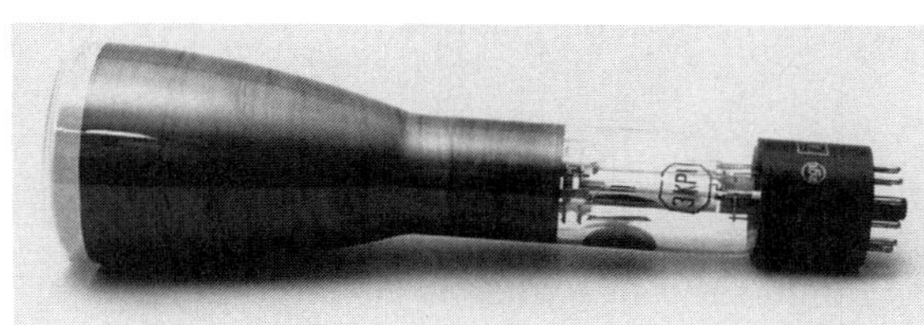

Figure 4.30 RCA 3KP1 manufactured in 1956. (From the author's collection.)

to all deflection plates. P7, P11 and even P16 phosphor for flying-spot scanner use were available. RCA and Sylvania were the principal manufacturers of the 3KP-. The 3KP1 was used in the RCA Model WO-79A oscilloscope and the 3KP4 is remembered chiefly for its use in the Pilot Model TV-37 television receiver of about 1949. The TV-37 is prized among collectors as a unique example of the early post-war commercial television period.

A short, three-inch diameter CRT employing the new 12-pin duodecal television base and known as the 3MP1 was registered in 1947 by General Electric. Only eight inches long, it was well-suited to compact oscilloscopes, such as the RCA WO-57A, Waterman Products S-11A and several others intended for television servicing. The 3MP1 also was manufactured by RCA, Sylvania and Waterman (Rayonic). Figure 4.31 illustrates the 3MP1.

Around 1950, a popular series of oscilloscope CRTs of relatively short length were developed by RCA. These included the 2BP1, 3RP1 and 5UP1 (Figures 4.32 through 4.34). All used the new 12-pin duodecal base and

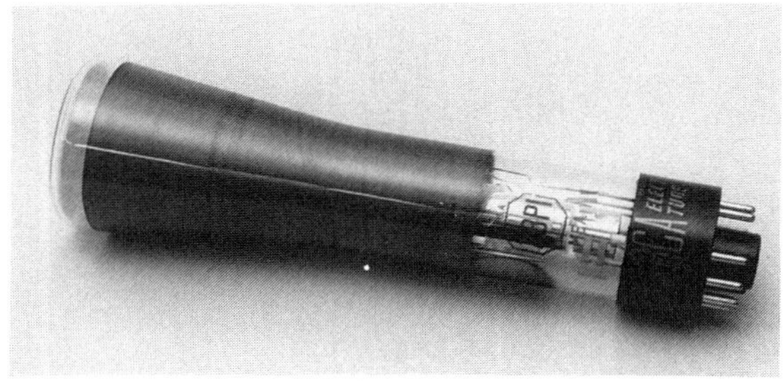

Figure 4.32 RCA 2BP1 manufactured in 1966. (From the author's collection.)

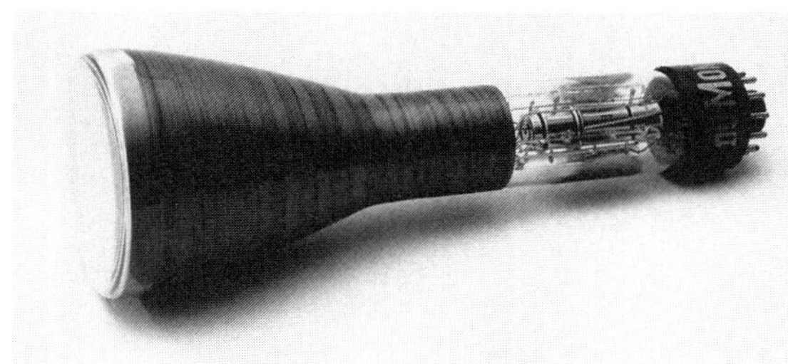

Figure 4.33 Du Mont 3RP1 manufactured in 1959. (From the author's collection.)

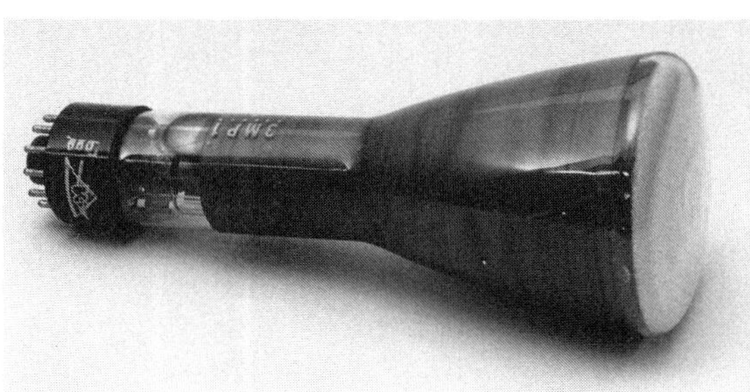

Figure 4.31 Waterman Products 3MP1 manufactured in 1956. (From the author's collection.)

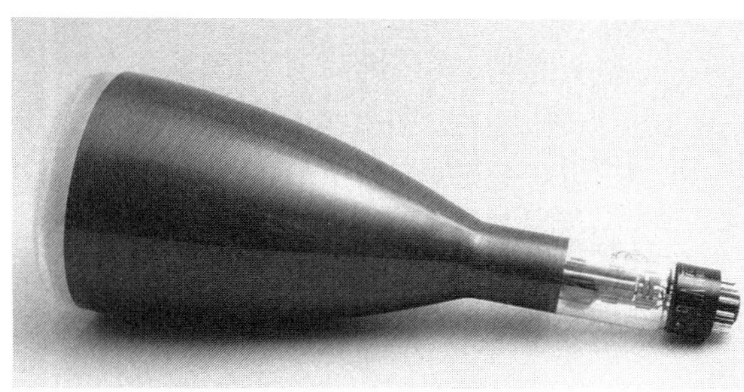

Figure 4.34 RCA 5UP1 manufactured in 1967. (From the author's collection.)

featured excellent spot size. They were used in many oscilloscopes intended for the television and industrial electronics service markets. Sylvania and General Electric also supplied these tubes and P7 and P11 phosphors were available in some cases. The tubes in this series were about the last of the curved face oscilloscope CRTs to be introduced with the exception of the Waterman 3AQP1 and 5DEP1 which were identical to the 3RP1 and 5UP1, respectively, except for higher deflection sensitivity. The 3AQP1 and 5DEP1 were introduced in 1958 and 1960, respectively. Du Mont developed a flat-faced version of the 3RP1 in 1950. The nomenclature departed from conventional practice since it was merely given an "-A" suffix to distinguish it from the original curved-face bulb.

A particularly interesting oscillographic CRT was announced circa 1953. This was the one-inch diameter 1CP1 (Figure 4.35) from Cossor Ltd. It was used mostly for monitoring of modulation of transmitters. The 1CP1 used an eight-pin loktal base. The loktal base, also called lock-in and loctal because of trademark restrictions, was developed by Sylvania Electric in 1938 for Philco Radio tubes. It used heavier lead wires which were the actual base pins. The metal "base" and keyway doubled as a ninth contact for the anode. The RCA 1EP1 (Figure 4.36) of 1956 was similar to the 1CP1 except for an 11-pin "unidekar" hard-pin base and semi-flat faceplate. P2 and P11 phosphors also were available for the 1EP.

Three seven-inch tubes for oscillography were developed in the late 1940s. Two of these, the 7GP1 and 7JP1 (Figure 4.37) were identical to the 7GP4 and 7JP4 developed for inexpensive television receivers except for

Figure 4.35 Cossor Ltd. 1CP1, manufactured circa 1956. (Courtesy of U.S. Army LABCOM, E. T. and D. Laboratory, Ft. Monmouth, N.J.)

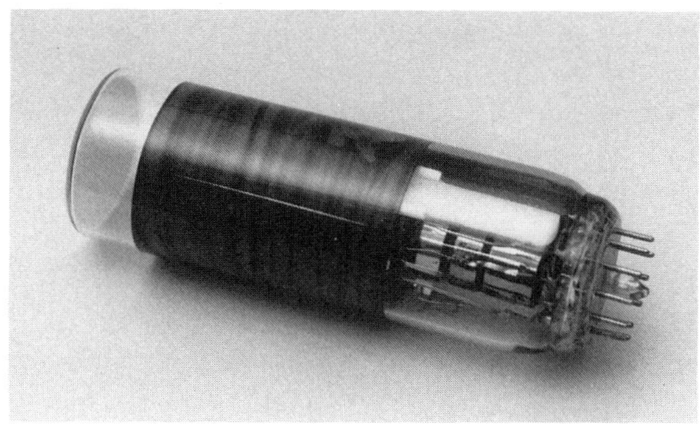

Figure 4.36 RCA 1EP2, circa late 1950s. (From the author's collection.)

Figure 4.37 Sylvania 7JP1 manufactured in 1949. (From the author's collection.)

their green P1 phosphor. These tubes were made primarily by Sylvania and RCA and were used in their own brand of oscilloscopes including the models 132 and 400 by the former and WO-56A by the latter. These tubes had relatively large spot size due to the need for enough beam current to produce a bright picture for television use since they were designed for fairly low accelerating voltages. RCA produced a modified version, the type 7VP1, in 1950 that was specifically for oscilloscopes and employed a beam limiting aperture for a very sharp trace. High brightness was not as much of a requirement for oscillography as it was for television so the trade-off of brightness for a smaller spot size was acceptable. Still later, RCA produced a version of the 7VP1 which was registered with EIA as the type 4490 in the industrial tube numbering system rather than the CRT numbering system. Several other CRTs were registered by RCA in the early 1960s with similar type numbers. The reason for the deviation from previous practice is unknown. The 4490 used a post-deflection accelerator and P31 phosphor, both of which contributed toward good brightness. All used the same seven-inch, curved-sided, blown-glass bulb and 14-pin diheptal base. Their length was quite short for their large screen size.

Two eight-inch tubes that were very similar to the seven-inch types also were produced. These were the National Union and Sylvania 8CP1 (Figure 4.38) which found limited use in television servicing oscilloscopes for a short time in the mid-1950s and the RCA 4491 in the early 1960s. Both were of the post-deflection accelerator type with bulbs that were scaled up versions of the seven-inch shape and employed the 14-pin diheptal base. The 8CP1 was also available

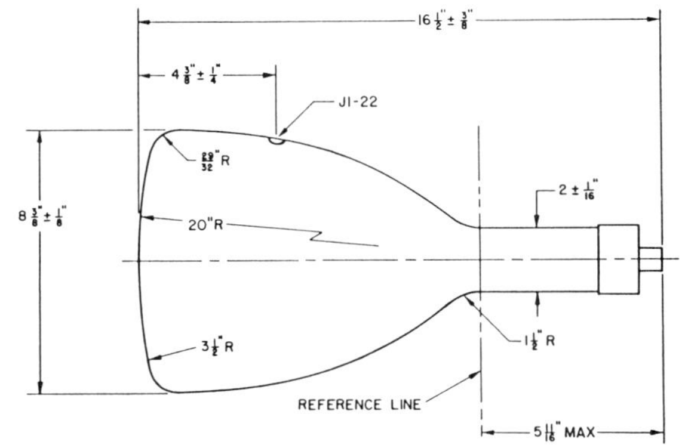

Figure 4.38 Sylvania 8CP1, circa late 1950s. (Courtesy of Philips Display Components Company.)

with P2, P4, P5, P7 and P11 phosphors. The RCA 4491 was cataloged with P31 phosphor.

4.6 FLAT-FACE CRTS

The first truly flat-face tube registered in the United States was the Du Mont 5RP1 (Figure 4.39) of 1945 described under multiband acceleration. It was soon followed by the Du Mont 5SP1 (Figure 4.40) dual-gun CRT. Both were available with P1, P2, P4, P7 and P11 phosphors. The flat-face tube was a signifi-

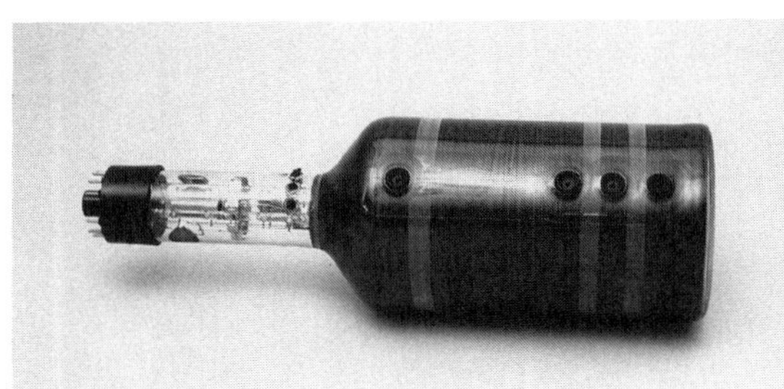

Figure 4.39 Du Mont 5RP11-A, manufactured circa early 1950s. (From the author's collection.)

cant step forward since it ushered in the era of precision measurement of amplitude and time from the screen of an oscilloscope. Until then, the oscilloscope's greatest value was for qualitative examination of waveforms rather than quantitative measurements. The flat-face tube combined with a flat plastic overlay called a graticule (having a scribed or printed scale) allowed measurement without much of the parallax and image distortion present with the older curved-face tubes.

The flat-face tube rapidly became the norm and by the mid-1950s was used for all oscilloscopes except those for the amateur experimenter and television serviceman. In 1952, three tubes were introduced which were among the most widely used throughout the 1950s. These were the Du Mont 3WP1 (Figure 4.41) and 5ADP1 (Figure 4.42) and the RCA 5ABP1 which was almost identical to the 5ADP1. All three were produced with a wide variety of phosphors and by most CRT manufacturers. The post-deflection accelerator 5ABP1 and 5ADP1 were improved replacements for the venerable 5CP1 and had improved deflection sensitivity and geometry to take advantage of the flat-face screen. Important oscilloscope applications for these tubes included the Tektronix portable Model

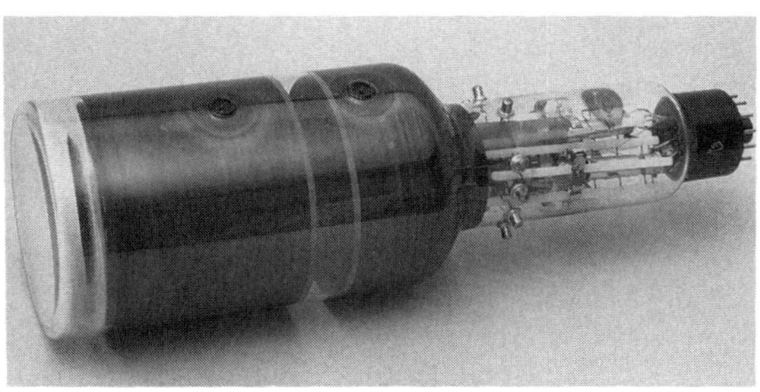

Figure 4.40 Du Mont 5SP7 manufactured in 1952. (From the author's collection.)

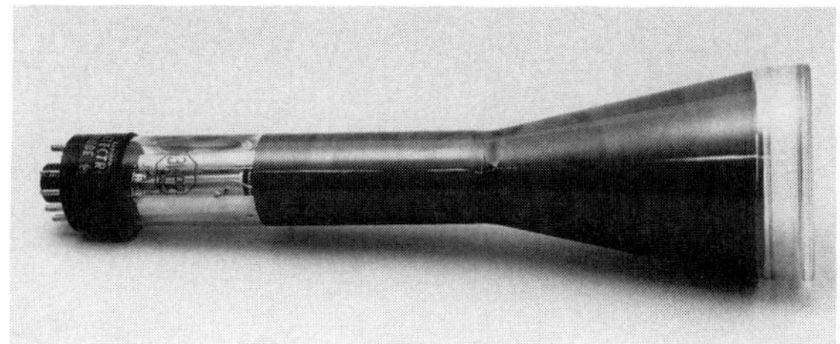

Figure 4.41 RCA 3WP2 manufactured in 1961. (From the author's collection.)

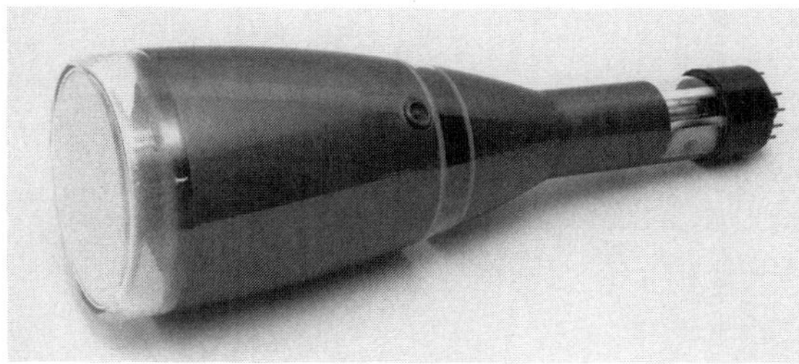

Figure 4.42 Du Mont 5ADP2 manufactured in 1956. (Courtesy of U.S. Army LABCOM, E. T. and D. Laboratory, Ft. Monmouth, N.J.)

310 for the mono-accelerator 3WP1 and the Du Mont 304-A for the 5ADP1. A tight tolerance version of the 5ADP- series was registered by Du Mont in 1960 as the 5ADP-A. The 5ADP-A was also aluminized for higher brightness.

4.7 MULTIBAND ACCELERATION

Du Mont Laboratories announced the previously mentioned five-inch diameter 5RP1 and 5RP11 multiband accelerator CRTs in July of 1945.[20–22] The 5RP- had several advanced features which made it particularly useful for viewing or photographically recording extremely high-speed, nonrecurring electrical phenomena which became their major application, and for the projection of conventional

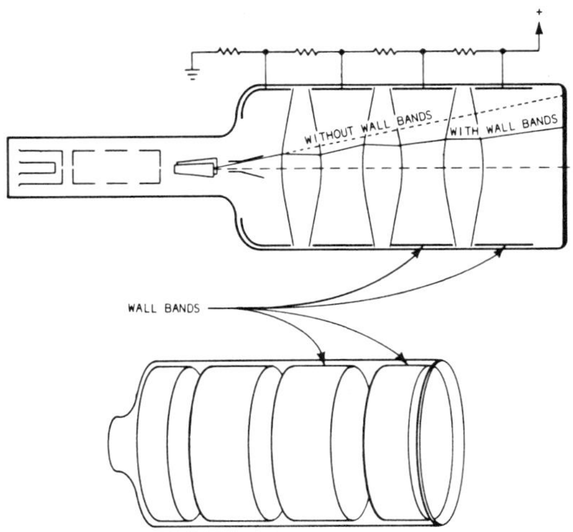

Figure 4.43 Multiband post acceleration.[44] *(Courtesy of Tektronix, Inc.)*

oscilloscope waveforms for group viewing. Instead of the previously used, single, post-acceleration electrode applied to the bulb interior near the screen, a series of separate accelerator electrode bands with insulating bands between were applied to a barrel-shaped bulb (Figure 4.43).

Post-deflection acceleration has the advantage of good deflection sensitivity due to the electron beam's lower energy in the deflection plate region while possessing the ability to increase the electron velocity with a higher voltage before reaching the screen to produce greater brightness. Practical limitations of this concept are reached, however, when the ratio of the final accelerator to anode voltage reaches about 2:1, due to increased astigmatism of the spot and geometric distortion of displayed patterns. The multiband acceleration principle uses a series of accelerator bands with each having a progressively higher voltage as it approaches the screen, thus providing a more gradual transition to the final acceleration voltage. Separate high-voltage connectors are provided through the bulb wall for each conductive band of *aquadag* or silver and an external resistive divider network provides the intermediate voltages between the anode and the final acceleration voltage. Figure 4.44 illustrates the evolution of the multiband concept.[23] The multiband accelerator CRT was used during the period from about 1945 to 1960, by which time the spiral accelerator described in Section 4.10 had largely displaced it.

The improved cylindrical bulb shape of the 5RP- combined with the smoother voltage gradient reduces penetration of the acceleration field into the deflection plate region and allows post-accelerator to second anode voltage ratios of up to 10:1. The 5RP- was used to replace the popular 5CP- series with few circuit changes except the additional 15,000 volt high voltage power supply and divider for the post-accelerator system. Green P1 phosphor was used for the type 5RP1 intended for visual observation of fast transients and projection of waveforms. Blue P11

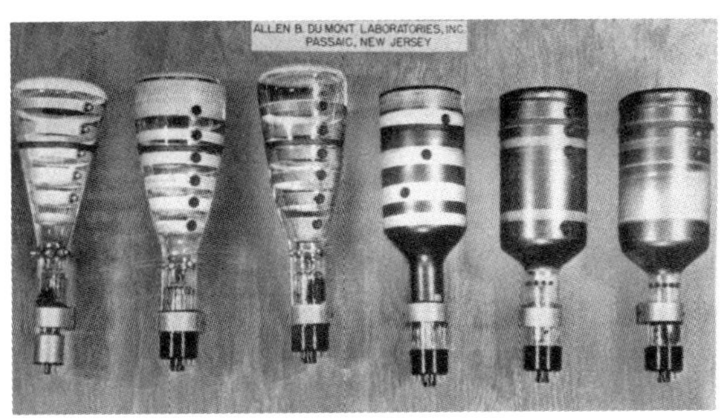

Figure 4.44 Evolution of multiband post-acceleration during the 1940s. (Courtesy of Hughes Aircraft Company.)

phosphor was supplied in the 5RP11 where photographic recording was desired. P2, P4, P5 and P7 phosphors also were cataloged for special purpose applications. Other features of the 5RP- included a flat faceplate to aid measurements made from the screen and deflection plate connections made through the neck to allow use at higher frequencies. The 5RP- was used in the Du Mont Model 247-A projection oscilloscope. An improved CRT, the 5RP-A, was used in the Du Mont Model 248-A, 250-H, 250-AH, 280, 281 and 281-A oscilloscopes beginning in 1947. Finally, a version with tightened tolerances was registered by Du Mont in 1959 as the type 5RP-B.

A similar appearing tube, the five-inch Du Mont 5XP- (Figure 4.45), was registered with RMA in 1948. The primary difference was a vertical deflection sensitivity three times greater than that of the 5RP- which especially enhanced its use at frequencies up to 200 megahertz (megacycles in those days).[24] Decreased vertical deflection plate width and closer spacing accounted for the improved performance. The trade-off was limited vertical scan which was as little as 1.25 inches when operated at 20,000 volts with a 10:1 PDA to anode voltage ratio. The maximum acceleration voltage rating was 25,500 volts. P1, P2, P7 and P11 phosphors were available depending on application. A tightened tolerance version was made available around 1953 as the type 5XP-A and about 1954, the 5XP-B which utilized a metalized (aluminized) screen. This benefited performance in three ways: Brightness was increased by reflecting the light ordinarily lost to the back of the tube; a direct return path to the anode was provided to prevent charging of the screen at high voltages; and light from the heater was prevented from reaching the camera and fogging the film in photographic applications requiring long exposure times.

The Du Mont 5AWP- was another similar tube introduced in 1956. It was basically similar to the 5RP- and 5XP- series and employed an aluminized screen and low capacitance vertical deflection plates with approximately four inches of useful scan. P1, P2, P7 and P11 phosphors were available.

Other five-inch multiband accelerator CRTs include the Du Mont developmental types K-1017 and K-1101. The K-1017 was produced for a high-speed oscilloscope developed by the MIT Radiation Laboratory in the mid 1940s and employed a coaxial deflection system which was capable of displaying 2,000 megahertz sine waves.[93] P11 phosphor was standard for the intended application of photographic recording of high frequency waveforms. The K-1101, circa 1951, was a similar tube except for greater length (approximately 21 inches) and higher voltage operation of up to 37,500 volts. P11 and P15 phosphors were standard for the K-1101. Note that the K-1017 and K-1101 were house numbers rather than RMA registered type numbers. House numbers were usually used for relatively low-demand, special-purpose CRTs. These CRTs achieved their wide bandwidth through the use of very short deflection plates at the expense of sensitivity.

The multiband accelerator principle also was applied to some dual-gun CRTs by Du Mont and Electronic Tube Corporation during the 1950s. These included the Du Mont type 5BDP- and the ETC type 5CTP-.

4.8 MULTIBEAM/MULTIGUN CRTS

Multibeam and multigun CRTs both provide a means to simultaneously display two or more

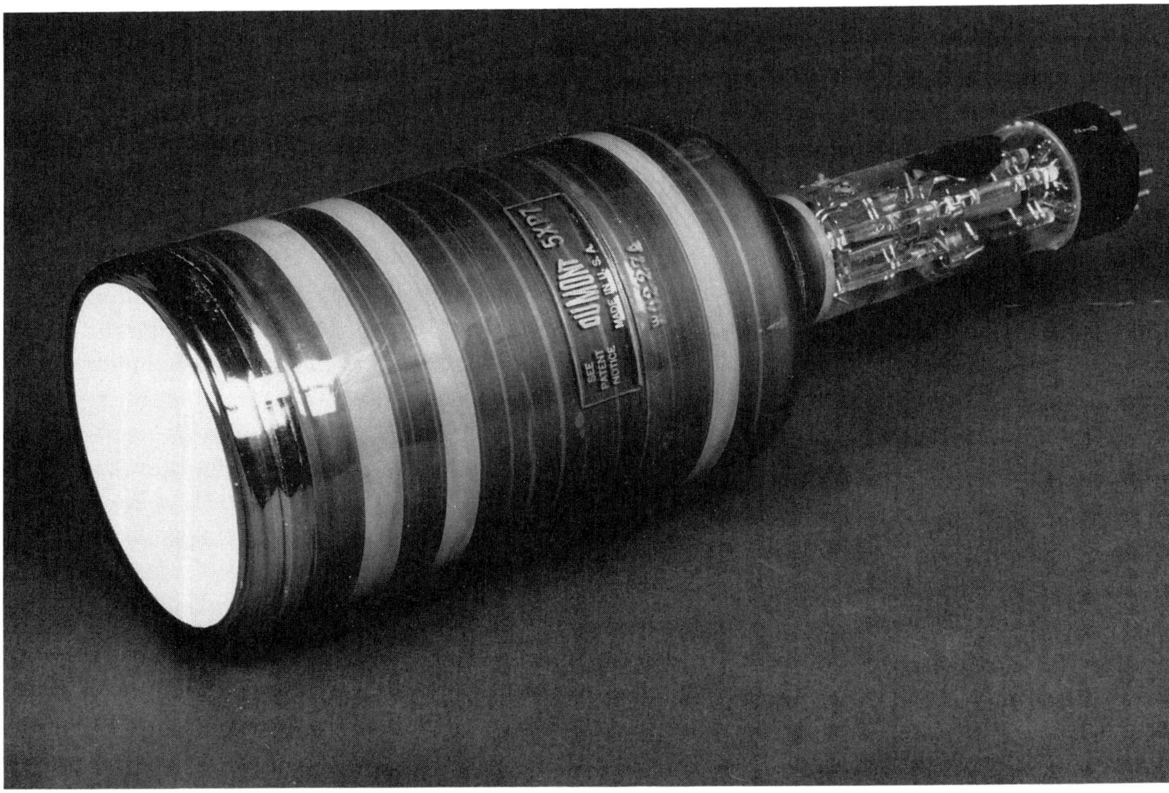

Figure 4.45 Du Mont 5XP7, manufactured in early 1950s. (Courtesy of U.S. Army LABCOM, E. T. and D. Laboratory, Ft. Monmouth, N.J.)

waveforms on the same screen without the use of an electronic switch circuit to alternately display the two input signals. CRTs containing as many as 10 guns within the same envelope have been produced commercially. Multibeam tubes share a common set of horizontal deflection plates and time-base circuit for comparison of two time-related events. These may be the input and output signals from an audio amplifier to evaluate distortion or phase shift or the voltage and current waveforms at the same point in a circuit. On the other hand, multigun CRTs have completely independent electron guns within the same envelope and each may have individual vertical amplifiers and time-bases capable of displaying totally unrelated waveforms. Early multibeam CRTs were announced in 1936 by Manfred von Ardenne in Germany[25] and Allen B. Du Mont reportedly made several in the United States in the same period. A variation of the multibeam tube was the *split-beam* CRT developed by the British company, Cossor, Ltd. in 1939.[26,27] The split-beam produced only one beam up to the deflection region where the beam was split into two beams by a splitter plate mounted between the first set of deflection plates (Figure 4.46). Two separate vertical deflection signals were applied to the individual plates of the set with the signals referenced to the common splitter plate.

The first commercially advertised multigun CRT in the United States was the Western Electric type 330 announced in January 1938.[28–30] Three independent beams and a seven-inch screen were featured. A five-volt heater, 1,000 to 5,000 volt accelerating voltage and 23-inch length were among its characteristics. Three versions were available: the 330A, 330B and 330C, differing only in phosphor type. These used P1, P2 and P5 phosphors, respectively, although at that time the "P" numbers had not yet been registered by RMA. The 330 series CRTs were not widely used and it remained for Du Mont to introduce the first commercially successful multibeam tube in the U.S.

In 1945, Du Mont announced the five-inch type 5SP- (refer back to Figure 4.40) double beam CRT which was the equivalent of two of the popular 5CP- CRTs within one envelope. The 5SP- used a large metal shield between the two deflection plate systems to reduce "crosstalk" or interaction. Deflection plate caps on the tube neck allowed enough contacts for all of the electrodes associated with two guns as well as reduced capacitance for higher frequency response.[31] P1, P2, P4, P7 and P11 phosphors were available as standard products. The 5SP-A tight tolerance version was developed in 1952 to 1953. The Du Mont Model 279 of 1947 and the Model 322 of 1952 were the principal oscilloscopes to use the 5SP-. The Electronic Tube Corporation (ETC), a company specializing in multigun CRTs, was another manufacturer of this tube.

During the 1950s, more new multigun CRTs were introduced by Du Mont and ETC, the most notable of which were the Du Mont post-deflection accelerator types 3ABP- and 5AFP- and the mono-accelerator 5ARP-. The 5AFP- (Figure 4.47) was used by Du Mont in the Model 322-A oscilloscope and the 5ARP- was used in their Model 333.

The multiband accelerator was combined

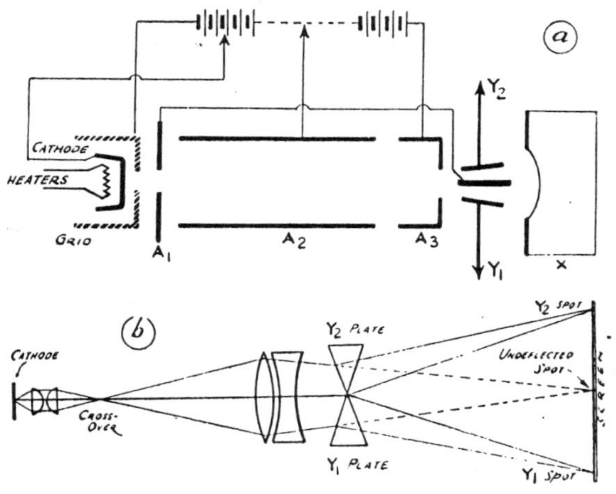

Figure 4.46 Cossor Ltd. split-beam CRT electron gun of 1939.[26]

Figure 4.47 Du Mont 5AFP- dual-gun CRT, circa late 1950s. (From the author's collection.)

with the dual gun in the Du Mont 5BDP- and the ETC 5CTP- in the mid 1950s. By the late 1950s, dual-trace oscilloscopes with delayed sweep had become commonplace, thus electronically eliminating much of the need for multibeam and multigun CRTs.

A few new tubes were developed by Tektronix for the specialized applications still requiring separate beams. These included the dual-beam, mono-accelerator type T502 (Figure 4.48) for the Model 502 oscilloscope in 1958, the dual-beam, spiral post-accelerator T551 (Figure 4.49) for the Model 551 in 1958, the dual-gun, spiral post-accelerator T555 (Figure 4.50) for the Model 555 in 1959, and the dual-gun, spiral post-accelerator T5560 (Figure 4.51) for the Model 556 in 1966. All were available with the customary oscilloscope phosphors of P1, P2, P7, P11 and P31 as well as several others on special order.

The award for the most guns squeezed into one tube probably should go to ETC for their 12-inch type 12Z10P- with 10 individual electron guns. Keeping everything working at the same time must have been a real challenge, especially considering the amount of support electronics needed to drive it. Despite the similarity of the type number to registered CRTs, this was another example of a house number.

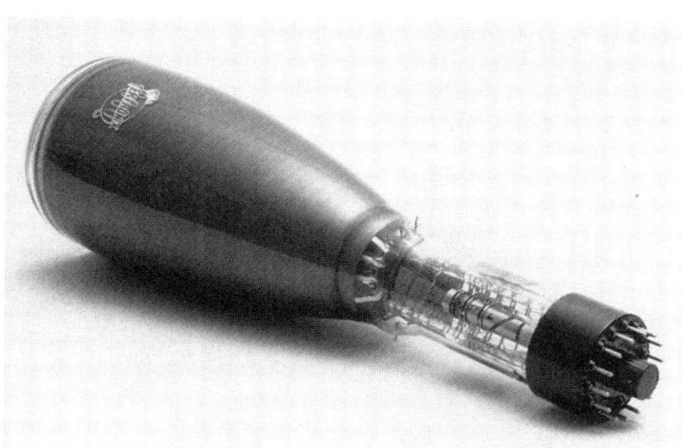

Figure 4.48 Tektronix T502 dual-beam, mono-accelerator CRT of 1958.

4.9 MONO-ACCELERATOR

In 1954, the pendulum swung from the post-deflection accelerator CRTs of the post-war years back toward the simpler acceleration method of the 1930s. This type was now termed the *mono-accelerator* CRT and a number of refinements were incorporated that resulted in much-improved performance.

The post-accelerator types suffered from several drawbacks, including pattern distortion, spot-defocusing near the screen edges and spurious background screen illumination due to secondary-electron emission from electron beam bombardment of the tube walls. The traditional reason for the use of the post-accelerator tube was to increase brightness through higher accelerating voltages while still maintaining good deflection sensitivity. These goals were met with the mono-accelerator principle by a combination of a redesigned deflection-plate structure and tightened tolerances. More stringent manufacturing tolerances permitted closer deflection plate spacing without beam intercept at deflection extremes. Longer deflection plates also increased deflection sensitivity without increasing capacitance significantly since the increased length was at the end of the plates where the spacing was widest. Hence the effect on capacitance was minimized. Accelerating voltages were then increased to about 2,500 to 3,000 volts to trade-off some of the newly increased deflection sensitivity for greater brightness. The result was a clean

Figure 4.49 Tektronix T551 dual-beam, spiral accelerator CRT of 1958. (Courtesy of Tektronix, Inc.)

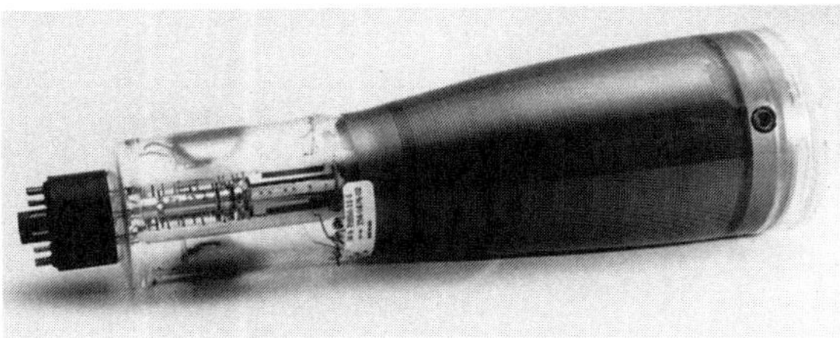

Figure 4.50 Tektronix T5550 dual-gun CRT (internal graticule version of T555).

display with freedom from post-acceleration pattern distortion, defocusing and background screen illumination.

Du Mont introduced the five-inch diameter mono-accelerator type 5AMP-, 5AQP- (Figure 4.52) and 5ARP- CRTs in 1954.[32] The 5AMP- was similar to the 5AQP- except that the deflection plate connections were made through the neck for higher frequency applications. Both also were eventually manufactured by Sylvania and Westinghouse. The 5ARP- was a dual-gun version. A short time later, the 5ATP- was announced. It was similar to the developmental type B1070 and was probably the highest voltage mono-accelerator oscilloscope CRT ever produced commercially. The 5ATP- was rated up to 10,000 volts, employed an aluminized screen, neck deflection plate connections and had limited vertical scan to maximize deflection sensitivity. All Du Mont mono-accelerator CRTs of the period were available with P1, P2, P7 and P11 phosphors. The type 5AQP- was used in the Du Mont 340 and 403 oscilloscopes as well as the Hewlett-Packard 120-A and 130-A. The 5AMP- was used in the Du Mont Model 323-A, the 5ARP- in their Model 333 and the 5ATP- in their Model 329-A. Tight-tolerance versions of the 5AMP- and 5AQP appeared with "A" suffixes in 1959 and aluminized versions with "B" suffixes in 1962 by Hewlett-Packard and 1960 by Thomas Electronics, respectively.

4.10 SPIRAL-ACCELERATOR

The spiral-accelerator was first suggested in Germany by E. Schwartz (in 1938) as a means to minimize compression of the beam in post-deflection acceleration systems but it was not developed into a practical device at that time. Howard Vollum, the founder of Tektronix, revived the concept again in 1952 to 1953[33] for the five-inch type T51 CRT for the 15 MHz Model 535 oscilloscope announced in late 1953. The idea behind the spiral-accelerator was to apply a gradually increasing accelerating voltage along the bulb's length to smoothly increase the electron beam's velocity, rather than the more abrupt steps of the multiband CRT (Figure 4.53). This would eliminate the beam distortions that occur at each transition to a higher voltage. Of course, inventing a means of painting a continuous helix on a curved surface, having the correct overall resistance and the ability to withstand tube processing temperatures was no mean feat. Joe Griffith was largely responsible for the successful development of the Tektronix CRT which became the key element to improved oscilloscope designs.

At first the T51 used the cylindrical bulb (Figure 4.54) of the 5XP-, but was soon changed to a tapered bulb (Figure 4.55) similar to that of the popular 5ABP- and 5ADP- tubes except for the anode button which was moved almost to the screen and changed from a small-ball type to a cavity connector. This made the entire bulb wall available for an uninterrupted spiral helix. Connection was made at the screen end of the helix to the anode connector, which also was connected to the

Figure 4.51 Tektronix T5560 dual-gun CRT of 1966. (Courtesy of Tektronix, Inc.)

aluminized screen. The gun end was connected to a thin, stiff wire neck pin connector that also was tied to an isolation shield between the two sets of deflection plates. Other neck pin connectors gave access to the vertical and horizontal deflection plates. Vertical deflection was limited to six centimeters for maximum sensitivity. A 10,000 volt accelerating potential assured a bright trace. The standard phosphor was P2, but a number of others were available on special order. The T51P2 was registered with EIA in 1957 as the type 5BGP2. The 5BGP2 was manufactured as a replacement tube by Du Mont, Sylvania, ETC and others to the consternation of Tektronix since the oscilloscopes would often not meet their original performance specifications with "foreign" CRTs.

Figure 4.52 Du Mont 5AQP11 mono-accelerator CRT manufactured in 1958. (From the author's collection.)

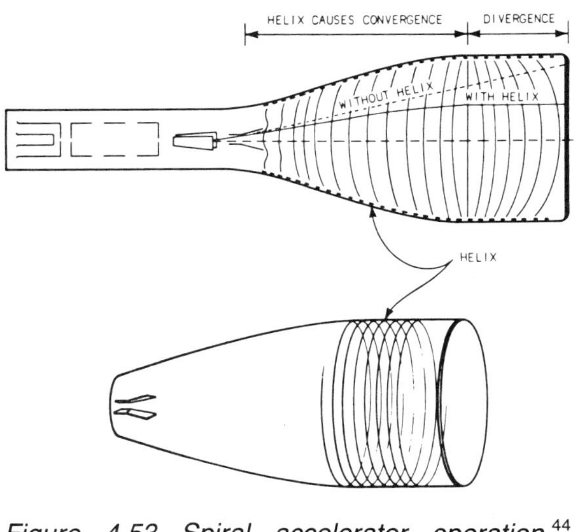

Figure 4.53 Spiral accelerator operation.[44] (Courtesy of Tektronix, Inc.)

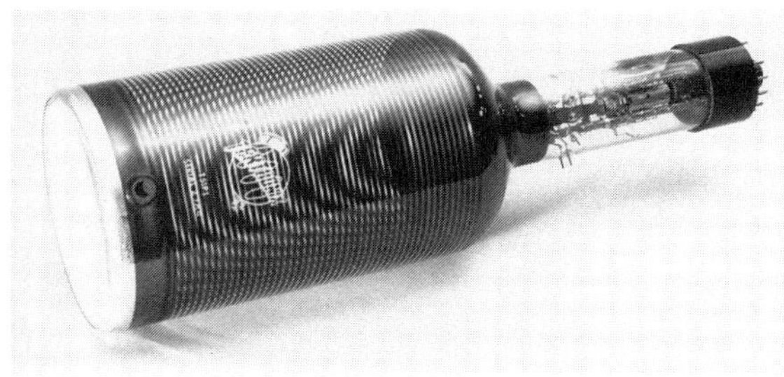

Figure 4.54 Tektronix T51 of early production (1954–1955). (Courtesy of U.S. Army LABCOM, E. T. and D. Laboratory, Ft. Monmouth, N.J.)

An entire series of spiral-accelerator CRTs from Tektronix, and later on other manufacturers, followed on the heels of the highly successful T51. Probably the most significant tube was the Tektronix five-inch T54 shown in Figure 4.56 (later registered as the 5BHP2 in 1957) with four centimeters of vertical deflection and used in the famous Model 545 oscilloscope in early 1955. A high-voltage version of the T51 for the 24,000 volt 517-A oscilloscope was the T54P11H, which replaced the multiband Du Mont 5XP11 used in the older Model 517. This improved tube design increased the usable scan to 4 × 8 cm. These tubes were soon followed by the T55 (5CBP2) for the Model 515 in 1955, the T52

Figure 4.55 Tektronix T51 of later production (1955 on). (Courtesy of Tektronix, Inc.)

(5CAP1) for the models 525 television oscilloscope and 570 vacuum tube curve tracer in 1957, the three-inch T317 (Figure 4.57) for the portable Model 317 oscilloscope and the five-inch dual-gun T555 used in the Model 555 (not portable!) in 1959. All of these Tektronix CRTs, and several additional ones which evolved, went through a series of three or four nomenclature changes over the years they were manufactured. This usually entailed adding extra digits to the type number. For example, the T31 progressed to T316 and

Figure 4.56 Tektronix T54 and electron guns (1956). (Courtesy of Tektronix, Inc.)

finally T3160, while the T51 became the T0510. Although P2 generally was standard, all common oscilloscope phosphors were usually available for these tubes, including P1, P2, P7, P11, and later, P31.

In 1962 Hewlett-Packard joined the laboratory oscilloscope fray in earnest with their Model 175-A oscilloscope[34] featuring their G205 CRT (Figure 4.58) with a 6 × 10 cm usable screen area, 12,000 volt acceleration, scan expansion mesh, 50 MHz vertical bandwidth, brighter P31 phosphor and an internal black graticule, all designed to take away market share from the 4 × 10 cm, 30 MHz Tektronix 545-A, which had long since become the industry standard. Not to be outdone, the folks in Beaverton responded in 1964 with a major instrument upgrade, the 545-B and 547, which featured the new T5470 spiral-accelerator CRT (Figure 4.59) with 6 × 10 cm scan, an *illuminated* internal CRT graticule, P31 phosphor and 50 MHz bandwidth which soon left the 175-A in the dust. The T5470 also had the best spot size of any oscilloscope CRT that this author has seen. This was a mixed blessing since H-P was promoting the fact that their phosphor screen could not be burned at any setting of the front panel controls, which was perfectly understandable in view of their fatter CRT spot size. Such was not the case with the T5470 with its nice crisp spot and resulting high beam current density which was not "burn-proof" like the H-P. The author spent many hours burning T5470 screens in an effort to find the phosphor and process with the highest resistance to burning.

The "Great CRT War" was well under way. The next advance was the H-P 180-A in 1966 with a *rectangular* 8 × 10 cm CRT, 50 MHz

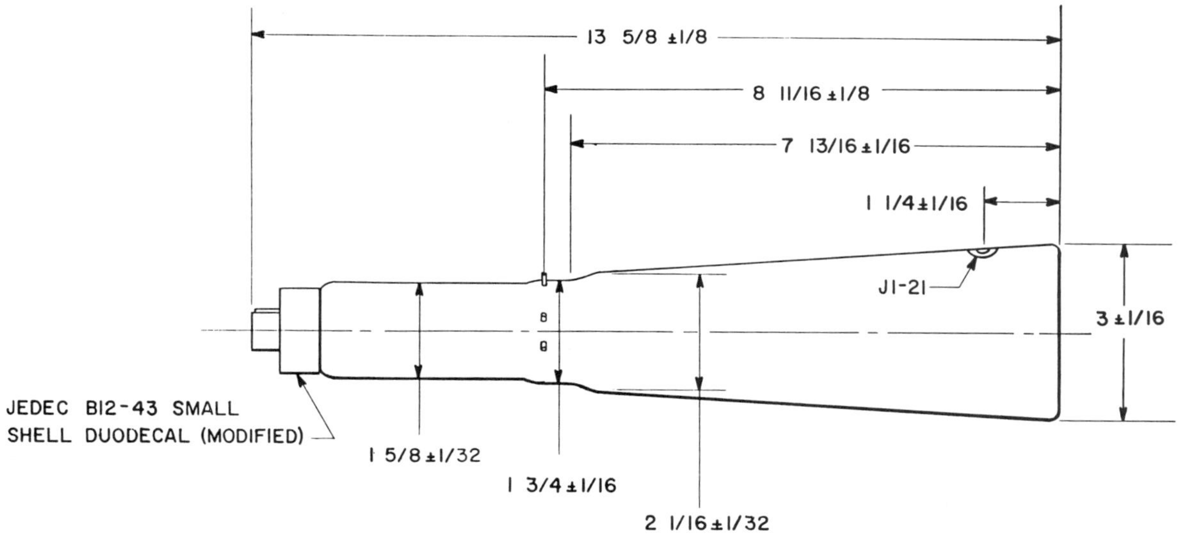

Figure 4.57 Tektronix T317 of 1959. (Courtesy of Tektronix, Inc.)

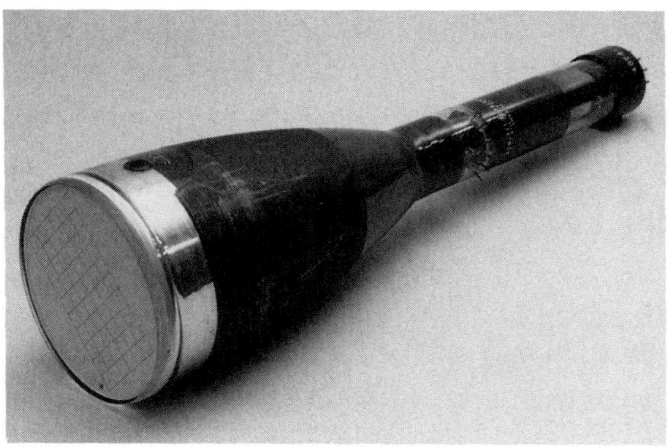

Figure 4.58 Hewlett-Packard G205 of 1962.

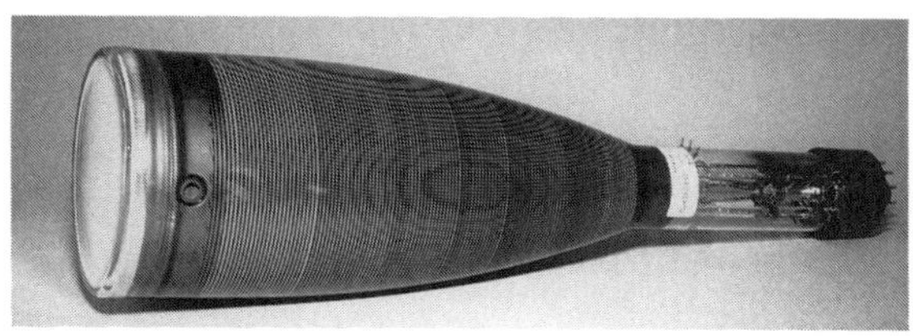

Figure 4.59 Tektronix T5470. (From the author's collection.)

bandwidth and a scan-expansion mesh instead of a spiral accelerator.[35] Tektronix responded with a volley of several new instruments, all with new CRTs. More about these later.

Many other companies in the United States and abroad produced spiral-accelerator CRTs during this period. The spiral accelerator was the standard for high-performance oscilloscope CRTs until the mid-1960s when scan expansion mesh CRTs became standard in the evolutionary process.

4.11 Distributed-Deflection

The bandwidth of oscilloscopes is limited by the capacitance of the deflection plates and the transit-time, or the time required for an electron to pass through the vertical deflection plate region. Neck-mounted deflection plate connections which had been used for years helped considerably by reducing the capacitance of the wiring to the deflection plates, but beyond about 50 MHz, the deflection plates themselves became too capacitive and transit times through the deflection plate region were long enough that phase reversals of the applied signal occurred during that interval.[36] Closer deflection plate spacing and longer plates which were used to increase the deflection sensitivity worked directly against bandwidth.

The traveling-wave principle appears in a patent application filed by A. V. Haeff in 1936, although the intended purpose was amplification and control of high frequencies rather than for CRTs.[37] In 1949, J. R. Pierce described a traveling-wave oscilloscope possessing a series of short deflection plates forming a terminated distributed capacitance delay line.[38] The use of a series of short deflection plates and delay line gave greater sensitivity than the very short deflection plates used in the earlier Du Mont K1017 high frequency CRT described in Section 4.7. Pierce's tube required a microscope for viewing the small waveform display, but had nearly flat frequency response up to 500 MHz. Several variations on this were reported in that period's literature by Lee in 1946,[39] Owaki et al in 1950,[40] Smith et al in 1952[41] and Germeshausen et al in 1957.[42] Only the latter appeared in a commercial product as the EG&G KR3 CRT, the first of its kind to be commercially available. It featured 2,000 MHz vertical bandwidth, but useful scan on the five-inch diameter tube was limited to only .4 × .6 inch.[33]

Tektronix applied the distributed-deflection concept to the type T581 CRT (similar in appearance to the T5470) for the 100 MHz models 581 and 585 oscilloscopes introduced in 1959.[43,44] A considerably more sophisticated distributed-deflection system was used by Tektronix in 1960 for their type T5190 CRT (Figure 4.60) employed in the Model 519 oscilloscope which had a bandwidth of one gigahertz.[45] The T5190 featured a spiral accelerator, 2 × 6 cm usable scan on a five-inch round flat-faceplate, P11 phosphor, .004-inch spot size, direct access (i.e., no vertical deflection amplifier) to the 125 ohm distributed deflection system and 24,000 volt acceleration-voltage. The distributed-deflection system had a number of internal tuning adjustments which were adjusted prior to sealing the gun to the bulb. The 519 oscilloscope was an important instrument for the 1960s nuclear weapons' testing programs and the tales of misadventures that they met with in that service were both colorful and plentiful.

Distributed-deflection recently has been used in a number of CRTs. Possibly the most noteworthy is the Tektronix T7100 (Figure 4.61) for the Model 7104 oscilloscope. Introduced in 1979, the T7100 CRT combined a distributed deflection system, box lens scan expansion and micro-channel plate to produce a real-time instrument with bandwidth in excess of one gigahertz.[46]

4.12 INTERNAL GRATICULES

The older blown-glass curved face CRTs made it very difficult to make accurate measurements from the screen due to parallax between the scale or graticule overlay because of the nonuniform distance between the graticule and the phosphor plane. The flat-faced tube greatly improved the situation, however

Figure 4.60 Tektronix T5190 one gigahertz CRT of 1960. (Courtesy of Tektronix, Inc.)

the spacing between the plane of the graticule (which usually was illuminated) and that of the phosphor (due to the thickness of the glass faceplate between) still resulted in parallax errors and the difficulty of always having the eye lined up in the same relative position for each measurement. Hewlett-Packard solved this problem in 1961 with their Model 120-B oscilloscope by forming the graticule internally on the glass faceplate before applying the phosphor screen (Figure 4.62). A finely ground black glass powder which was fused to the glass faceplate was used to withstand the acid rinses and high-temperature baking associated with CRT processing.[47] This graticule's only drawback was the difficulty of simultaneously photographing the graticule and display. The black lines would not show unless back-lit with "flood" electrons or unless the entire screen was illuminated with ultraviolet light to silhouette the graticule. Both techniques were used by Hewlett-Packard and had the side benefit of "prefogging" the film for higher photographic writing speeds. In 1964, Tektronix introduced the illuminated, titanium dioxide, photo-deposited internal graticule which allowed the use of conventional incandescent edge-lighting, similar to that used for external graticules. A Plexiglas™ combination light-pipe and implosion shield was laminated to the CRT faceplate to conduct the light to the graticule markings. Etched graticules were also investigated in about 1965. In the late 1960s the use of a separate faceplate frit-sealed to the funnel allowed a considerable improvement in the visual appearance of the graticule by lending itself well to the application of screen-printed, illuminated graticules using colored frit applied to the inner glass surface. This technique remains the process in use today.

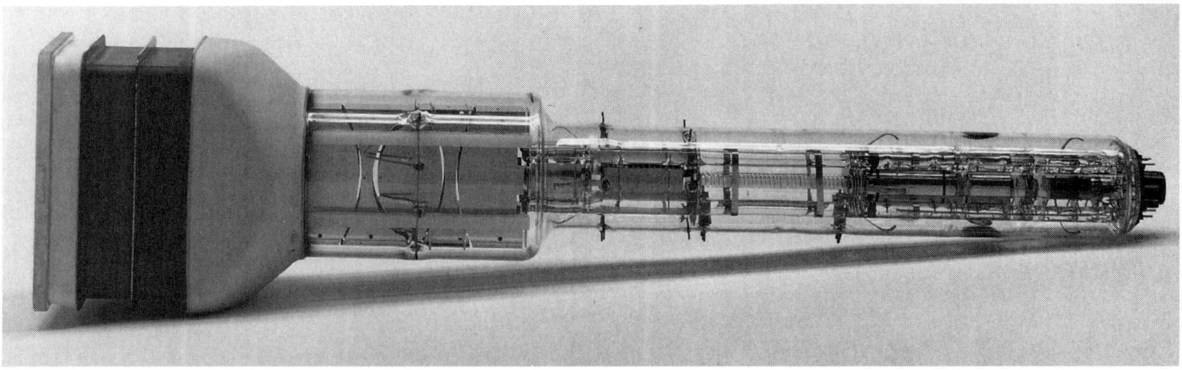

Figure 4.61 Tektronix T7100 one gigahertz CRT of 1979. (Courtesy of Tektronix, Inc.)

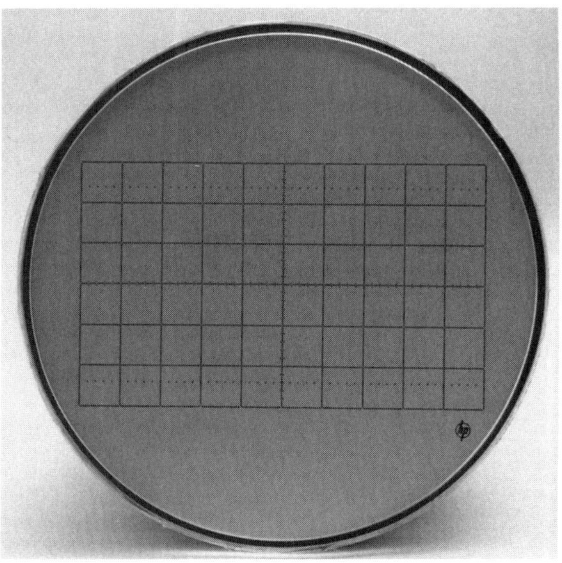

Figure 4.62 Hewlett-Packard black line internal graticule as used for their G205 CRT.

4.13 RECTANGULAR SCREENS

At about the same time that rectangular screens were displacing round screens in television receivers, the first rectangular CRT for oscillography appeared. This was the Waterman Products three-inch type 3SP1 (Figure 4.63) of about 1950. The primary advantage

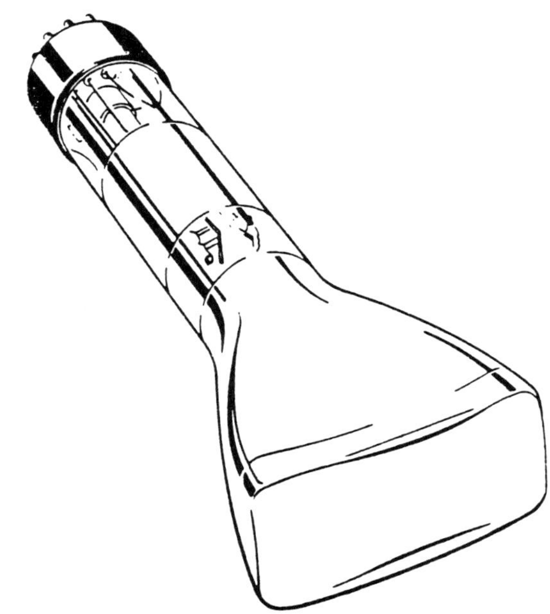

Figure 4.63 Waterman 3SP1 rectangular screen CRT, circa 1950. (Courtesy of Electronic Industries Association.)

was the saving of space in portable oscilloscopes and Waterman used it to advantage in their Model S-15A. Since its curved face had a height of only 1.5 inches and a width of three inches, they were able to stack two tubes vertically in little more than the space required for a conventional three-inch round CRT. The result was a dual-gun oscilloscope of sorts although the beams obviously could not be superimposed for direct comparisons. The 3SP1 used an electron gun borrowed from the 3RP1 with a new glass envelope and was available with P1, P2, P4, P5, P7 or P11 phosphors. Waterman was a specialized manufacturer in that they produced only three-inch CRTs and oscilloscopes through the 1950s, with the exception of their first "Pocketscope" of 1946 which employed a two-inch 2AP1 CRT purchased from RCA.

The General Electric GL-3UP1 shown in Figure 4.64 (GE used the "GL" prefix to denote their tubes but this was not included in the EIA registration designation) was announced in 1951. It had a screen width of only two inches and a height of 1.5 inches but had a three-inch designation. The 3UP1 used the gun from the 3MP1 and shared its short length. Because of that short length, deflection sensitivity of the 3UP1 was not spectacular and the author is unaware of any product actually using it.

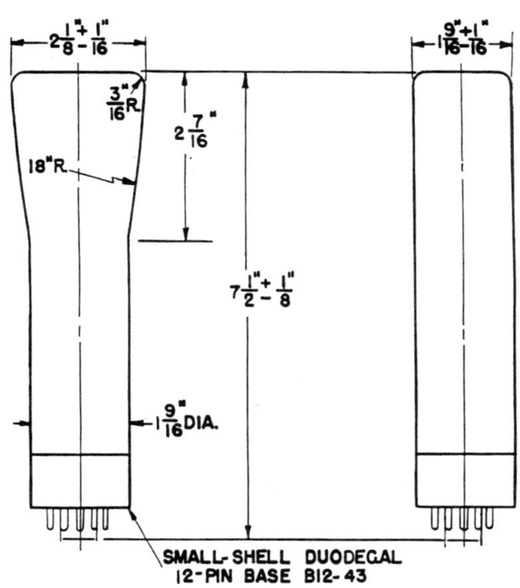

Figure 4.64 General Electric 3UP1 of 1951. (Courtesy of Electronic Industries Association.)

Waterman also developed two unusual rectangular CRTs during the early 1950s, the 3XP1 (Figure 4.65) and 3YP1 differing primarily in length and hence deflection sensitivity. The lengths were 8-7/8 and seven inches, respectively, and the screens were 1-1/2 × 3 inches. Both used the eight-pin Loktal base used for radio tubes and the second anode contact was made through the metal base key and shell, the only tubes of their kind except for the Du Mont 3AYP1 of 1959 which had an improved electron gun and was designed as a replacement for the 3XP1. The 3XP1 was used in the Waterman Model S-6A oscilloscope and also was manufactured by Du Mont.

The rectangular CRT offered two advantages. First, it made smaller oscilloscopes possible, which became a matter of greater importance in the 1960s as solid-state electronics displaced the vacuum tube. Secondly, scan often had been limited in newer CRTs as a trade-off for bandwidth and deflection sensitivity. High-sensitivity round tubes had large wasted areas at the top and bottom of the screen which could not be reached by the electron beam. Other than a few tubes of relatively low volume, however, the rectangular oscilloscope CRT did not displace the round tube overnight as it had in television. It was not until the early 1960s that it began to be

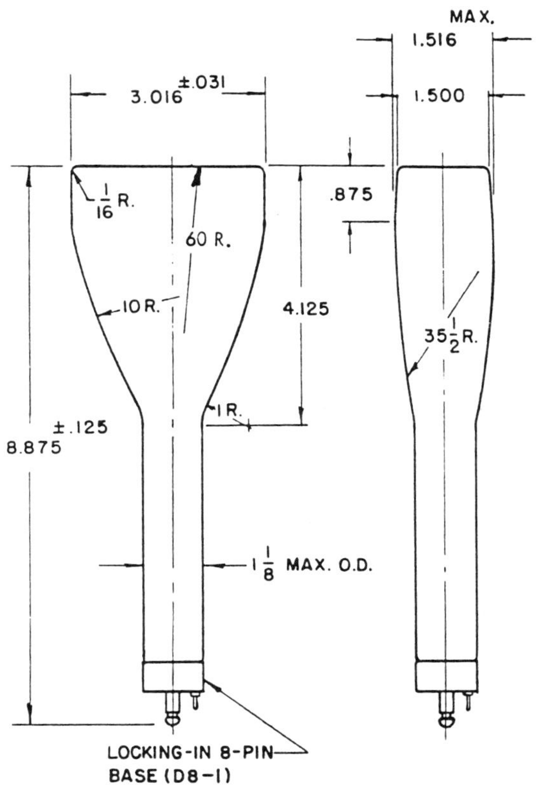

Figure 4.65 Du Mont version of 3XP1 originally registered by Waterman in 1952. (Courtesy of Electronic Industries Association.)

used in higher-performance instruments. It was used for the five-inch CRT (the most widely used laboratory oscilloscope screen size) in 1960 as the Du Mont type 5BXP- and in 1962, it was used in the Tektronix non-registered T503RS, which later evolved to the T5031 and T5032 (Figure 4.66). Soon after, Tektronix began to use the rectangular screen in almost all new instrument designs, even those using a spiral accelerator. This was made possible by use of a cylindrical portion of the funnel containing the helix gradually changing to a rectangular shape near the screen.

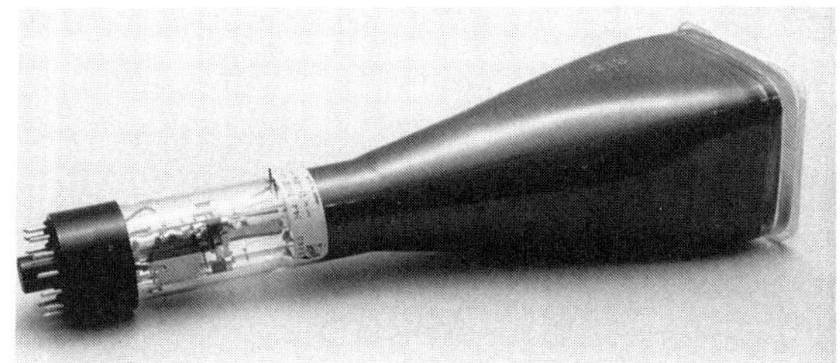

Figure 4.66 Tektronix T5032 8 × 10 cm rectangular CRT (late production).

Tubes of this configuration included the 6 × 10 cm T6470 (Figure 4.67) for the Model 647 oscilloscope, the T4220 (Figure 4.68) for the Model 422 and the T4530 for the Model 453. All of these later changed from all-glass envelopes to ceramic funnel versions during the late 1960s. Today it is a rarity to encounter a round screen CRT in a modern oscilloscope.

4.14 Ceramic envelopes

Ceramic CRT funnels are a unique Tektronix innovation. They were first developed in 1961 by William Wilbanks et al[48,49] and were applied with particular success in 1963 to the five-inch rectangular T5640 storage tube (Figure 4.69). Actually, the first ceramic CRT was the rectangular T5610 (Figure 4.70) in-

Figure 4.67 Tektronix T6470 6 × 10 cm rectangular CRT of 1964.

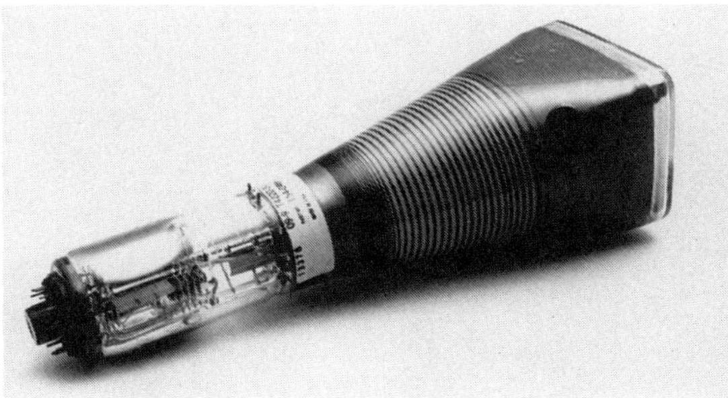

Figure 4.68 Tektronix T4220 of 1965, the first of several high-performance CRTs for portable oscilloscopes.

troduced the year before for their Model 561 A oscilloscope. Several reasons have been expressed for the use of ceramic rather than glass for the funnel. The benefits include utilization of excess capacity of the Tektronix ceramic plant used to manufacture ceramic terminal strips which were being made obsolete as printed circuits came into use; lower cost; ability to rapidly and relatively inexpensively produce new shapes and sizes; ease of making the multiple electrical contacts necessary for storage tubes; ability to fabricate designs that are impossible with glass; high sheer and tension strengths; and lessening the dependence on outside glass vendors and their occasional labor disputes. In all probability it was the combination of several of these factors that prompted the use of ceramic envelopes. They have been produced in many sizes and shapes ranging from the two-inch T2110 (Figure 4.71) to the 11-inch T6110 (Figure 4.72) and many of the best selling Tektronix CRTs were converted to ceramic envelopes. The use of ceramic was practical for CRT production runs of just a few hundred funnels per year, whereas the glass companies tended to require a few hundred thousand per year to justify tooling.

4.15 Scan-Expansion

By the early 1960s, the high-voltage, post-accelerator CRT had reached a high degree of refinement with the spiral accelerator and distributed-deflection systems. The transistor was beginning to displace the vacuum tube for oscilloscope circuitry. Further advancements were needed to improve the deflection sensitivity of the CRT to better match the lower output voltage swing possible with transistorized deflection amplifiers and to permit their use for the ever-increasing demands for wider bandwidth oscilloscopes. Electron-optical scan-expansion techniques were the key to the next wave of major oscilloscope improvements.

The concept of an electrically conductive mesh between the low- and high-voltage regions of a PDA CRT dates back to Scheller in 1920 although at that time the mesh was placed directly behind the screen. In 1939 Rogowski and Theilen used a mesh near the deflection plates to form an electron-optical lens for magnification or expansion instead of the compression usually associated with post-deflection acceleration and to prevent the accelerating fields from entering the deflection plate region.[50,51] The British Marconi Ltd. is reported to have been one of the first companies to have used the principle commercially.[52]

Figure 4.73 illustrates the effect of the scan expansion mesh on a deflected electron beam.[44,45,96] The gains in improved deflection sensitivity and the ability to use larger amounts of screen area without the deflection plates intercepting the beam were tempered by the larger spot size which also was magnified

Figure 4.69 Tektronix T5640 storage tube, late production with hard-pin base instead of over-capped type. (Courtesy of Tektronix, Inc.)

proportionally to the deflection expansion, some loss of beam current due to interception by the mesh and ghosting of the displayed waveform by secondary electrons emitted from the mesh.

The first domestic use of scan expansion was in the Hewlett-Packard G205 CRT used in their Model 175-A oscilloscope of 1962.[34] That was the opening skirmish of the previously mentioned Great CRT War between Hewlett-Packard and Tektronix. This CRT had 6 × 10 cm scan area, 50 MHz bandwidth and an internal graticule. It was designed to present a substantial improvement over the popular Tek 545-A oscilloscope with its 4 × 10 cm scan, 30 MHz bandwidth and external graticule. The first Tektronix use of scan expansion was in the 1963 all solid-state Model 647 employing the rectangular 6 × 10 cm T6470 CRT using a frame grid (Figure 4.67).[52] The frame grid is similar to a scan-expansion mesh except that it is cylindrical rather than spherical. This gave expansion in the vertical axis only. The 647 never achieved large volume sales, thus it remained for the hybrid Model 547 using the five-inch round T5470

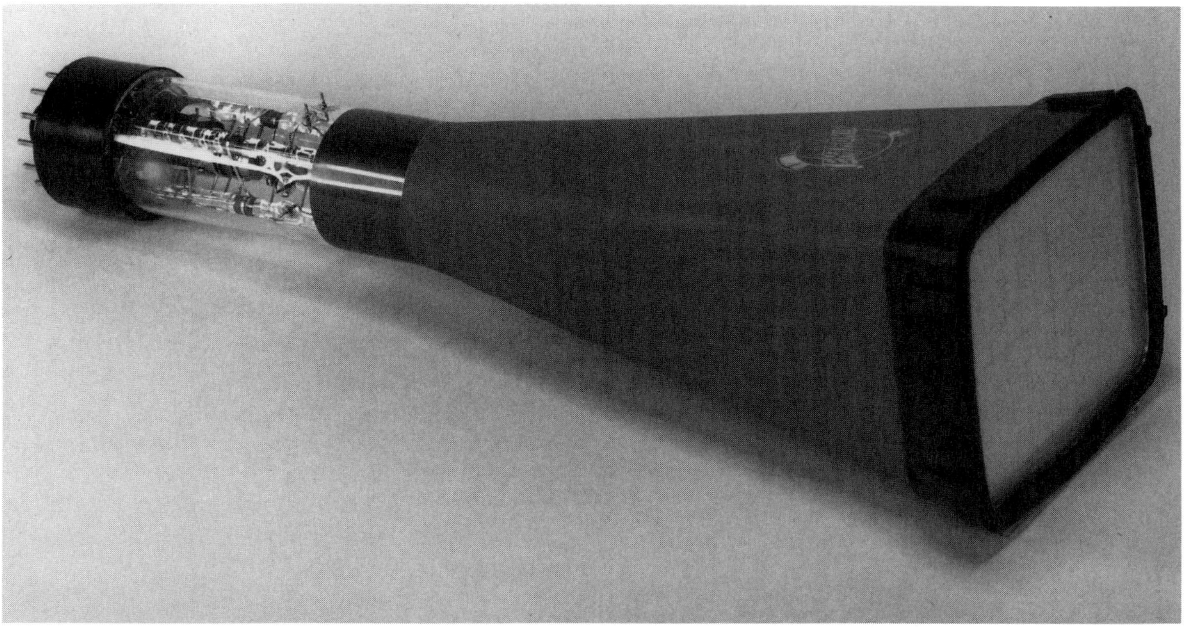

Figure 4.70 Tektronix T5610 ceramic envelope CRT of 1963. (Courtesy of Tektronix, Inc.)

CRT introduced in 1964 to regain the lead. The 547 matched the performance of the H-P 175-A oscilloscope without resorting to a scan-expansion mesh, therefore it possessed a far better spot size.

H-P next countered in 1966 with the Model 180-A using the scan-expansion mesh to achieve a shortened rectangular CRT with 8 × 10 cm scan and 50 MHz bandwidth in an all solid-state oscilloscope which was considerably smaller and lighter than previous instruments.[53] The 180 series of instruments did have a noticeable impact on Tektronix scope sales and were H-P's most significant oscilloscopes.

Also in 1966, Tektronix announced the first in what was to become a highly successful series of high-performance portable oscilloscopes using scan-expansion mesh CRTs. This was the Model 453 using the four-inch rectangular T4530 CRT, also with 50 MHz bandwidth. The 150 MHz Model 454 built around the distributed deflection T4540 (Figure 4.74) expansion-mesh CRT in 1967 proved that performance equal to the best of laboratory instruments could be packed into a highly portable oscilloscope eminently suited to computer field service applications, as well as the design bench.

In 1970 the HP183-A oscilloscope with 250 MHz bandwidth used the next scan expansion CRT (Figure 4.75) in the oscilloscope sweepstakes.[54,55] Progress did not stop here as the Tektronix models 465, 475 and 485 proved in 1973 with their larger screens while retaining the highly portable configuration. These were specified at 100, 200 and 350 MHz and used the T4650 (Figure 4.76), T4750

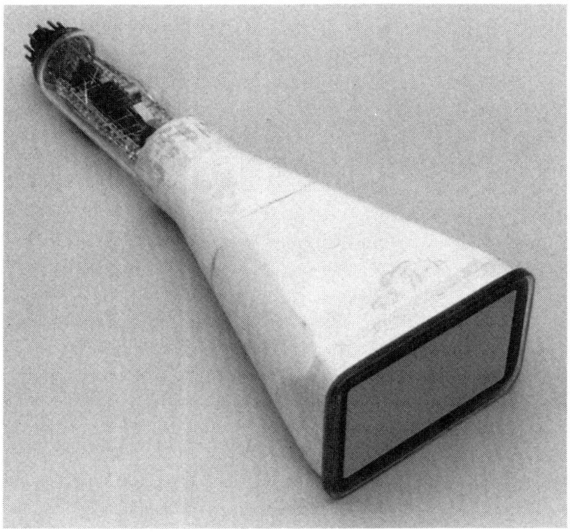

Figure 4.71 Tektronix T2110 two-inch ceramic envelope CRT for hand-held portable oscilloscopes.

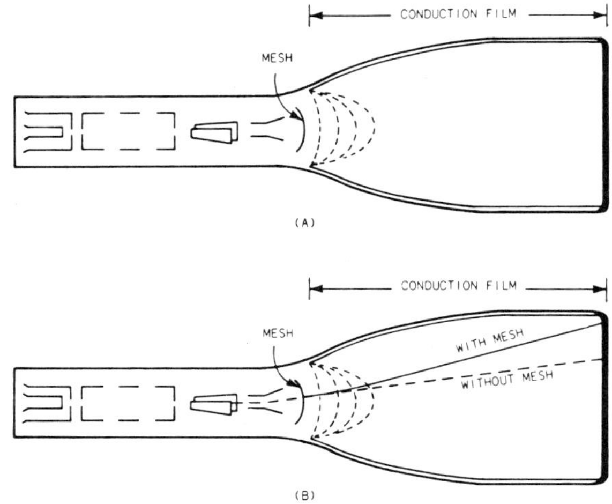

Figure 4.73 Scan expansion mesh operation showing increase in deflection sensitivity.[44] *(Courtesy of Tektronix, Inc.)*

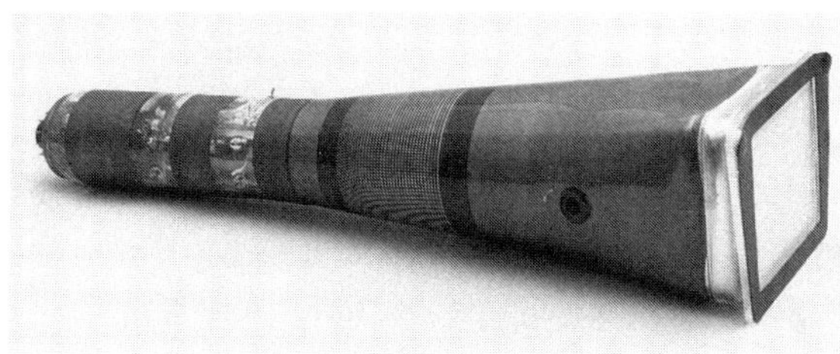

Figure 4.74 Tektronix T4540 CRT, early production.

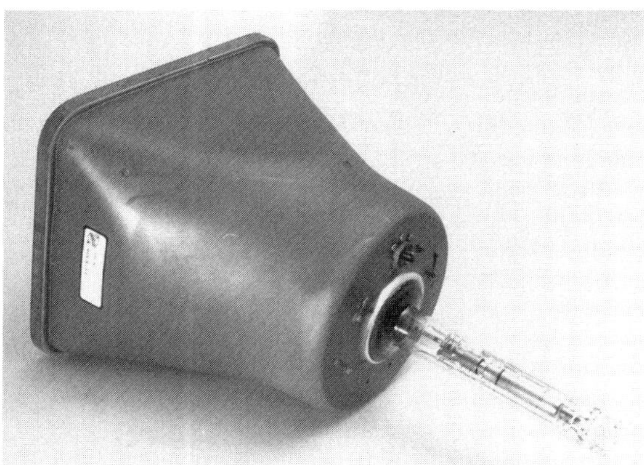

Figure 4.72 Tektronix T6110 11-inch ceramic envelope storage tube with electromagnetic deflection.

and T4850 CRTs respectively. The first two were actually five-inch rectangular tubes despite their type numbers which would imply four inches. Initially, the first digit of Tektronix tube numbers indicated the size. Later, the type number often was based on the instrument model number, the 400 series being the high-performance portable line.

A particularly unusual scan expansion CRT was developed in 1967 by Hewlett-Packard for their 20 Mhz Model 1300A X-Y display

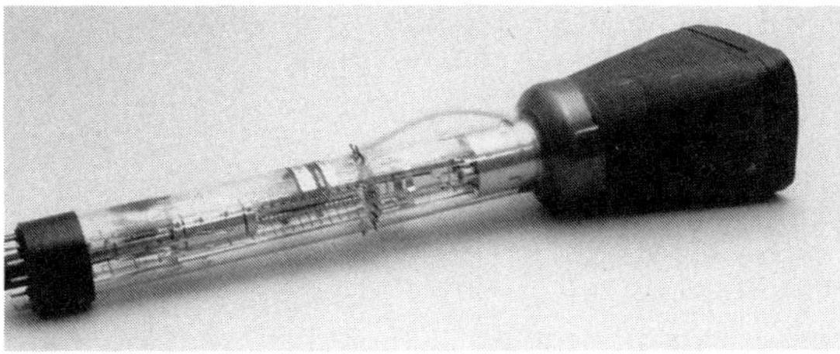

Figure 4.75 Hewlett-Packard CRT for the Model 183-A oscilloscope.

(Figure 4.77). The 14-inch diagonal screen had an 8 × 10 inch useful area and was held to 17-3/4 inches in length using a scan expansion mesh and 110-degree television bulb.[56]

The drawbacks of a domed mesh, interception of beam current, focus degradation and secondary emission, led to the development of a box lens in 1977 by Odenthal[57–59] of Tektronix and which was based on a concept by Klemperer[60] and El-Kareh.[61] The box lens is divided into sections having curved gaps to form a series of electron-optical lenses as illustrated in Figure 4.78. These have different magnifications for the vertical and horizontal axes and geometry adjustment may be made by adjusting the voltages on each element. The box lens was used in the T7100 CRT for the state-of-the-art Model 7104 1 GHz oscilloscope announced in 1979.

Another approach was taken by Amperex for their type D14-360 CRT in 1982 to eliminate the problems associated with the scan-expansion mesh.[62] They used permanently magnetized wire rings to form lenses for expansion, beam centering, orthogonality correction and beam shaping. These rings were magnetized from outside the finished tube using techniques developed for convergence correction for color television picture tubes more than three years earlier.

The current state-of-the-art scan-expansion method is the meshless scan expansion (MSE) system by B. Janko of Tektronix in 1977.[63–69] The complexity of this highly refined electron gun with three quadrupole lenses and the "elkhorn" scan expansion lens system is shown in Figure 4.79. It is used in the

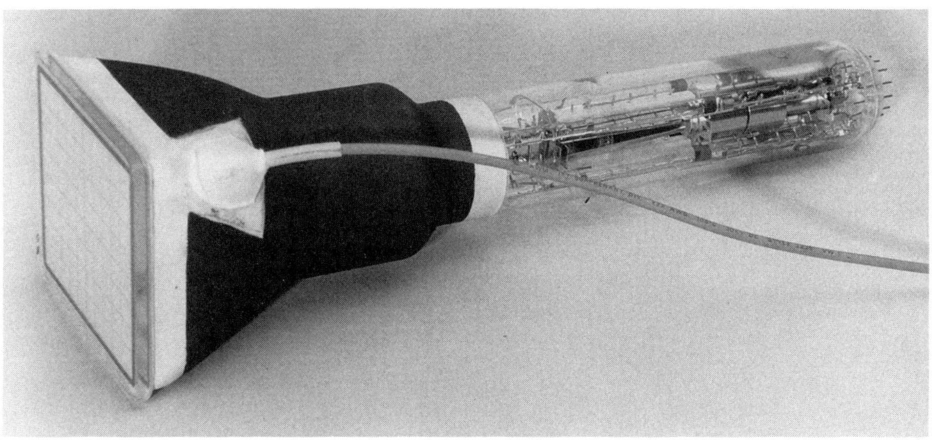

Figure 4.76 Tektronix T4650 ceramic envelope CRT.

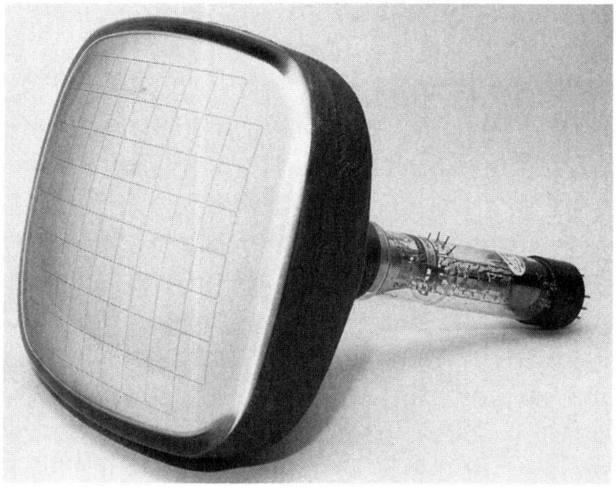

Figure 4.77 Hewlett-Packard short 14-inch electrostatic deflection CRT. (From the author's collection.)

T2440 CRT for the Model 2465 oscilloscope and other high-performance instruments. Advantages include shorter overall tube length, large PDA ratio of 8:1 for a bright trace, 3X expansion in the vertical and 4.5X expansion in the horizontal, plus rugged lightweight design.

4.16 MICROCHANNEL PLATE CRTS

The microchannel plate (MCP) is another CRT development that was used to achieve state-of-the-art performance. Its purpose is to increase the beam current, hence the brightness of high-speed waveforms. The MCP consists of a large array of microscopic conductive glass channels similar in construction to a fiber-optic faceplate and is mounted just behind the screen. The impinging electron beam causes secondary emission and electron multiplication from the interior channel walls as the electrons are accelerated through it toward the screen which has an applied high positive voltage (Figure 4.80).[44,70,95]

The MCP was first applied commercially to the Tektronix T7100 CRT for the Model 7104 1 gigahertz oscilloscope in 1979, the fastest real-time oscilloscope in the world.[46,71] Later it was used for the T2467 CRT in the

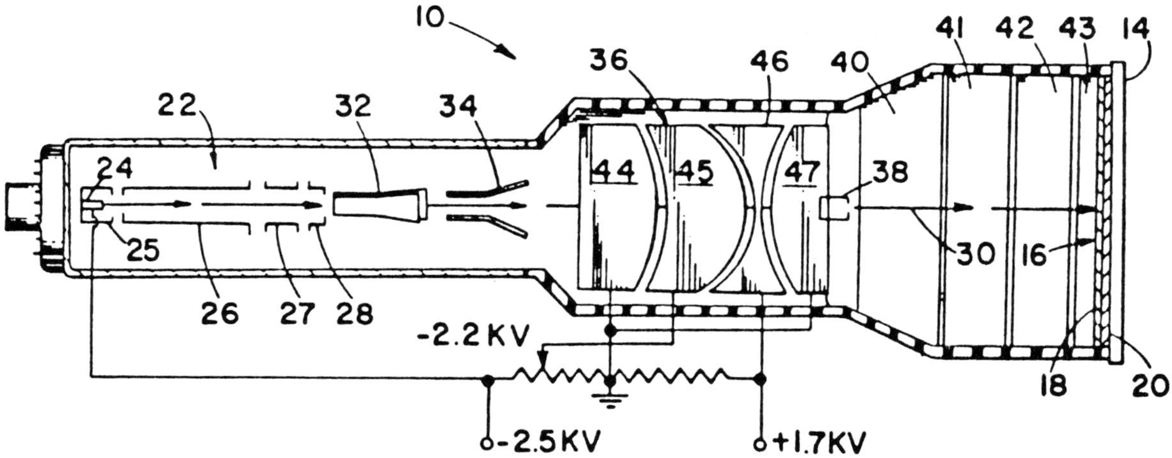

Figure 4.78 Box lens CRT (elements 44 through 47).[59]

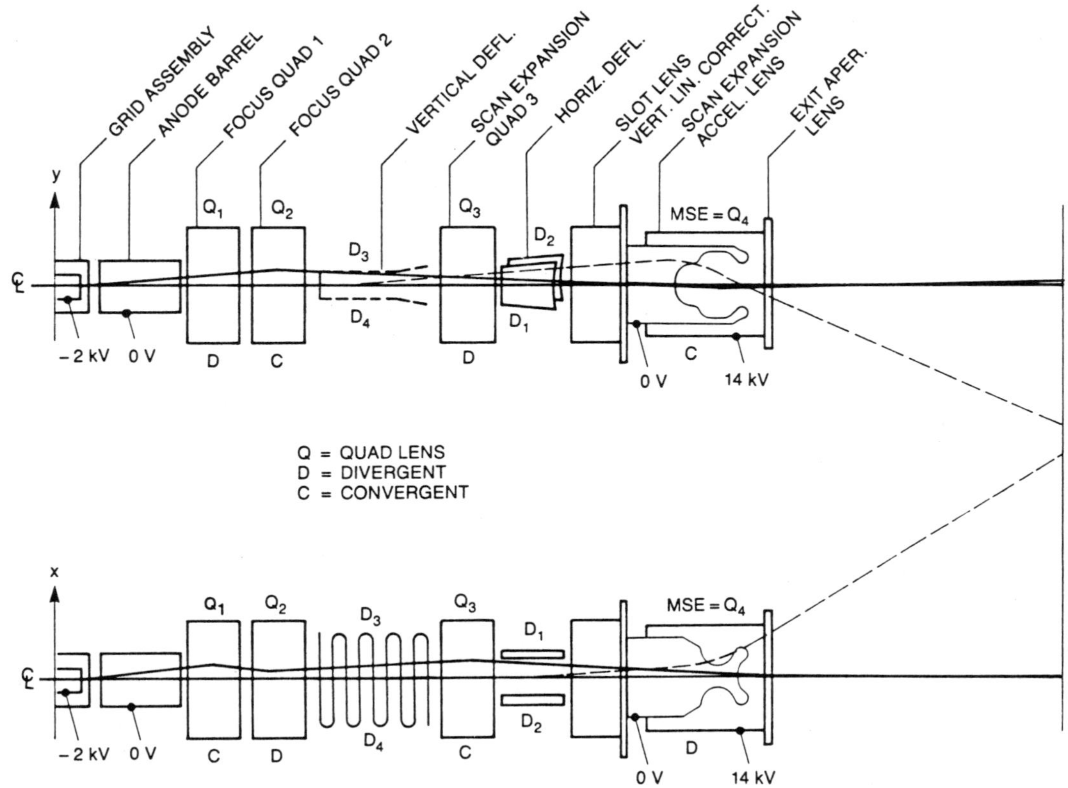

Figure 4.79 Meshless scan-expansion (MSE) lens CRT in vertical axis (top) and horizontal axis (bottom). (Courtesy of Tektronix, Inc.)

350 MHz portable Model 2467 oscilloscope in 1986.[72]

It is worth noting that all of the preceding oscilloscope advances were largely due to improvements in cathode-ray technology.

4.17 Storage Tubes

By the mid-1950s, the electrostatically deflected CRT was capable of excellent results in the field of recurrent and high-frequency waveform measurement. Very low-frequency waveforms and transient phenomena, however, were another matter. Long-persistence phosphors, such as P7 and P12, developed during World War II for radar were a partial solution to the problem of retaining these fleeting patterns long enough for interpretation, however subdued lighting and higher accelerating voltages also were required in order to view the low-level persistence beyond about one second. The direct-view storage cathode-ray tube provided a means of viewing waveforms for one minute or more at the expense of greater CRT and associated circuit complexity and cost. Generally, the storage tube is more prone to screen damage

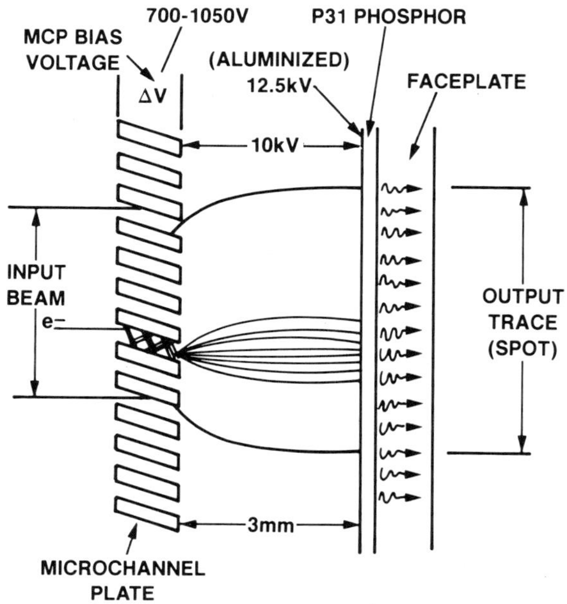

Figure 4.80 Microchannel plate for multiplication of beam current by secondary emission.[44] *(Courtesy of Tektronix, Inc.)*

from misuse so care is required to prevent formation of poor storage areas.

Secondary electron emission, the primary mechanism for electrical charge storage, was first identified in cathode-ray devices in 1902 by L. Austin and H. Starke.[73,74] Until the advent of the storage tube, it usually was more of a hindrance than a help to cathode-ray tube operation. Secondary emission caused charging of insulating surfaces within the tube which created pattern distortion and a phenomena known as *sticking potential*. This phenomena presented a limit which prevented the accelerating voltage from being increased for greater brightness. This was because the secondary emission ratio, which is depended upon to maintain the insulated screen at the accelerating voltage, drops below unity at some voltage and allows the screen to charge negatively due to the electron beam's deposit of negative charges.

Secondary emission is a function of the material being bombarded by the electrons and the electrons' acceleration voltage. Figure 4.81 illustrates an example of a curve of secondary emission ratio versus electron accelerating voltage. As acceleration voltage is increased, more secondary electrons are ejected from the material up to a point at which the quantity begins to decrease again due to trapping of the electrons at the greater depths to which they penetrate. The point where the secondary emission ratio is one, or unity, is significant because electrons arriving at either higher or lower energies will cause the insulating surface to charge away from that point—comparable to balancing a marble on a convex surface. A conductive backing electrode beneath the insulating material will introduce an upper voltage limit beyond which the material may not charge. By employing a writing gun and a flood gun, each at a different accelerating voltage relative to the backing electrode, different points on the secondary emission curve will be used simultaneously, thus permitting storage as will be described.

Cathode-ray storage tubes for oscillographic applications differ from those used in raster scan displays. Halftone rendition of images is not required, therefore the bi-stable storage, as described in 1947 by Andrew Haeff of the Naval Research Laboratory,[75] was simpler and entirely adequate for the purpose.

The Hughes Products type 6498 Memotron (1956) was the first CRT designed specifically for storage of oscilloscope waveforms and was used in their models 103 and 104 storage oscilloscopes. The Memotron uses

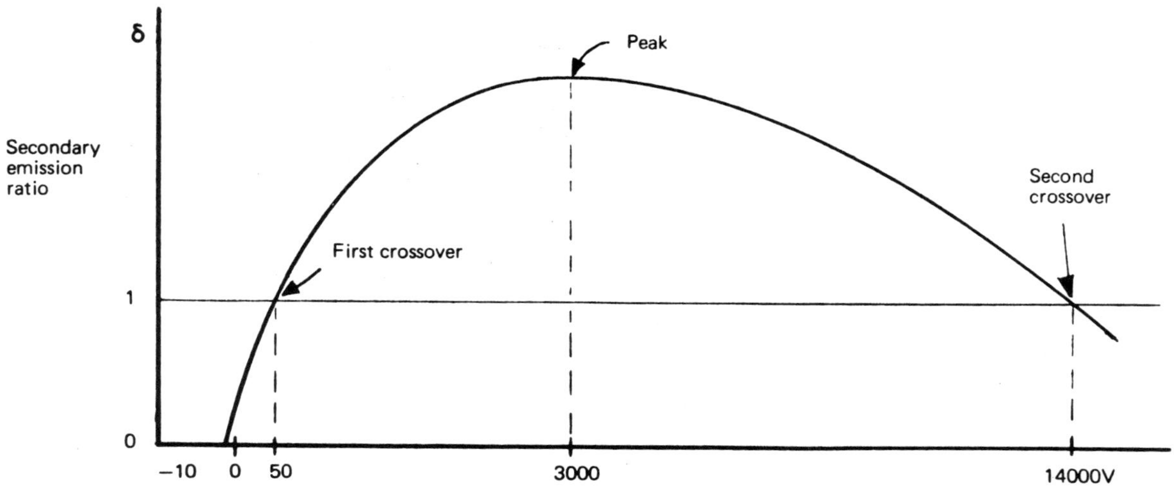

Figure 4.81 Secondary emission yield curve showing ratio of secondary to primary electrons versus beam landing energy.[44] *(Courtesy of Tektronix, Inc.)*

a storage mesh coated with an insulated secondary electron emitting material, such as magnesium oxide, and is located behind the screen. The storage mesh is operated at a voltage relative to the writing gun (similar to a conventional electrostatic deflection electron gun) such that the secondary emission ratio is greater than one. That is, more electrons leave the mesh than the writing beam deposits; the net effect is to leave a positive charge in the areas struck by the beam.

A *flood* electron gun operating at a low voltage relative to the storage mesh uniformly floods the entire mesh with *collimated* or parallel-traveling electrons. Initially, the entire mesh has a slightly negative charge and electrons from the flood gun are repelled back toward a slightly positive collector mesh. The result is a completely dark screen. Where the mesh is *written* with a positive charge, the electrons are accelerated through it toward the phosphor screen, which is at a considerably higher voltage, thus producing a bright image of the stored pattern on the screen. Because any ions generated within the large volume of the tube tend to limit viewing time, an ion repeller mesh, operated at an appropriate voltage to prevent ions from reaching the storage mesh, also is included.[17,76,77]

The next major innovation was the Anderson bi-stable, direct-view, storage tube developed by Robert Anderson of Tektronix during the late 1950s and first introduced commercially in the Model 564 storage oscilloscope in 1963.[44,78–85] This tube, the five-inch rectangular T5640, greatly simplified the construction since the phosphor screen did double duty as the charge storage mechanism. Willemite, the long-popular phosphor for low voltage cathode-ray tubes known as P1 (composed of zinc orthosilicate with a manganese activator), was an ideal storage material due to its secondary emission properties and ability to emit light efficiently when struck by the low-energy flood gun electrons. No meshes were used, resulting in a rugged, relatively low-cost tube.

The phosphor *target* is deposited on a flat glass faceplate having a transparent electrically conductive coating of tin oxide to provide electrical contact (Figure 4.82). Neck-mounted flood guns, whose cathodes are typically 150 to 250 volts (the tube's *operating point*) below the target's conductive backing where the secondary emission ratio is less than unity, cause the phosphor (which is an insulator) to charge negatively to the flood-gun cathode voltage as electrons are received. This condition is known as the *ready-to-write* mode.

The writing beam, whose cathode is about 3,500 volts below the target, has considerably higher energy and the secondary emission ratio of the target at that voltage is greater than unity, thus causing the phosphor to charge to the more positive level of the conductive backing in areas struck by the beam. This voltage increase on the phosphor raises it above the point where the secondary emission ratio is greater than unity for the flood guns and they maintain it at that voltage, producing light from these areas after removal of excitation from the writing beam.

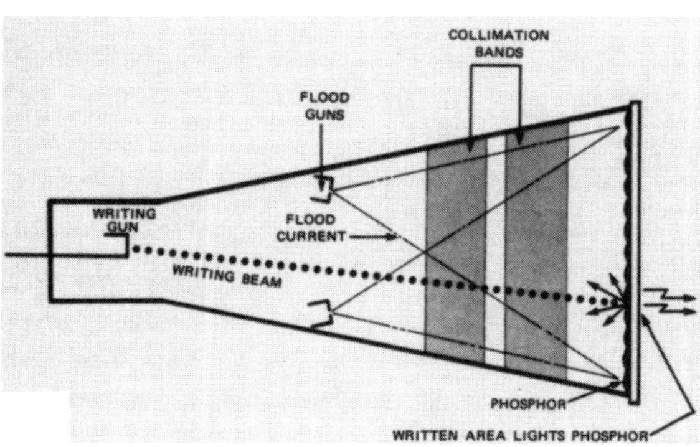

Figure 4.82 Schematic of direct view bi-stable storage tube. (Courtesy of Tektronix, Inc.)

Erasure of the patterns is accomplished by momentarily pulsing the target backplate voltage more positive to a point that the secondary emission ratio is greater than the one for the flood guns, which then raise the entire phosphor surface to the backing voltage in a condition called *fade positive*. The backplate then is suddenly dropped to the flood-gun cathode voltage and allowed to recover gradually to the operating point. Since the target is at a level at which the secondary emission ratio is less than one, the flood guns maintain the phosphor surface at their cathode voltage during the recovery period and a clean dark screen in the ready-to-write condition is presented.

Many improvements were made in the Anderson tube in subsequent years. Most changes involved the materials and processes for target structure and included refined target preparation, greater isolation of the phosphor in dot patterns and improved backing layers where the conductive collector was in the form of "islands" protruding through the phosphor layer (as in the Tektronix T5490 CRT [Figure 4.83] used in the Model 549 oscilloscope (1966). Target *writing speeds,* the ability to

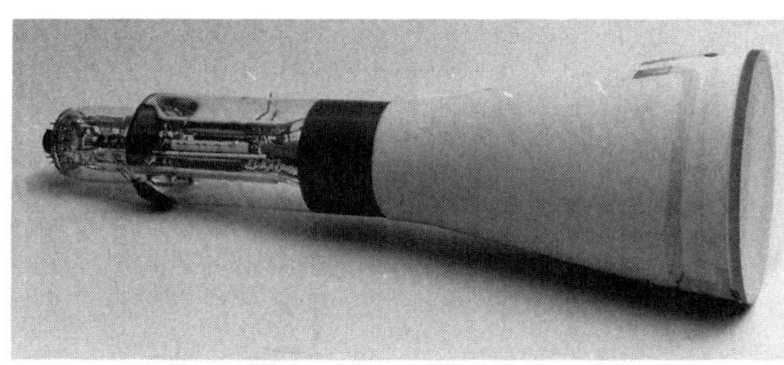

Figure 4.83 Tektronix T5490 wideband storage tube of 1966.

store very fast transient waveforms, were improved by the addition of better secondary emitters, e.g., magnesium oxide, although this cost some brightness degradation. In fact, low storage brightness was the main fault of the Anderson tube since the flood guns operated at such a low acceleration voltage and only 10 to 20 foot-lamberts were possible without appreciably shortening life. Nevertheless, many storage oscilloscopes and other displays were produced using these devices and are just now being replaced by digital storage techniques which permit a much simpler conventional CRT to be used.

The third important advance in direct-view storage CRTs for oscillography was the Hewlett-Packard five-inch, round, variable persistence tube introduced in 1965 in their Model 141A oscilloscope. Actually, the CRT was partially manufactured by Westinghouse Electric, who had considerable experience in manufacturing storage CRTs, with electron guns supplied by H-P.[86] Their 100 MHz Model 181A announced in 1968 used a rectangular version of the variable persistence CRT.[87] These CRTs could be used in any of three operating modes. The first was as a conventional CRT where waveforms of medium- to high-repetition rate would be viewed. The second was as a storage tube where single events would be recorded and viewed for periods up to one hour. The final and most innovative mode was the variable persistence mode where the trace could be made to fade out at some time after writing similar to normal long-persistence phosphors, except that the waveform could be retained for up to one minute. This was particularly useful where slow sweeps were used with recurring waveforms. By adjusting erase pulses applied to the storage mesh, the trace on the screen would fade from view just as the next waveform was written, providing continuous viewing. An added feature was the lack of an erase flash as found in other storage tubes. The tube construction was similar to that of the Memotron except for the absence of an ion repeller mesh. The main differences were in operating voltages and use of pulsed variable persistence erasure. Hewlett-Packard further improved the variable persistence storage tube in 1978 with an added high-voltage mesh and electron optical crossover lens between the electron gun and storage mesh, thus increasing the stored writing speed.[88] Additional variable persistence storage CRTs were later produced during the 1970s by Thomson CSF in France, Amperex in the Netherlands and English Electric Valve Co. Ltd. Figure 4.84 illustrates a rectangular H-P storage tube.

The final step in the development of the storage CRT for oscilloscopes was the T7410 transfer storage tube developed by Tektronix and first used in 1972 in the Model 7623 storage oscilloscope. The transfer storage tube had something for everyone: conventional nonstore, halftone storage, bi-stable storage and high-speed, bi-stable operation. Variable persistence also was available in the halftone mode. The added high-speed, bi-stable mesh (Figure 4.85) was the principal difference be-

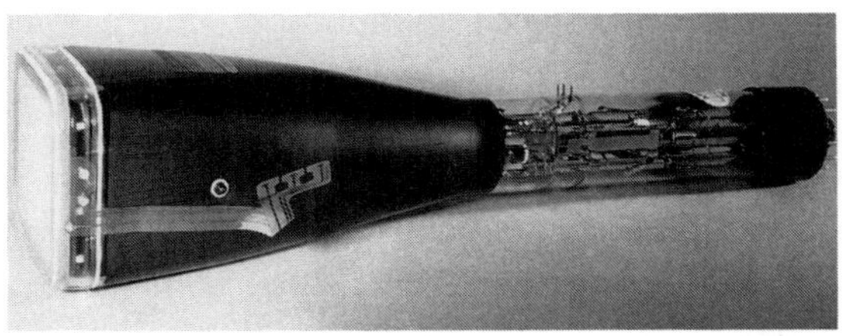

Figure 4.84 Hewlett-Packard storage tube (type unknown).

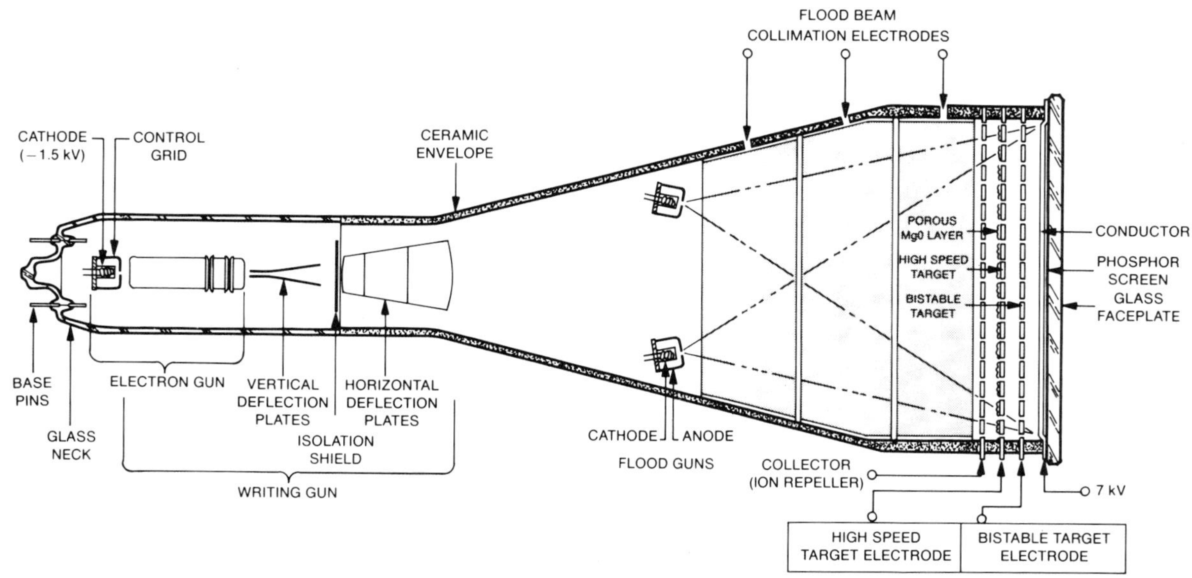

Figure 4.85 Tektronix transfer storage tube of 1973.[44] (Courtesy of Tektronix, Inc.)

tween it and the previous Hughes and Hewlett-Packard designs. This mesh overcame the problem of shortened viewing time that accompanied higher bi-stable storage writing speeds by first storing the high-speed waveform on the high-speed mesh and transferring it to the low-speed mesh for long viewing times.[44,94]

4.18 Digital Storage

Digital storage or digitizing oscilloscopes are not an actual CRT development, but are included herein because of their impact on CRTs used in these products. In recent years, the cost of digital memory for waveform storage has dropped dramatically while capacity and speed have increased greatly. It is now more economical to add digital storage to an oscilloscope than it is to use a storage cathode-ray tube. Digital storage reduces the requirements for CRT performance considerably, representing a cost savings. Further processing of the stored waveform data by the oscilloscope or an external computer also is possible for functions such as averaging, fast Fourier transform, peak measurements, comparison and hardcopy. An additional benefit is the ability to record higher speed waveforms. Several companies currently manufacture digital storage oscilloscopes with performance up to and beyond 300 MHz in a portable instrument.

The CRT requirement for wideband deflection is eased greatly since the digital circuitry handles the high-frequency signals and delivers a low-frequency stored waveform to the CRT. Even low-cost magnetically deflected CRTs are used with the stored waveforms converted to a raster scan format. Color CRTs also may be used for coding and identification of information. With all of these advantages, it is anticipated that the majority of future oscilloscope sales will be digital storage models.

4.19 THE WORLD BEYOND ONE GIGAHERTZ

Beyond one gigahertz, scan conversion tubes are used rather than conventional electrostatic CRTs. The scan converter tube is not strictly a cathode-ray tube per se, but performs the function of storing the acquired waveform and reading it out for viewing on a television type raster-scanned CRT having a larger and brighter display.

One early type of scan converter tube was the Tektronix T6011 (1969). This development was intended to overcome two of the main limitations of their direct-view bi-stable storage CRT at the time: lack of brightness and adequate display size. Fundamentally, the scan converter tube was a storage oscilloscope CRT optimized for electrical readout. Direct viewing of the stored image was also provided in this particular scan converter since the tube employed a phosphor storage target. Direct viewing was only of secondary importance, therefore the target was optimized to maximize the electrical output signal from the transparent conductive layer between the phosphor layer and glass faceplate. The waveform, or other information, was written by the electrostatically deflected electron gun. The same gun was used to read the stored image by raster-scanning across the stored charge patterns. Variations in beam landing energy of the read beam caused by the stored charge pattern resulted in an electrical current of varying amounts collected by the conductive layer on the faceplate. This signal was amplified by a video amplifier and fed to one or more standard television monitors scanned in unison with the read raster. The T6011 was useful as high as 10 megahertz.[89,90]

Scan converters came into their own for nuclear research requiring bandwidths beyond one gigahertz. The Tektronix T7912 scan converter CRT (Figure 4.86) (1981) was capable of recording 2.5 gigahertz signals. A "write" electron gun with traveling wave deflection system deposited charge patterns on a thin silicon wafer. These patterns were read by a low velocity "read" electron beam from a gun mounted in the opposite end of the tube scanning the back surface of the silicon wafer.[44]

EG&G has supplied three gigahertz scan conversion units for government nuclear applications. A seven gigahertz tube using an electron gun, a traveling wave deflector, a

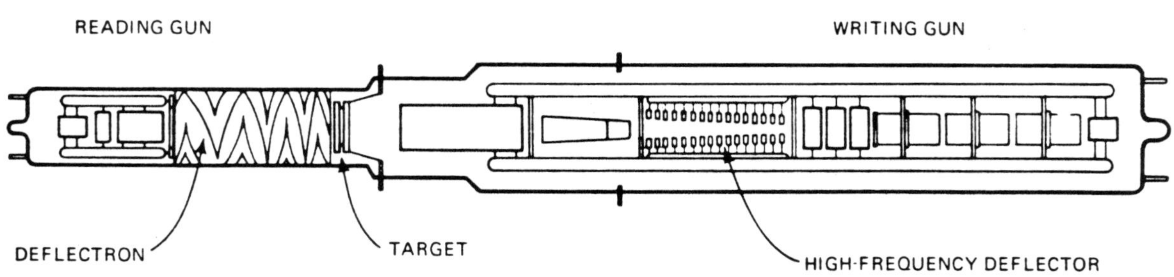

Figure 4.86 Tektronix T7912 scan converter tube of 1981.[44] (Courtesy of Tektronix, Inc.)

channel plate electron multiplier and an external conventional television camera tube to pick up the small image for display on a larger monitor were described by C. Loty of France in 1983.[91] This tube was used by Tektronix for their 7250 transient digitizer. A Tektronix-developed lower-cost scan converter tube, the T7950, was recently introduced for the Model SCD-5000 transient digitizer capable of recording and displaying 4.5 gigahertz signals. Other instruments using these and similar scan converter tubes have been developed in the USSR and France for recording single-event transients in the four to seven gigahertz realm.[92] Scan conversion techniques appear practical up to at least 10 gigahertz where signal handling with any degree of fidelity becomes increasingly difficult.

5

Monochrome Television Picture Tubes

5.1 Background

Up until 1940, the use of the cathode-ray tube in television was confined mostly to developmental work with commercial operation just beginning in the late 1930s. Von Ardenne demonstrated all electronic television in Germany in 1929. Zworykin was the first to use the cathode-ray tube in all-electronic television in the United States the same year while at Westinghouse Electric. He used a magnetically deflected tube which was one of the earliest high-vacuum CRTs.[1–3] Zworykin made further refinements to the CRT in the course of television research at RCA[4,5] especially the use of black conductive aquadag on the internal glass walls of the funnel to provide an anode electrode and to absorb reflected light from the screen to increase picture contrast.[6]

Until about 1935, most CRTs had screen diameters in the range of three to five inches. Television developments of the period pointed up the desirability of larger pictures for viewing by more than one or two people, thus the trend developed toward larger screens which persists to this day. The electromagnetically deflected CRT progressed to the commercial RCA type 903 and 1800 nine-inch tubes and RCA 1801 five-inch tube available for experimenters in 1935 to 1937.[7–9] The 1800 and 1801 employed yellow phosphor for a more pleasing picture rather than the previously used green phosphors. Fernseh A. G. of Germany had the largest tube of the period, a 25-inch which was used with the 441-line interlaced scanning just being developed in 1937. Both Fernseh A. G. and Telefunken were demonstrating projection systems that same year. The former used a hard glass five-inch bulb with magnetic deflection and focus. The latter was similar except for use of electrostatic focusing.[10] In 1939, RCA introduced the 12-inch type 1803P4 (Figure 5.1) which was soon registered with RMA as the 12AP4 and the nine-inch 1804P4 (Figure 5.2) which became the 9AP4. Both were used in several commercial television receivers introduced at the time of New York's 1939 World's Fair, which marked the start of regular television broadcasting in the United States. These tubes also were produced by National Union. Other magnetically deflected pre-war picture tubes included the Philco 6AP4 and 10AP4, RCA 7AP4 and 9BP4, and National Union 9CP4 and 12CP4 registered around 1940. Baird

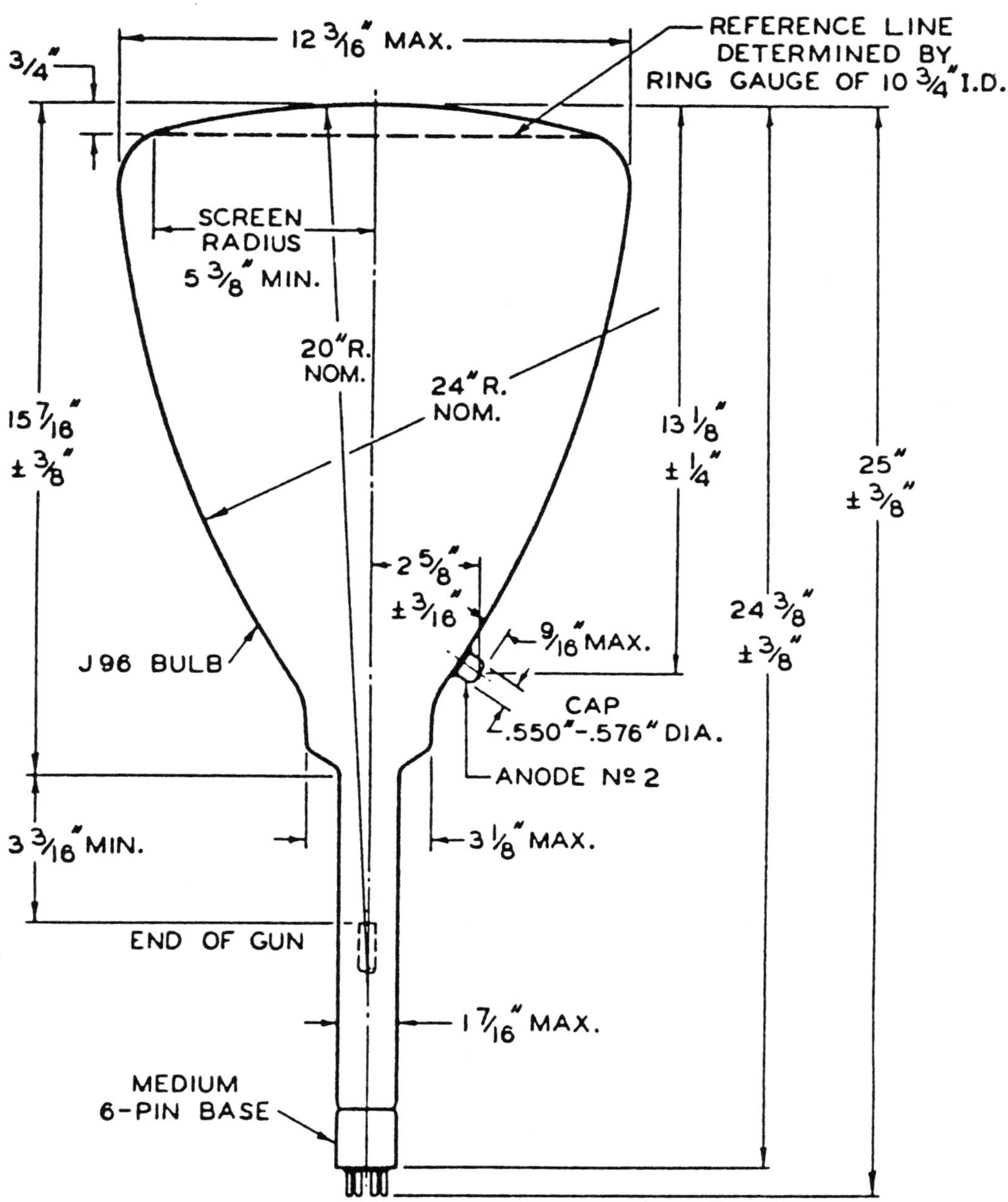

Figure 5.1 RCA 1803P4/12AP4 picture tube of 1939. (Courtesy of Electronic Industries Association.)

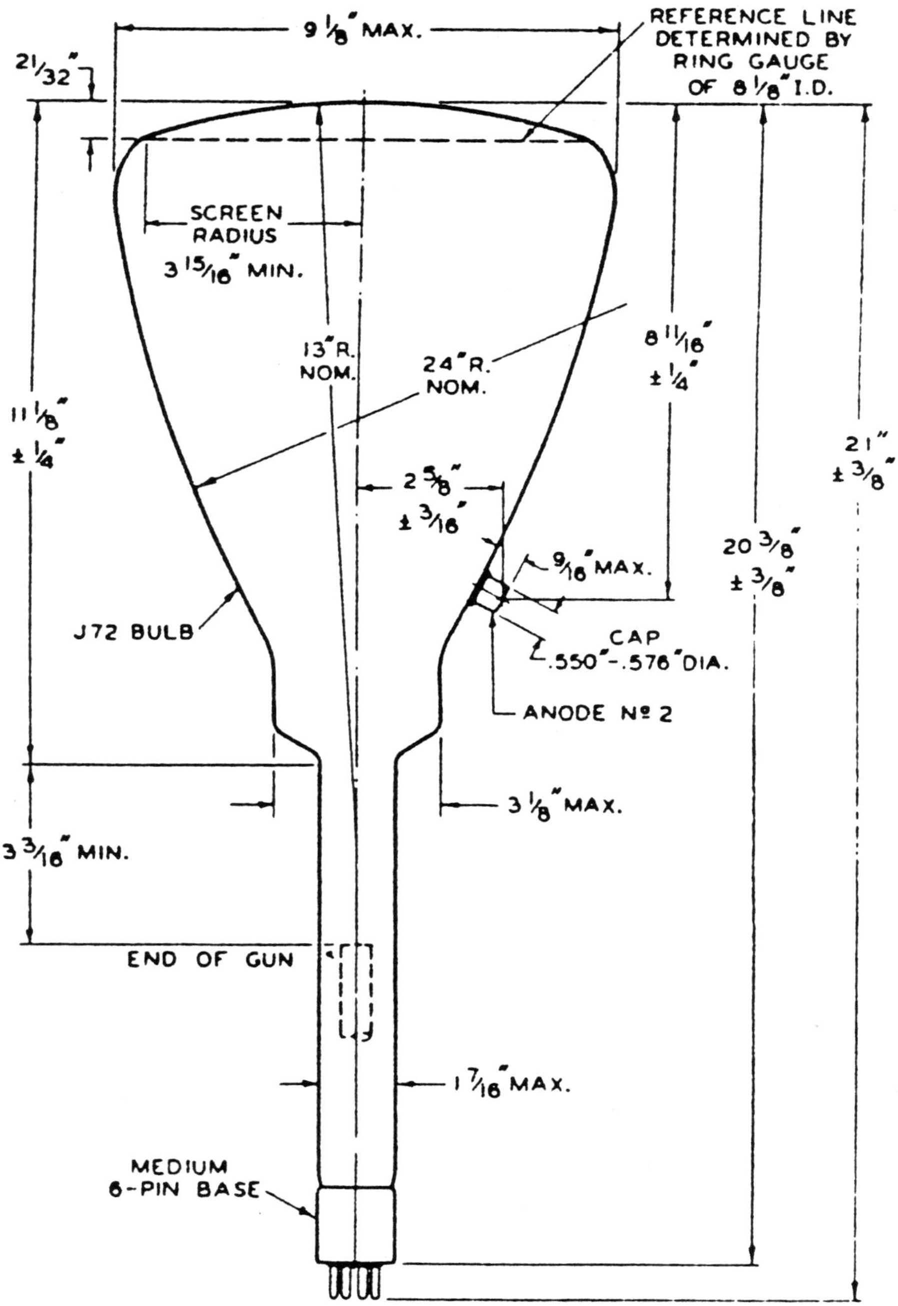

Figure 5.2 RCA 1804P4/9AP4 picture tube of 1939. (Courtesy of Electronic Industries Association.)

Television Ltd. in England was producing large magnetically deflected tubes in 12- (Figure 5.3), 15- and 22-inch sizes under the trade name Cathovisors.[11,12] The 1803, 1804 and Cathovisors all used white phosphors later registered with RMA as P4 and which became and remains the universally used monochrome television phosphor.

Electrostatically deflected CRTs experienced a similar evolution during the 1930s with Du Mont and Cossor championing them in the U.S. and England, respectively. In 1937, Dr. Du Mont visited Europe and saw large-screen electrostatic CRTs made by A. C. Cossor, Ltd. On his return to the U.S. he asked Corning Glass to make large bulbs. Corning's idea of a large bulb was nine inches, but at Du Mont's insistence they were able to produce strain-free 14-inch bulbs and later, 20-inch bulbs.

Du Mont introduced their Model 180, the "Clifton," 14-inch, electrostatic deflection television receiver at the 1939 New York World's Fair and, in 1940, the Model 195 television receiver using a 20-inch electrostatic CRT (Figure 5.4). Reportedly only 25 of the 20-inch receivers were manufactured. These large screen tubes were marvels of their time considering the difficulty of obtaining strain- and defect-free glass capable of withstanding the thousands of pounds of atmospheric pressure on them resulting from their high vacuum and large surface area. A further complication was the high temperature required for baking during processing to remove contaminants. Despite their large size, they were able to withstand five atmospheres of pressure.[13]

Most electrostatically deflected television receivers of the era were of the low-cost, small-screen type with five-inch CRTs (types 1802 and 1805) being used almost exclusively for complete receivers, kits and "attachments" to convert existing console radios. General Electric, RCA, Meissner,

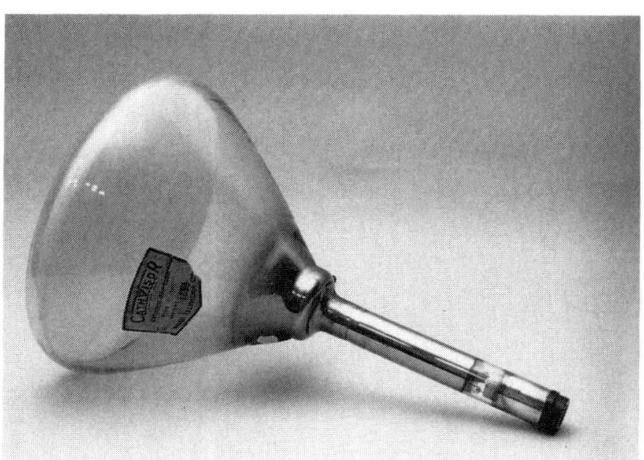

Figure 5.3 Baird 12MW2 Cathovisor of 1939. (Courtesy of C. E. "Sonny" Clutter.)

Figure 5.4 Du Mont 14- and 20-inch electrostatic deflection CRTs, circa 1940. (Allen B. Du Mont Collection, National Museum of American History, Smithsonian Institution.)

Andrea and Westinghouse all introduced five-inch receivers in 1939.

An unusual electrostatically deflected and focused television CRT was developed by C. J. Davisson with the assistance of C. J. Calbick at Western Electric.[70–72]

This CRT was constructed for use in the 1937 demonstration of a broadband coaxial cable link between New York and Philadelphia. To evaluate the capability of the link for television, a CRT having the best possible reproduction of video signals was desired. The result was a CRT that was probably the longest all-glass CRT ever built, measuring about five feet in length. The screen was 12 inches in diameter. Physically large electron gun elements permitted precise assembly and alignment.

Two new concepts were introduced with this tube: the use of deflection Z axis modulation[70] and contrast improvement through use of a separate glass panel fastened one to two inches in front of the screen with an airtight gasket. The space was filled with Nujol, a fluid having an index of refraction matching that of glass. This spread the light from the CRT spot, which is reflected by the glass' surface over a larger area, thus improving contrast.[73] Further improvement was proposed by adding a dark dye to the fluid to produce an effect similar to "filterglass" gray faceplates, which were introduced over 10 years later.

A continually pumped laboratory CRT in a similar vein was a 31-inch metal shell CRT by RCA also produced in 1937.[74] Electrostatic deflection also was used for this tube. The intent was to investigate large-screen television with a direct-view CRT instead of a projection tube. A beam current of eight milliamperes was used instead of the usual one-half milliamp.

World War II put a temporary halt to the fledgling home television industry. The interruption gave ample time to reexamine the question of transmission standards which had evolved during the late 1930s to the 441-line interlaced scanning system de facto standard of 1939. The Federal Communications Commission would not authorize commercial television broadcasting until the industry agreed on one set of standards. This led to the formation of the National Television Standards Committee (NTSC) in July of 1940, which formalized the present 525-line system that was approved by the FCC in July 1941. The war delayed any further growth of the television industry once a standard finally had been approved.[14–17]

5.2 WARTIME TELEVISION

Military television applications predominated during the period of 1942 through 1945. These included reconnaissance, pilotless aircraft and guided bombs. Cathode-ray tubes for image viewing were similar to the radar and oscilloscope tubes of that period. P1 green and P4 white screens were used instead of the long-persistence P7 screens common to radar. Electrostatic tubes such as the five-inch 5BP4 and 5CP4 probably had rather limited wartime television usage. The seven-inch, magnetically deflected 7CP1 and 7CP4 announced by RCA in 1943 are the most commonly encountered. The SCR-549 and SCR-550 airborne television systems, BC-1213 receiver and BC-1214 monitor all used the 7CP1 (Figure 5.5) instead of the pre-war RCA type 7AP4 monitor CRT since the green phosphor gave better image contrast in high ambient light conditions, especially when used

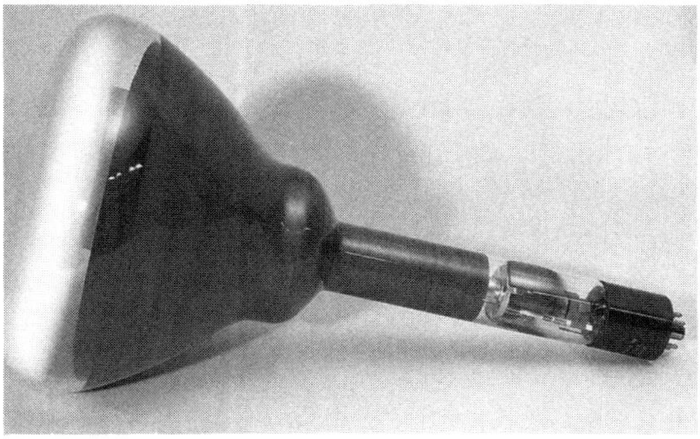

Figure 5.5 RCA 7CP1 manufactured in 1957. (From the author's collection.)

with a green light filter in front of the screen.[18] The 7CP1 also had an anode voltage rating double that of the 7AP4 because of its bulb anode connection instead of through the base. The increased anode voltage resulted in far brighter pictures and allowed operation at high altitude by removing the high voltage anode connection from the base.

5.3 EARLY POSTWAR—ELECTROSTATIC DEFLECTION

Electrostatic deflection was used briefly in the late 1940s for low cost and portable television receivers. Screen sizes ranged from three to 10 inches with seven inches being most common. Most were of blown glass, round screen bulbs with large radii of screen curvature. These represented some of the last uses of blown glass bulbs except for replacement oscilloscope CRTs.

The first postwar electrostatic CRT, the 7EP4 (Figure 5.6), was from Allen B. Du Mont Laboratories who were the strongest promoter of electrostatic deflection during the pre-war years. The 7EP4 was their last at-

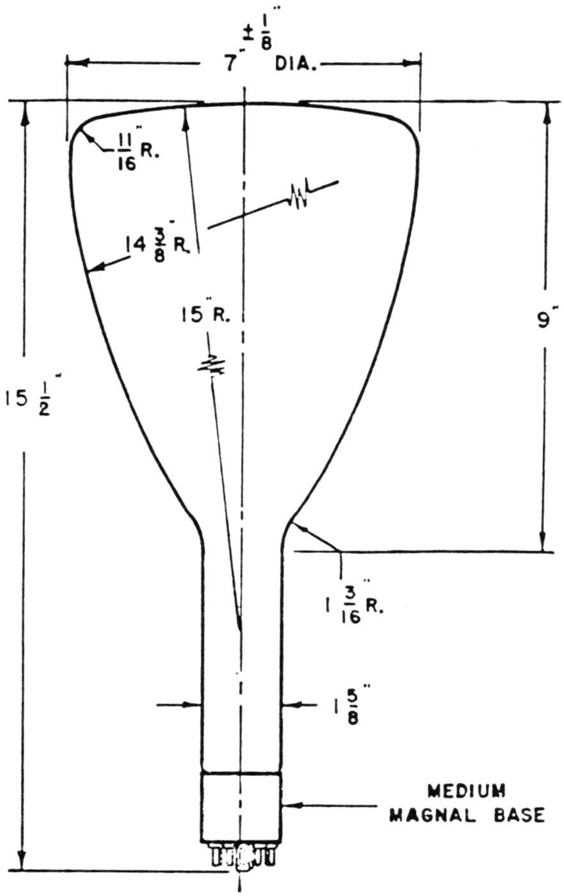

Figure 5.6 Du Mont 7EP4 electrostatic deflection picture tube of 1945. (Courtesy of Jacob Ruiter.)

tempt in 1945 and henceforth they confined the use of electrostatic deflection to oscilloscopes and related equipment. Du Mont continued to lead the industry in large screen sizes as they had during the 1930s. The seven-inch 7EP4 was not widely used and the author has not yet determined which manufacturers, if any, used it in commercial receivers. Its 11-pin magnal base with no separate anode connection limited it to an anode voltage of about 3,000 volts which would not produce a very bright picture.

RCA introduced the 7GP4 in early 1946. It used the larger 14-pin diheptal base and was rated at 4,000 volts. Sylvania Electric also produced the 7GP4. It was soon superseded by the RCA 7JP4 (Figure 5.7) in 1947 which became the most widely used electrostatically deflected CRT. The anode rating of 6,000 volts was the primary improvement over the type 7GP4. The seven-inch type 7JP4 was produced by RCA, Sylvania and National Union and was used in most portable models of the period. These included the popular Motorola VT-71 as well as models by Hallicrafters, National, Admiral, Belmont, Coronado, Crosley, Teletone, Airline, Automatic Radio, Sentinel, Emerson, Philco, Belmont and Raytheon from 1947 to 1950.

Two large-screen electrostatic CRTs were introduced in 1947 which allowed a receiver chassis designed for the 7JP4 to be used for a larger receiver as well. These were the 10-inch North American Philips 10GP4 and Sylvania 10HP4. Both were limited to 5,000 volts accelerating voltage which was inadequate for a 10-inch screen.[19] The 10GP4 and 10HP4 were similar except for bulb shape. The 10HP4 (Figure 5.8) was used in the Belmont Model 22AX22 receiver. No other usage has been identified for the 10GP4 and 10HP4.

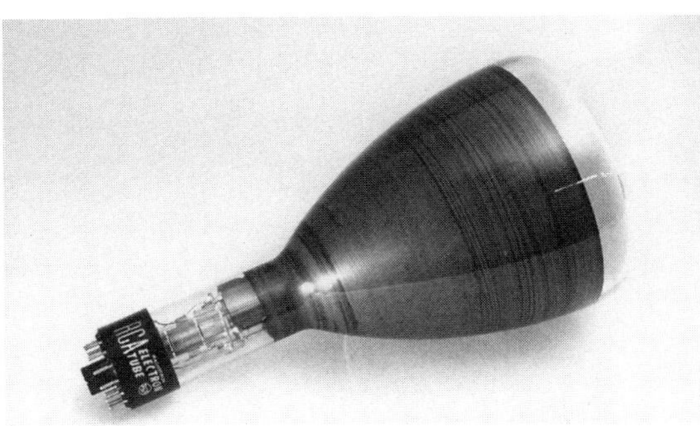

Figure 5.7 RCA 7JP4 electrostatic deflection picture tube manufactured in 1948. (From the author's collection.)

The three-inch RCA 3KP4 (Figure 5.9) registered in 1948 was used in the Pilot Model TV-37 "Candid" receiver of 1949. Except for the P4 white phosphor, the 3KP4 was otherwise identical to the 3KP1 oscilloscope CRT. This model receiver is highly prized by collectors of early television sets and 3KP4 CRTs for replacement purposes command a premium price.

Motorola also made a version of the seven-inch VT71 receiver using the National Union eight-inch diameter 8BP4 in 1949. Similar to the 7JP4 except in screen size, the 8BP4 (Figure 5.10) was the last electrostatically deflected CRT to be introduced for television and by 1950 the use of electrostatically deflected television picture tubes was dead.

5.4 EARLY POSTWAR—MAGNETIC DEFLECTION

1946 brought forth the first magnetically deflected picture tubes in what was to became a deluge of new tubes by 1950. The issue of electrostatic versus electromagnetic deflection was settled by 1950, the deciding factor being the advantages of magnetic deflection for larger screen sizes.

In 1947, magnetically deflected tubes were beginning to be manufactured with bulbs of hand-pressed glass rather than blown glass, as was the custom for electrostatic CRTs.[19,77]

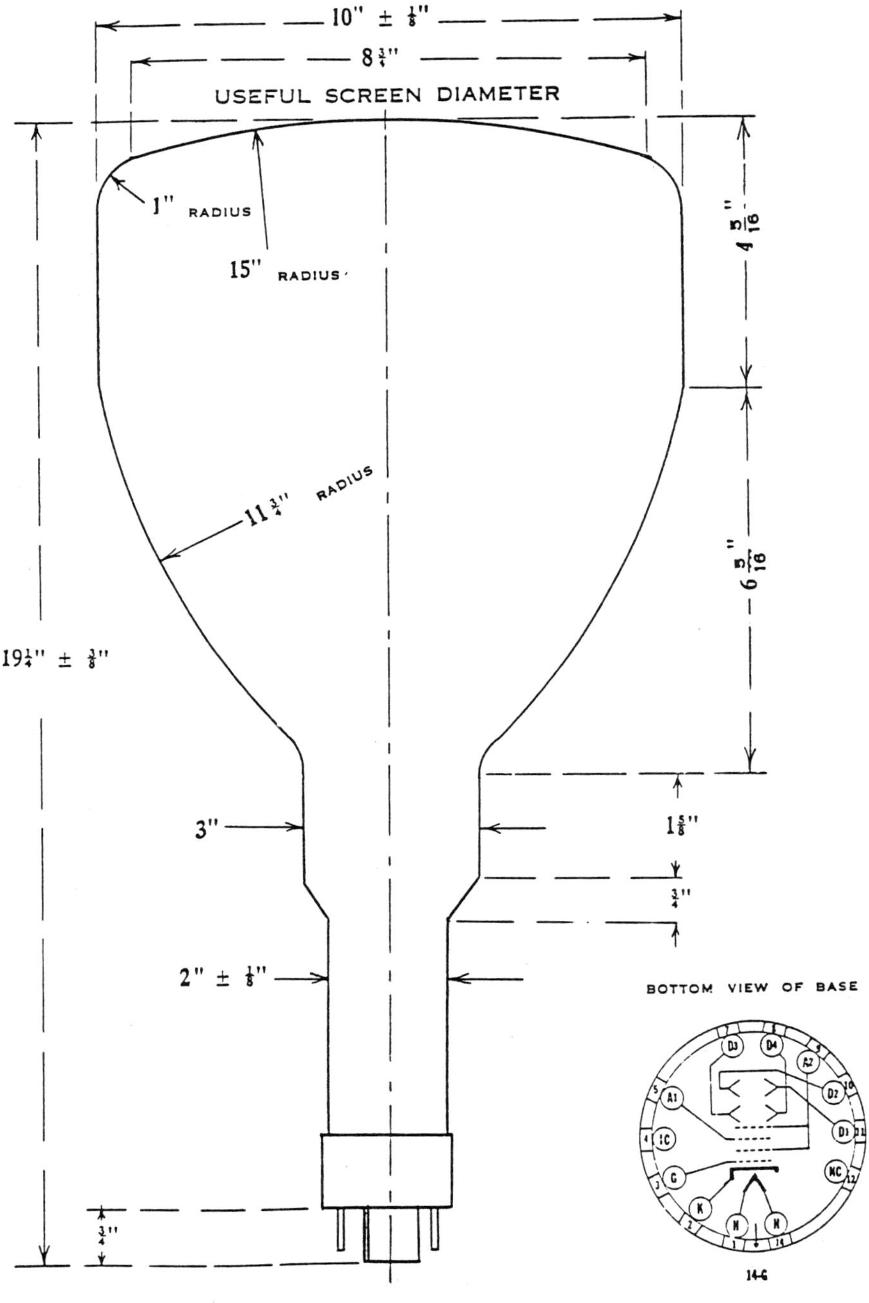

Figure 5.8 Sylvania 10HP4 electrostatic deflection picture tube of 1947. (Courtesy of Electronic Industries Association.)

Figure 5.9 RCA 3KP4 electrostatic deflection picture tube manufactured in 1948. (Courtesy of James Richardson.)

Figure 5.10 National Union 8BP4 electrostatic deflection picture tube manufactured in 1949. (From the author's collection.)

The most notable differences were a much flatter screen and a corresponding increase in weight due to the greater thickness of glass required for flatter screens to safely withstand atmospheric pressure.

The first electromagnetic-deflection picture tube registered with RMA in early 1946 was also the largest until 1950. This was the Du Mont 20-inch diameter 20BP4 (Figure 5.11). The 20BP4 was the first tube to use the new 12-pin duodecal base of which only five to seven pins usually are present. Almost all subsequent picture tubes introduced up until the mid-1950s used this base.[77] The anode contact for the 20BP4 was still of the pre-war medium cap type similar to that of receiving tubes. Only limited quantities of the 20BP4 were ever produced and as far as is known, only Du Mont manufactured receivers using it, referring to them as "deluxe class."

The 20BP4 was followed immediately by

Figure 5.11 Du Mont 20BP4 of 1946. (Courtesy of National Museum of American History, Smithsonian Institution.)

the registration of the 7HP4 and 10CP4 by North American Philips (Norelco) and the 5TP4, 7DP4 and 10BP4 by RCA. The Norelco tubes were never widely used. Both were magnetically focused and used the small-ball anode contact carried over from wartime radar and oscilloscope CRTs, but they were the first registered types to use a conductive aquadag (dag) coating on the funnel's outside surface. The coating was connected to the receiver chassis ground and, in conjunction with the internal anode coating, served as an additional filter capacitor for the high-voltage power supply. The external dag coating soon became standard for most picture tubes introduced thereafter.

The RCA 5TP4 was a projection CRT and will be discussed in Section 5.4. The 7DP4 (Figure 5.12) was a high-voltage electrostatic focus tube and was used only in the RCA Model 621TS table model TV receiver.[20] The 10BP4 (Figure 5.13), however, was another story. The widely licensed and copied RCA Model 630TS[20,21] designed around the 10BP4 was *the* classic television chassis in the early years of television. The 10BP4 became the best-selling picture tube of the 1946 to 1950

Figure 5.12 RCA 7DP4 of 1946. (Courtesy of C. E. "Sonny" Clutter.)

Figure 5.13 RCA Victor 10BP4 manufactured in 1949. (From the author's collection.)

period with almost all major manufacturers marketing receivers using it in both the table model and console categories. All picture tube manufacturers except Du Mont, made the 10BP4 or a close equivalent for the OEM (Original Equipment Manufacturer) and replacement markets. The accelerating voltage was rated at 10,000 volts, which was more than sufficient for a bright, sharp picture on its 10-inch diameter screen.

The 10BP4 featured a pressed glass faceplate and funnel, magnetic focus, the new cavity-type anode connector,[77] a seven-pin duodecal base, external conductive coating and use of an external, neck-mounted, double-magnet ion-trap. The latter consisted of a pair of either permanent magnets or electromagnets to cause any negative ions generated by collisions of the electron beam with residual gas atoms to be deflected sufficiently that they were intercepted by apertures in the gun. This prevented the heavy ions from reaching the screen and causing deadening of the phosphor in the screen's center. The resulting discoloration was known as an *ion burn* and was one of the primary drawbacks that had plagued early magnetically deflected cathode-ray tubes. A number of 10BP4s and receivers using them have survived and are now in the hands of collectors.

Du Mont, ever the promoter of large screen CRTs, next introduced the 15-inch diameter 15AP4 (Figure 5.14) in January 1947. It used the small-ball anode contact but was otherwise similar in construction to the 20BP4 with no external conductive coating or ion trap. As with the 20BP4, the 15AP4 was never made in any substantial quantities.

Two other relatively obscure 15-inch tubes were introduced in the 1946 to 1948 period. These were the Zetka Laboratories 15CP4 and Du Mont 15DP4, which were basically similar to the 15AP4 except for the use of an ion trap. All of the above mentioned, 15-inch tubes used hand-pressed bulbs in one of two alternate shapes.

Several picture tubes similar to the 10BP4 were announced between 1946 and 1949. Most were interchangeable with the 10BP4 with only minor mechanical and/or electrical differences. These included the General Electric 10DP4, Du Mont 10EP4, Rauland 10FP4 and 10RP4 and Sylvania 10MP4, in addition to the previously described Norelco 10CP4. The 10DP4 and 10RP4 used electrostatic focus with the former requiring a high focus voltage (about 30 percent of the anode voltage) and, the latter, a low voltage focus type requiring less than about 2 percent of the anode voltage. The 10DP4, 10EP4, 10FP4 and 10RP4 all featured aluminized (also called metalized or metal-backed) screens which was a major advance and was described by Bachman[22] in 1945 and Epstein and Pensak in 1946.[23] The 10FP4 was the most widely used aluminized 10-inch tube. Aluminizing solved four problems in one sweeping advance: (1) A doubling of brightness by reflecting light forward that would be otherwise wasted to the rear, (2) contrast enhancement by preventing scattered light reflected from the funnel walls from washing out black areas of the picture, (3) elimination of "sticking," which limited acceleration voltage to about 10,000 volts and (4) elimination of ion burns or the need for an ion trap because the heavy ions could not penetrate the aluminum film while the lighter electrons could.[77] Also during this period the glass bulbs went from hand pressing to automatic pressing as the television boom took hold.

In 1948, Du Mont registered a 12-inch diameter tube, the 12JP4 (Figure 5.15) using the same bulb used in the 12DP7 radar tube except that a small-ball anode connector was substituted for the medium cap type. The popular 12KP4 registered by GE and 12LP4 (Figure 5.16) by Sylvania soon followed. These were quite similar tubes except that the 12KP4 was aluminized. The Du Mont 12QP4 and 12RP4, Tel-O-Tube 12TP4, Sylvania 12VP4, Thomas 12YP4 and Sheldon 12ZP4 were variations on the 12-inch theme with the 12YP4 having the first automatic focusing electron gun and the 12ZP4 having an aluminized screen. Minor differences in ion trapping, length, external conductive coating and anode contacts (and probably manufacturers' egos) prompted the different registrations of otherwise similar tubes.

Philco-Lansdale produced a one-of-a-kind oddity designated as the 12WP4 (Figure 5.17) in late 1949 with the purpose of low cost and light weight. Several features that were unique for the time included the use of small neck diameter (7/8 inch), to reduce the deflection power requirements by placing deflection coils closer to the electron beam axis; a nine-pin

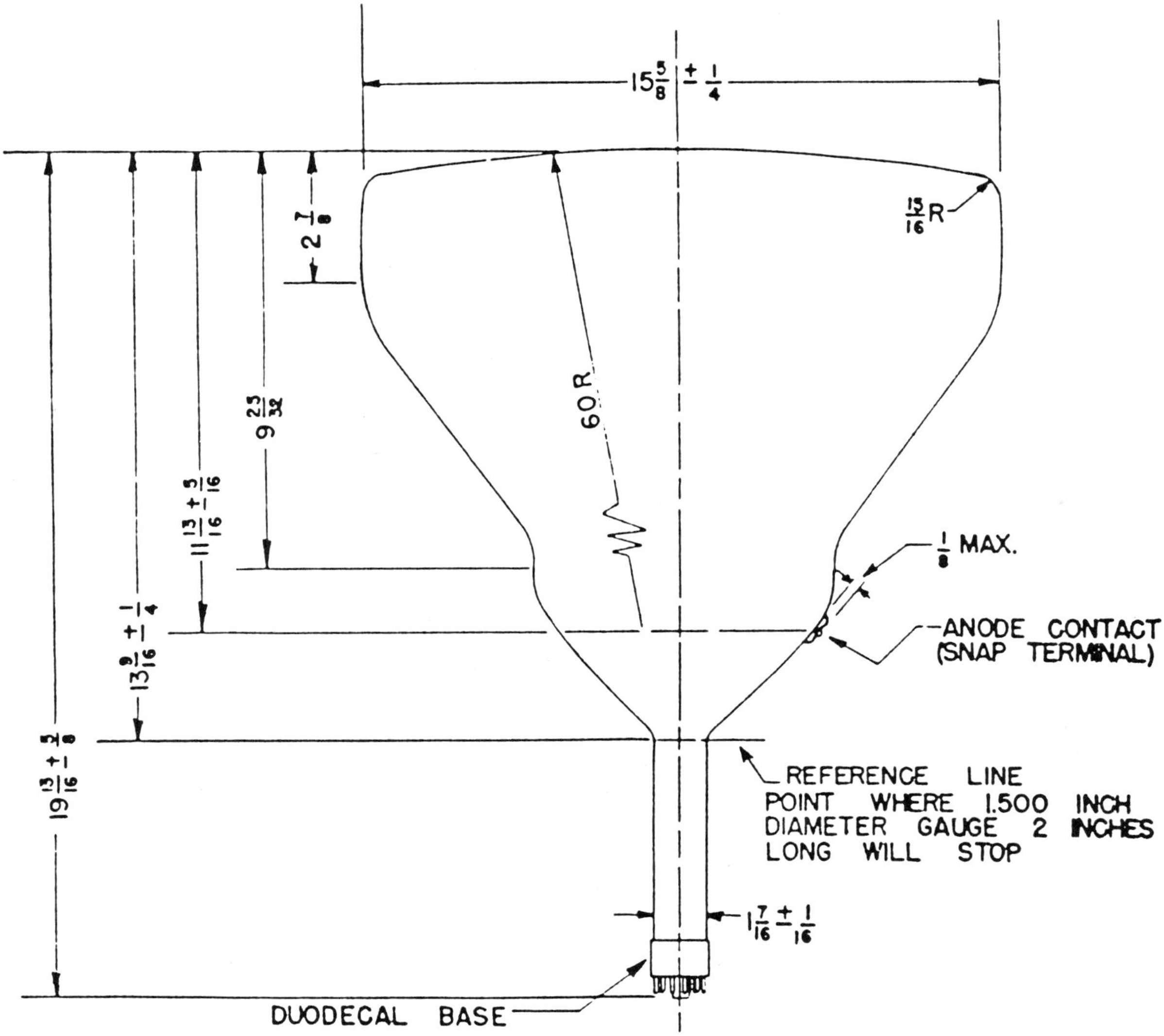

Figure 5.14 Du Mont 15AP4 of 1947. (Courtesy of Electronic Industries Association.)

miniature glass base similar to that used on receiving tubes; a triode electron gun; a large cavity anode connector; and exhaust tubulation through the funnel rather than through the base stem. The 12WP4 was used only in television receivers manufactured by Philco.

The years 1948 and 1949 marked the beginning of mass-produced, large-screen picture tubes and television receivers, first with the metal funnel RCA 16AP4 (to be described later) and followed shortly thereafter by a number of 16-inch diameter all-glass

Figure 5.15 Du Mont 12JP4 of 1948. (Courtesy of Jerry Talbott.)

Figure 5.16 RCA 12LP4-A manufactured in 1950. (Courtesy of James Richardson.)

Figure 5.17 Philco 12WP4 manufactured in 1954. Note the unusual small diameter neck, miniature receiving tube base, exhaust tip seal on bulb and large diameter cavity anode connector. (Courtesy of C. E. "Sonny" Clutter.)

picture tubes which rapidly displaced the earlier 10- and 12-inch tubes in sales volume. At about the same time larger screens were becoming popular, two other trends developed. The first trend was the introduction of gray faceplate glass to improve picture contrast in room light. The second trend was the use of wider deflection angles, thus decreased overall tube length which allowed a minimum cabinet depth. The 50- to 56-degree deflection angle that had been standard in the first postwar picture tubes first gave way to 60 degrees, then 70, 90, 110 and finally 114 degrees. Substantial savings in materials and shipping costs plus improved cabinet styling were the results.

The first all-glass 16-incher was the 16CP4

registered by Tel-O-Tube Corporation of America (early 1949) immediately followed by the Zetka 16DP4; the Du Mont 16FP4; the Sheldon 16HP4, 16LP4, 16MP4 and 16SP4; the National Union 16JP4; the Tel-O-Tube 16VP4 and 16WP4 (Figure 5.18); and the Philco-Lansdale 16ZP4. The new 60- and 70-degree deflection angles and gray glass faceplates added new dimensions to the manufacturers' ability to register slightly different new picture tubes. One variable, however, had disappeared temporarily. Electrostatic focus had lost out to magnetic focus for these new tubes. The ion trap, external conductive coating and anode connector variations remained, though. The 16DP4 and 16LP4 were the most widely used by receiver manufacturers although the 16WP4 also was fairly common in Philco receivers. Note that for a period of time, tubes that were originally registered with clear glass faceplates received an "-A" suffix when manufactured with

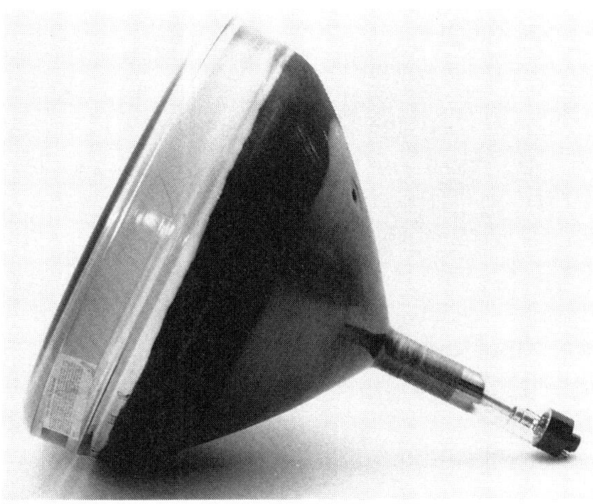

Figure 5.18 Philco 16WP4-A circa 1950. Note its short length made possible by 70-degree deflection. (Courtesy of Jerry Talbott.)

gray glass faceplates for contrast enhancement. Examples include the 10BP4-A, 12LP4-A and 16LP4-A. At the same time, the change was made from double- to single-magnet ion traps.

Three 19-inch diameter all-glass picture tubes were registered in the period of late 1949 to early 1950. They were the 66-degree deflection Sheldon 19DP4 and Tel-O-Tube 19FP4 and 19GP4. By that time the metal funnel 19AP4 by Du Mont had established a firm lead by virtue of its significantly lighter weight, thus the 19-inch all-glass tube never achieved any great commercial success.

5.5 PROJECTION CRTS FOR THE HOME

Projection television systems using three- to five-inch diameter cathode-ray tubes were briefly employed in the post-war period to produce pictures larger than could be produced by large direct-view picture tubes with their size limitations imposed by glass fabrication technology. Projection television has led a sporadic existence since its initial development during the late 1930s in Germany by Fernseh A. G.[24] and Dr. Knoll of Telefunken,[10] in England by Baird and Scophony[25,26] and in the United States by RCA.[27–30]

RCA registered the five-inch diameter type 5TP4 projection CRT (Figure 5.19) in 1946 and marketed the model 648PTK and 8PCS41 console projection sets in 1947 using the 5TP4 in a Schmidt optical system to produce 15 × 20-inch pictures.[20] A silicate P4 white phosphor was used for its ability to withstand the very high beam power density without saturation. A total of 27,000 volts was used for acceleration with high voltage electrostatic focus. Aluminization of the screen was used to improve brightness and prevent phosphor charging. Otherwise, the 5TP4 was similar in most respects to other picture tubes in early post-war production. This probably was the first commercially available projection CRT. General Electric, Emerson and United States Television receivers also utilized the 5TP4. The 5TP4 CRT was manufactured by both RCA and General Electric.

Philco also produced a 15 × 20 inch projection receiver, the Model 48-2500, in 1948 using their TP-400 four-inch CRT.[31,32] The TP-400 and the later TP-400A (Figure 5.20) used an eight-pin octal base derived from radar CRTs. The acceleration voltage was less than that of the RCA 5TP4, only 20,000 volts, and depended on a Micro-Lens projection screen to concentrate the light toward the viewer to improve brightness.

The remaining post-war projection CRT was the three-inch 3NP4 (Figure 5.21) by North American Philips (mid-1948). The 3NP4 was the heart of the Norelco Protelgram projection system which could be used to convert

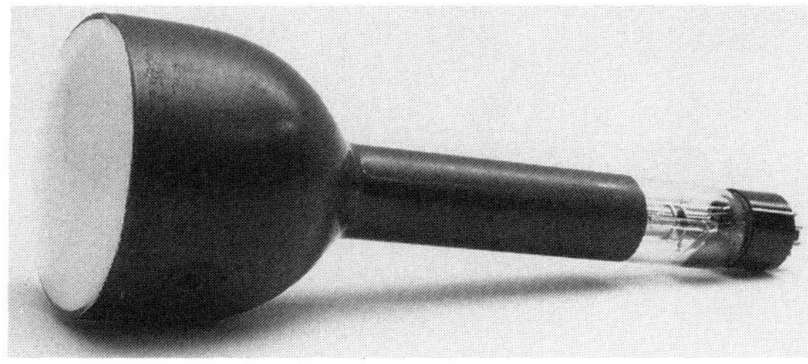

Figure 5.20 Philco TP-400A projection tube, manufactured circa 1950. (Courtesy of C. E. "Sonny" Clutter.)

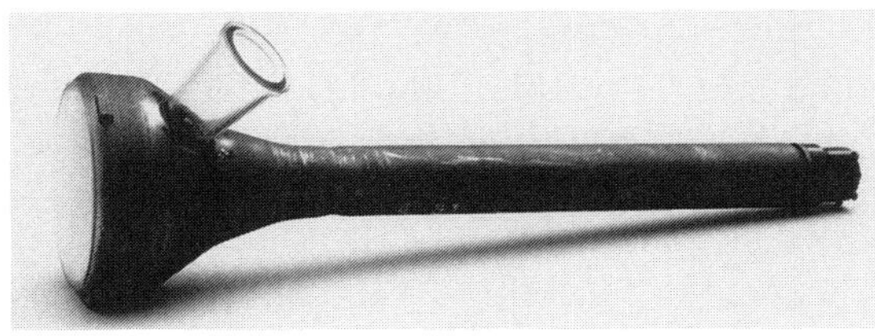

Figure 5.21 North American Philips (Norelco) 3NP4 of 1948. (Courtesy of C. E. "Sonny" Clutter.)

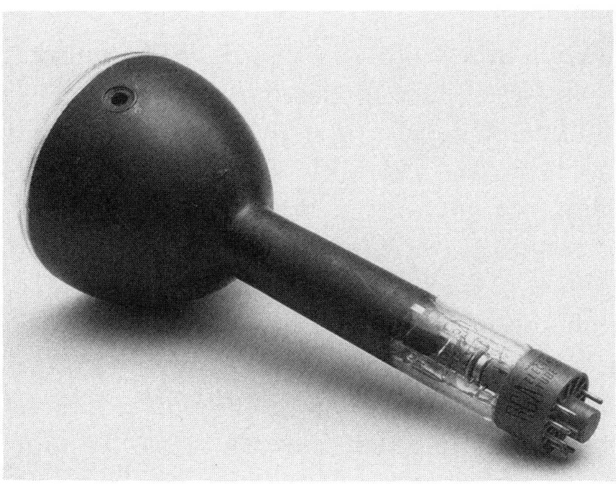

Figure 5.19 RCA 5TP4 projection kinescope manufactured in 1954. (Courtesy of C. E. "Sonny" Clutter.)

conventional direct-view television receivers to large screens.[33–35] The 3NP4 appears to have been designed and manufactured in Europe judging by the five-pin radial contact base and tapered glass cup surrounding the small ball anode connector which operated at 25,000 volts.

By 1950, home use of monochrome projection television systems with their dim, albeit large, pictures was temporarily doomed by the advent of large, direct-view picture tubes, particularly the 21-inch rectangular tubes. The only exception was the two-inch round Philco 2EP4 CRT (Figure 5.22) developed in 1959 for their "Safari" model portable television receiver. While not strictly

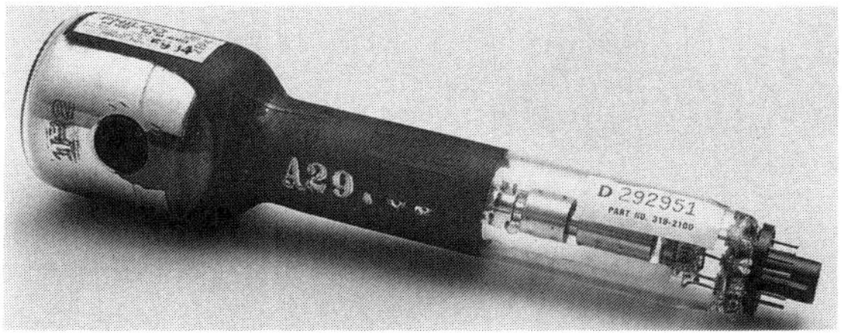

Figure 5.22 Philco 2EP4 picture tube for optically magnified portable television receiver manufactured in 1959. (Courtesy of C. E. "Sonny" Clutter.)

a projection tube per se, the 2EP4 produced an optically magnified picture through use of reflective optics to achieve the same net result, a picture many times larger than the CRT's screen size. The Safari was an all-transistorized receiver capable of operating from rechargeable batteries. The 2EP4 operated at an accelerating voltage of about 9,000 volts and used low-voltage electrostatic focus.

The home projection television concept became dormant until revived for home color television in 1972 by Henry Kloss and his Advent Corporation. (Projection color cathode-ray tubes are discussed in Chapter 6.)

5.6 PROJECTION CRTS FOR THE THEATER

Television projection in the theater began with the mechanical/optical scanning Scophony system in England in 1939.[26,36,37] The first theater projection television system using a cathode-ray tube was shown at the New Yorker theater in New York City in 1941 by RCA.[38,39] A 7-1/2-inch CRT operating at 65,000 volts was employed to produce a 15 × 20-foot picture. World War II interrupted the further development of projection CRTs in England and the United States for entertainment purposes.

Meanwhile, the Eidophor light valve television system was undergoing development in Switzerland. Dr. F. Fischer of Swiss Federal Institute of Technology had applied for a patent on the fluid film light valve in 1939.[40,41,75] Light valve CRTs employ a thin film of liquid in place of a conventional phosphor. The film deforms where struck by the electron beam. The CRT envelope is constructed to allow light from a high-intensity arc lamp to be projected through the film. A Schlieren optical system projects the picture formed on the film onto a large motion picture screen. In 1944 the Eidophor concept was demonstrated by Dr. Fischer with discouraging results. A second model under development at the time of his death in 1947 was completed by his associates in 1948 with better results, and a third version (1950) met all essential objectives.

J. Donal and I. Langmuir of RCA also explored the light valve prior to 1940.[42,43] After 1950, General Electric, Hughes Aircraft, Gretag and others devoted considerable effort to light valve development, although by then the emphasis had shifted from theater use to industrial and military applications. Many approaches to light valve technology evolved from this work including demountable tubes with replaceable fluid films as a solution to life problems, proprietary fluids with improved vacuum compatibility for sealed off tubes and full-color displays.[41,44–50,78,79]

Theater projection systems using conventional CRTs briefly received renewed interest as home projection systems were dying out around 1950. The RCA models TLS-86 and TLS-87 systems used the 5TP4 CRT with a modified Schmidt optical system to produce pictures up to 7 × 9 feet. RCA developed the seven-inch type 7NP4 projection CRT in 1950 which was capable of producing 15 × 20-foot pictures at a throw distance of 60 feet.[51] An acceleration voltage of 75,000 volts was used with forced air cooling of the faceplate and the glass bulb was corrugated to provide longer leakage paths for the high voltage. A 14-pin diheptal base was used for adequate isolation of the approximately 17,000 volt focus potential which was higher than the total accelerating voltage of most CRTs. The deflection angle was limited to 35 degrees to increase bulb length for better insulation, reduce deflection power requirements and improve focus in the picture corners.

A very similar tube, the RCA 7WP4 (Figure 5.23), was developed circa 1952 for an 80-foot throw distance. The principal difference in the two tubes was the inside radius of the faceplate's curvature.

A smaller tube, the RCA 5AZP4 (Figure 5.24) with a five-inch screen, was developed later in the 1950s for a 6 × 8-foot picture. Only one-half of the anode and focus voltage of the 7NP4 was required and, except for the integral anode lead wire, the 5AZP4 closely resembled the older 5TP4. The RCA type 4820-A from about 1972 was identical to the 5AZP4, except for a fine-grain rare-earth white phosphor screen optimized for good resolution and high brightness.

During the 1970s and 1980s, monochrome projection CRTs were used principally in military and industrial applications and color television. During this period, RCA produced a number of projection tubes in limited quantities for these markets using house numbers rather than registering them through EIA. This continued until the sale of RCA to General Electric in 1987 when RCA's Lancaster, Pennsylvania, special-purpose tube operation was spun off as Burle Industries

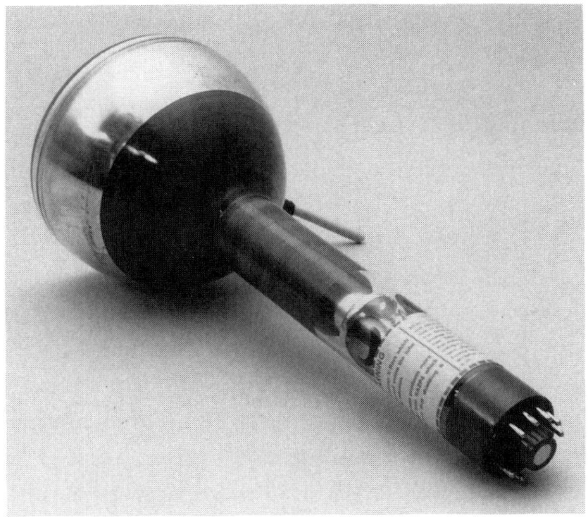

Figure 5.24 RCA 5AZP4 projection tube, circa 1955. Note permanently attached anode lead. (Courtesy of C. E. "Sonny" Clutter.)

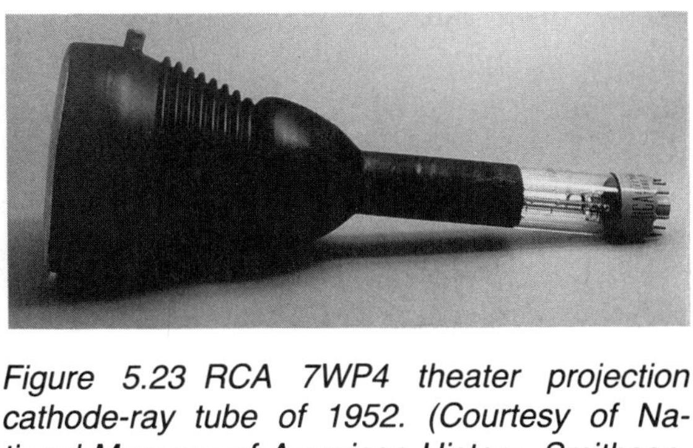

Figure 5.23 RCA 7WP4 theater projection cathode-ray tube of 1952. (Courtesy of National Museum of American History, Smithsonian Institution.)

(founded by former executives from the RCA plant).

5.7 METAL CONE TUBES

The metal envelope picture tube's origins date back to the RCA 913 one-inch oscilloscope CRT of 1936. Of course, scaling the glass-to-metal sealing techniques for a one-inch faceplate up to those required for a 16-inch and larger faceplate involved a concerted development effort to establish high-speed production on automatic machinery. The metal-cone picture tube achieved a large, relatively flat-faced tube with a savings in weight and bulk while strength was increased. The anticipated cost savings, however, never materialized.

RCA announced the development of the directly viewed 16-inch diameter type 16AP4 (Figures 5.25 and 5.26) to television receiver manufacturers in January 1948.[52,53] The 16AP4 employed a chrome-iron spun cone chosen for its match of coefficient of expansion to suitable faceplate and neck glass, adherence

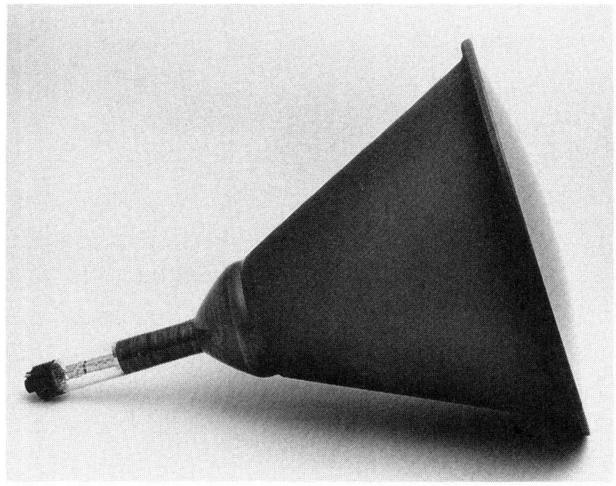

Figure 5.26 16AP4 metal-cone kinescope of 1948. (Courtesy of Jerry Talbott.)

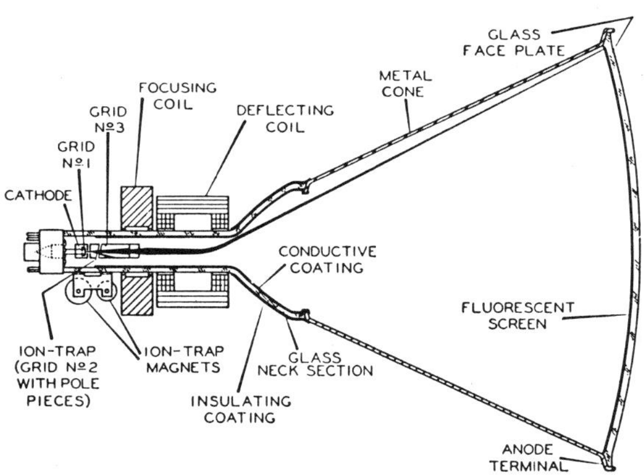

Figure 5.25 RCA 16AP4 metal-cone kinescope cross section. (Courtesy of RCA/Thomson.)

of glass at the seals and vacuum integrity. High-quality window glass, sagged or pressed to a slight curvature, was used for the faceplate. Because of some difference in thermal expansion properties between the faceplate and metal cone, the faceplate was under compression from the edge. Figure 5.27 illustrates the glass to metal seal.

Glass under compression has very high strength, thus it could be made quite thin (3/16 inch) and still withstand the atmospheric pressure of 1-1/2 tons on a 16-inch faceplate. Testing to three times that amount ensured safety under normal conditions. Conventional pressed glass faceplates required approximately three to four times that thickness. The flared glass neck sealed with another glass to metal seal to the small end of the cone provided adequate insulation between the deflection yoke and the metal cone which operated at up to 14,000 volts. High voltage contact was made to the lip at the screen and the tube was mounted to the cab-

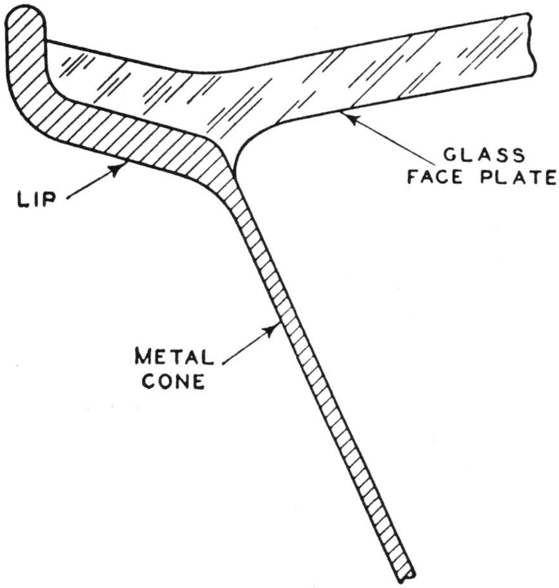

Figure 5.27 Glass-to-metal seal cross section as used for the 16AP4. (Courtesy of RCA/ Thomson.)

inet with an insulated plastic assembly. Often the entire cone was enclosed in a loose fitting polyethylene sleeve to provide a degree of high voltage protection to the serviceman during receiver servicing. The 16AP4 had a deflection angle of 53 degrees and magnetic focusing similar to that of many of the glass picture tubes at that time; it was over 22 inches long. Despite its length, the 16AP4 enjoyed a brief popularity and was a major milestone in the picture tube's development.

An eight-inch metal-cone tube, the 8AP4 (Figure 5.28), was introduced by General Electric in 1949. The 8AP4 was simply a scaled down version of the 16AP4 designed for portable or small table model receivers, such as one by Arvin. One minor difference in the 8AP4 was its use of a simplified triode

Figure 5.28 General Electric 8AP4 metal-cone picture tube manufactured in 1950. (Courtesy of James Richardson.)

electron gun. Small screens were losing favor by then and relatively few receivers were manufactured with the 8AP4.

The next step in metal-cone picture tubes was the 19AP4 (Figure 5.29) by Du Mont

Figure 5.29 Advance Television Tube Inc. 19AP4 circa 1950. (Courtesy of Jerry Talbott.)

(1949). This 19-inch diameter tube was similar to the 16AP4, but used a wider deflection angle of 66 degrees to shorten the overall length to about one inch less than that of the smaller screen 16AP4. The 19AP4 was used in many receivers, especially by RCA and Du Mont and progressed through a series of improved versions using gray glass, aluminizing and frosted faceplates to improve viewability. These were identified by suffixes "A" through "D."

Rauland also registered two metal-cone tubes in 1949. Their 12-inch 12UP4 was another scaled-down version of the 16AP4 while their 16EP4 was a shortened version of the 16AP4 using a 60-degree deflection angle, resulting in an overall length savings of over two inches. Neither tube had a significant impact on television receiver designs. Although it would be logical to assume that they found application in receivers by Rauland's parent, Zenith Radio.

The metal-cone RCA 16GP4 (Figure 5.30) (late 1949) was a different matter. Another increase in deflection angle, this time to 70 degrees, shaved over four inches off the overall length of the 16AP4 and reduced the weight even further.[54] A smaller cabinet, reduced shipping and materials cost, and improved aesthetics were the immediate result. The 16GP4 was used in many RCA television sets (as well as other brands) and went through a series of screen improvements and suffixes from "A" to "C."

The 16GP4 was followed by another advance to a 22-inch screen size by the Rauland 22AP4 (1950) and to 24 inches in the GE 24AP4 (1951). Both were scaled-up versions of the 16GP4 and retained the length-reducing 70-degree deflection angle. Rauland also made a similar 24-inch tube, the 24BP4 in

Figure 5.30 RCA 16GP4 metal-cone 70-degree deflection picture tube manufactured in 1950. (Courtesy of Jerry Talbott.)

1951 using low-voltage electrostatic focus. None of these tubes achieved great commercial success but were another step in the progression of larger screen sizes.

Paralleling the development of large-screen rectangular glass picture tubes (described in section 5.8) was the introduction of rectangular metal-cone tubes. These began with the 17-inch RCA 17CP4 (Figure 5.31) in 1950 which used magnetic deflection and focus. The following year, 1951, brought several new 17- and 21-inch rectangular metal-cone tubes, including the 21AP4 with magnetic focus, 17GP4[55] and 21DP4 with high voltage electrostatic focus, and the 17TP4 and 21MP4 with low-voltage electrostatic focus. All of these tubes were from RCA—the leading proponent of metal tubes. Most RCA television receivers of that period used one of these tubes, but other manufacturers were slow to adopt

Figure 5.31 RCA 17CP4 rectangular metal-cone picture tube manufactured in 1958. (From the author's collection.)

them. The reason for the resurgence of electrostatic focus in the 1950 to 1951 period was the critical shortage of cobalt used in permanent magnet focus assemblies during the Korean War. Electrostatic focus, its viability proven during that period, has remained the dominant method to this day due to advantages in minimizing cost and weight.

The last frontier for the metal-cone picture tube was the giant screen. For many years Du Mont held the size record with their 30-inch round metal 30BP4 (Figures 5.32 and 5.33) of 1950 to 1951, which was used in their Royal Sovereign model deluxe television console. The 30BP4 used 90-degree deflection to keep the length within reason and employed magnetic focusing. Du Mont soon encountered cathode life problems because getters could be used only in the neck due to the unaluminized screen. Any getters used in the forward portion of the tube would leave barium patches on the phosphor, resulting in nonuniform brightness across the screen. Neck-mounted getters were inadequate to maintain a good vacuum with the large volume and internal surface area of such a large picture tube. A last-time build of 500 tubes was made for replacement purposes and the 30BP4 was discontinued after only one to two years of production. A conversion kit to allow replacement with a 27-inch rectangular glass picture tube also was produced to allow future replacements to be made.[56]

Rauland announced the metal-cone rectangular 27AP4 in 1952 and RCA produced the similar 27MP4 with an aluminized screen in 1953. Both tubes had 90-degree deflection angles, but the 27AP4 used low-voltage electrostatic focus, in contrast to the 27MP4's reversion to magnetic focus. By the early 1950s, large-screen glass tubes were being produced in volume at lower cost than the metal cone tubes and the metal cone CRT was relegated to special-purpose applications. Some metal cone tubes continue to be manufactured for replacement purposes for radar and air traffic control applications.

5.8 RECTANGULAR SCREENS

It was apparent from the early days of television that a rectangular picture displayed on a round screen was not a particularly good fit (sort of a square peg in a round hole). Either the picture could be expanded to the full width of the tube, which clipped the picture corners, or the picture corners could be kept within the viewable screen area and the remainder of the screen, which was unused, masked to the viewer. The latter method wasted a large area of the screen. Zenith made its early receivers with the full round screen area visible and expanded the height to fill

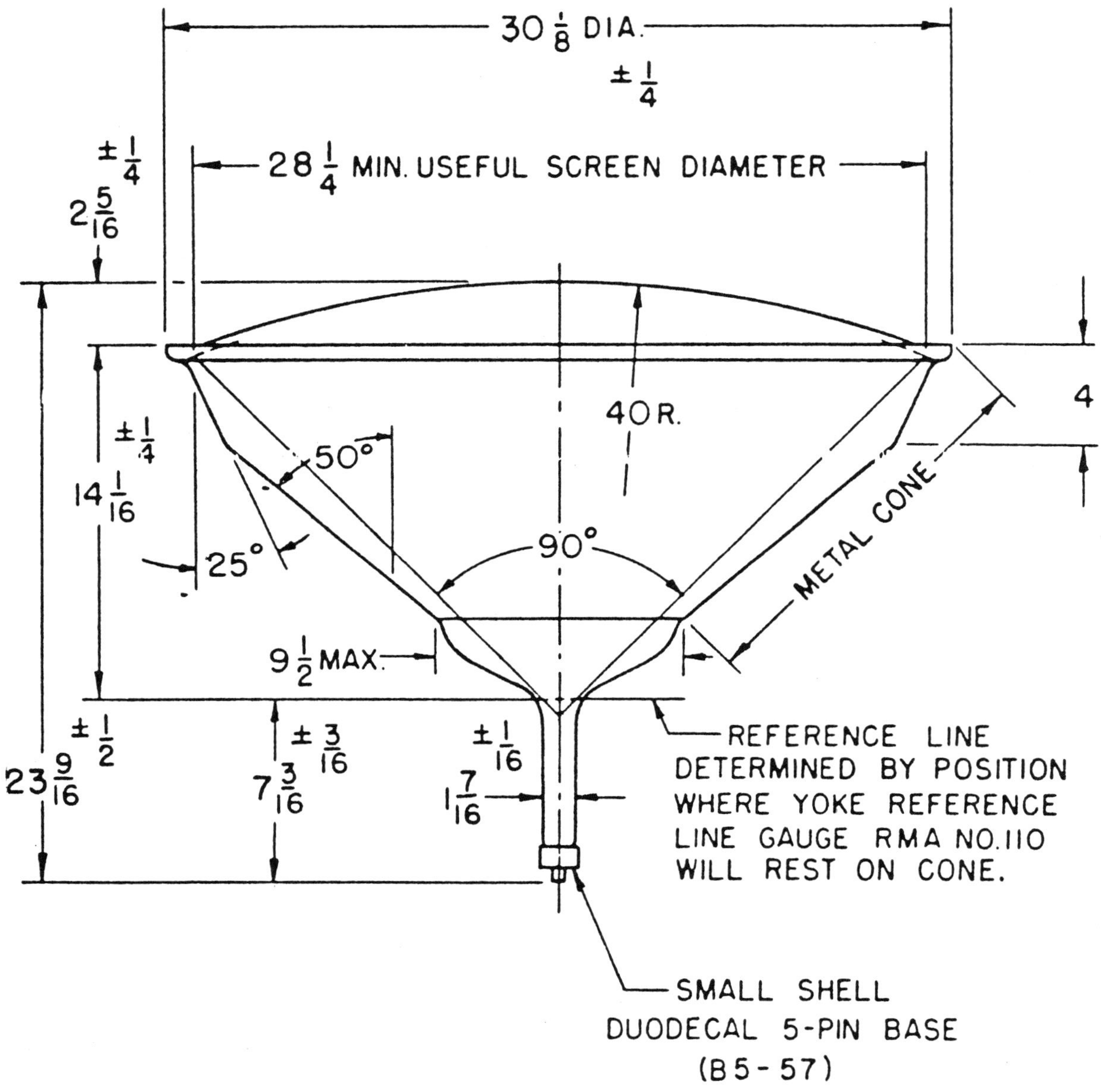

Figure 5.32 Du Mont 30BP4 metal-cone 90-degree deflection picture tube of 1951. (Courtesy of Electronic Industries Association.)

Figure 5.33 Photo of John McQueen sealing faceplate to metal cone for Du Mont 30BP4. (Courtesy of John McQueen of Southwest Vacuum Devices, Inc.)

the tube. The result was that not only were the corners lost, but the sides of the 4:3 aspect ratio picture as well. Additional picture area also was lost in all cases since it was necessary to overscan the mask area by about 10 percent to ensure that no black borders would become visible due to picture shrinkage at low power line voltage, or as tubes and other components aged.

Pre-war German television receivers were described in 1939 using rectangular picture tubes which were strikingly similar in size and appearance to the first postwar American rectangular tubes.[24,57] These receivers were termed Standard Receivers and were designed jointly by five German television firms, Fernseh A. G., Telefunken, Lorenz, Loewe and TeKaDe. The tubes were described as having 19.5 × 22.5 cm screens with a square (*sic*) flat picture. A 27.5 × 31.5 cm version, roughly equivalent to a 14-inch tube, also was reported at private showings by Fernseh. The tubes were described as short and judging from published photos the deflection angle appears to have been approximately 70 degrees as were the first American rectangular picture tubes 10 years later.

The first series of rectangular picture tubes in the United States was launched by Hytron with the 16-inch 16RP4 registered in November 1949. Several similarly shaped tubes having very flat sides and a 70-degree deflection angle were part of this group. General Electric introduced the nearly identical 16KP4 (Figure 5.34) and 16TP4 shortly after the

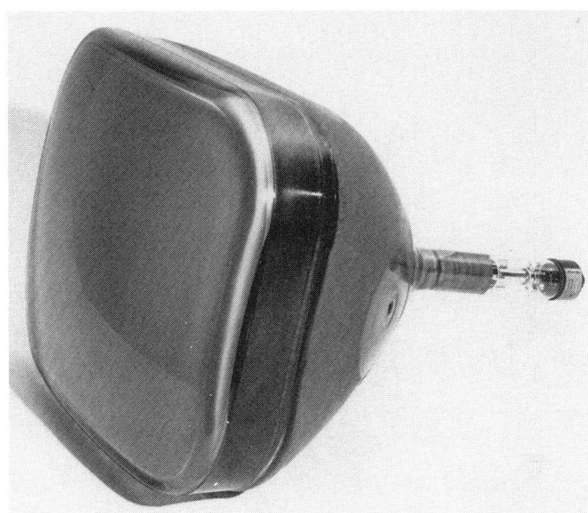

Figure 5.34 Westinghouse 16KP4-A manufactured in 1956. (Courtesy of James Richardson.)

16RP4. All three of these tubes were quite successful and were manufactured by many CRT manufacturers as OEM and replacement tubes. Other similar tubes (except for lack of external conductive coating and other minor differences) were the Vacuum Tube Products 16QP4 and Tel-O-Tube 16UP4 and 16XP4 (1949 to 1950).

Two other screen sizes using the squarish bulb shape of the 16-inch series appeared in 1950. The first, a 14-inch series included the firstborn Sheldon 14BP4 followed by the General Electric 14CP4 (Figure 5.35), Tel-O-Tube 14DP4 and Hytron 14EP4. Only the 14DP4 showed any degree of originality in its lack of external conductive coating and the need for a double-magnet ion-trap. Electrostatic focus (developed because of cobalt shortage during the Korean conflict) appeared in the 14-inch series in 1951 with the otherwise similar RCA 14GP4 using high-voltage focus and then the National Video

Figure 5.35 Sylvania 14CP4, manufactured circa 1951. (Courtesy of James Richardson.)

14HP4 and General Electric 14QP4 with low-voltage focus. The entire series was popular for small table model and portable receivers. The other series consisted of the obscure 19-inch Sheldon 19EP4, Hytron 19JP4 and 19QP4 from the 1950 to 1952 period. The first two tubes used magnetic focus while the 19QP4 used low-voltage electrostatic focus. Otherwise these tubes were simply larger versions of the 14- and 16-inch series.

The rectangular picture tube displaced the round tube almost overnight and by 1950 almost all new picture tube registrations were for rectangular tubes. In mid-1950 a new series of rectangular tubes appeared that set the style for some time to come. These began with 17-inch diagonal glass tubes having more curved sides than the 14-, 16- and 19-inch tubes previously introduced. The result was greater tube wall strength with a slight loss of picture in the corners. The first of these tubes registered, the eminently successful 17BP4 (Figure 5.36) by Sylvania, spawned the usual registering of similar tubes by other manufacturers, as well as electrostatic focus versions and other variants. These tubes included the 17AP4 by Du Mont, 17FP4 by Sylvania, 17HP4 by Rauland, 17JP4 by Philco-Lansdale and the General Electric 17RP4.

Twenty- and 21-inch rectangular picture tubes entered the scene in great profusion in the 1950 to 1951 period and the 21-inch screen rapidly became the dominant screen size for not only console models but some table model receivers as well. The first tube was the magnetically focused Hytron 20CP4 followed by the Sylvania 20DP4 and the electrostatically focused Hytron 20FP4, RCA 20GP4, Rauland 20HP4, Du Mont 20JP4, General Electric 20LP4 and RCA 20MP4. Several 21-inch

Figure 5.36 National Video 17BP4 manufactured in 1951. Note the more rounded screen sides than those previously used for rectangular picture tubes. (Courtesy of James Richardson.)

glass tubes with conventional spherically shaped faceplates and 70-degree deflection angle also were produced including the National Video Corporation 21WP4 and 21XP4, Westinghouse 21YP4 and Sheldon 21ZP4.

Another screen style, the cylindrical faceplate, emerged in 17- and 21-inch picture tubes in 1951. The faceplate was shaped like a slice out of the side of a cylinder with the screen flat in the vertical plane and curved in the horizontal. This screen style was an attempt to eliminate the curvature in the vertical direction which tends to reflect ambient light toward the viewer regardless of the height of the light source above the floor. By tilting the cylindrical picture tube's face slightly forward, reflections were directed downward, away from the viewer's line of sight. Cylindrical faceplate picture tubes began with the magnetically focused National Union 21EP4 and 17QP4 and their electrostatically focused counterparts, the 21FP4 and 17LP4. Others in the series included the 17SP4, 17UP4 and 21KP4 by Thomas Electronics, 17VP4 and 21JP4 by General Electric, and 17YP4 (Figure 5.37) by Philco-Lansdale. By the mid-1950s, the cylindrical faceplate had fallen by the wayside and it was not until Sony revived it for their highly successful Trinitron color picture tube (see Chapter 6) that it was used again commercially.

It should be noted that by the time the rectangular picture tube was introduced, gray glass faceplates had become universal for contrast enhancement, thus all of the rectangular glass picture tubes employed them from the start. Still, variations of the rectangular tubes were produced with contrast and brightness improvements including etched (or frosted) faceplates, aluminized screens and combinations of the two. Usually these variations were identified with suffixes from A through D. There was no consistency in suffixes be-

Figure 5.37 Sylvania 17YP4 manufactured in 1955. Note the cylindrical faceplate. (Courtesy of James Richardson.)

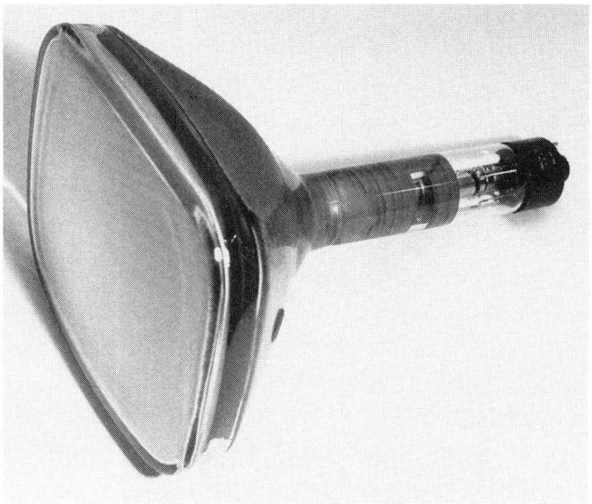

Figure 5.38 RCA 8DP4 employing 90-degree deflection to greatly shorten the funnel. Example manufactured in 1956. (Courtesy of James Richardson.)

tween differing tube types, they were simply issued on a first-come basis and it cannot be safely inferred that a 20CP4-A possesses the same feature configuration as a 17KP4-A. Differences in external conductive coatings and ion-trap magnets also were sometimes identified by the same suffixes.

Until 1953, all rectangular glass picture tubes had 70-degree deflection angles which required fairly deep cabinets, especially for the larger screen sizes. Weight was becoming appreciable as screen sizes (24 pounds for the 21-inch) continued to increase. 1953 marked the first usage of 90-degree deflection angles by Westinghouse with their 21ALP4 with electrostatic focus and the similar but magnetically focused 21AMP4. Many 90-degree picture tubes followed in screen sizes of eight, 10, 14, 17, 21, 24 and 27 inches. These picture tubes continued to use the now-standard duodecal 12-pin base and various combinations of magnetic and electrostatic focus and aluminized screens.

The 90-degree deflection RCA 8DP4 shown in Figure 5.38 was the basis of an extremely compact portable television introduced in 1956. Although not truly portable in the battery-operated sense, it was nonetheless the smallest receiver of the early television years and had a cabinet only slightly larger than the picture tube dimensions despite its vacuum tube circuitry.

General Electric registered the 9QP4 (Figure 5.39) in 1958 for a similar television receiver model of their own manufacture. The 9QP4 was designed for minimum manufacturing cost since by that time the novelty of television had worn off and prices were under increasing competitive pressure. An inexpensive one-piece blown-glass bulb was used for the 9QP4. The anode connection was

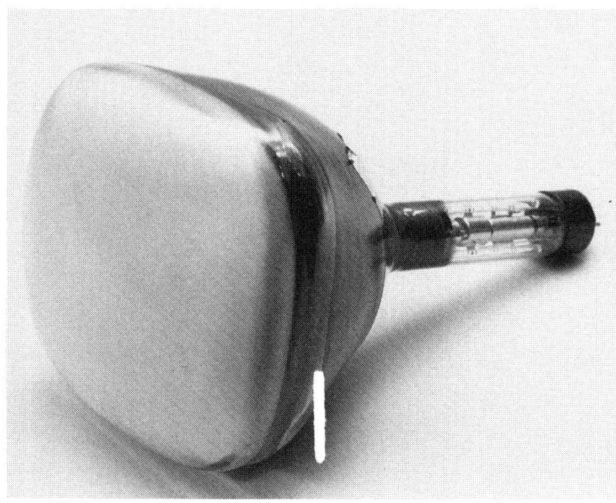

Figure 5.39 General Electric 9QP4 low-cost picture tube of 1958. (Courtesy of James Richardson.)

made through the base, rather than requiring a conventional anode button to be sealed into the bulb. The 9QP4 reverted back to a 70-degree deflection angle, thus cutting deflection circuit costs at the expense of slightly more length.

Another change in picture tubes occurred circa 1957 as a result of increased cost-consciousness. There was an introduction of lower current heaters followed by a variety of heater voltages, currents and controlled warm-up times designed specifically for "series string" operation with all of the other tubes' heaters in the receiver.[58] Until that time, 6.3 volt, 0.6 amp heaters operating from the receiver's power transformer had been standardized since about 1939. Elimination of the power transformer represented a significant cost and weight savings. Series string operation had been common for many years for radios, and increasing use of it was being made for television. Of course, the new heater ratings further increased the number of new picture tube registrations.

The introduction of 110-degree deflection in several 17- and 21-inch screen sizes in 1957 (Figure 5.40) became widely used to reduce cabinet depth and costs during the waning years of black and white television. Smaller neck diameters of 1-1/8-inch diameter or less were used to reduce deflection power requirements which had been steadily climbing as wider deflection angles evolved. Several new bases were designed for the smaller neck diameters. Such bases were characterized by the use of stiff wires passed through the glass. These wires functioned as the tube pins, were similar to miniature receiving tubes and were referred to as *hard pin stems*. This eliminated the necessity of a cemented base and soldering operation, although a small plastic guide was fastened over the pins to provide indexing with the socket and some protection to the more fragile pins and their glass-to-metal seals.

Figure 5.40 Tung-Sol 21EAP4 picture tube having 110-degree deflection and small neck for reduced cabinet depth, circa 1960. (Courtesy of Jerry Talbott.)

The beginning of integral implosion protection in 1959 was a positive addition to many new picture tubes. By permanently affixing the implosion protection to the picture tube, the previously used cabinet-mounted safety glass or plastic implosion protection and its mounting hardware were eliminated with four advantages to the manufacturer and customer: (1) cost savings, (2) prevention of dirt (attracted by the high anode voltage) collecting between the tube face and inside of the safety glass, (3) reduction in ambient light reflections from three surfaces to only one and (4) shortening of the cabinet depth by the amount of the customary spacing between the tube face and implosion shield. Of course, the tube's replacement cost was higher, but this had become relatively insignificant since tube life had been steadily improving as advanced manufacturing methods came into use and lower-cost picture tube rebuilding became more common.

The first implosion protection technique was the laminated or bonded-shield construction. This utilized a glass panel having the same curvature as the faceplate's outside surface. The panel and faceplate were spaced slightly and the gap filled with a clear resin. Upon curing, the result was a laminated sandwich slightly thicker than the similar safety glass used for automobile windshields.

Other integral implosion protection methods include Shelbond,™ Kimcode,™ T-band and molded covers. (Shelbond™ is a trademark of Corning Glass Works and consists of a resin-filled one-piece contoured metal frame bonded around the critical rim area of the tube face. Kimcode™ is a trademark of Owens-Illinois and uses a formed metal rim held in place by a steel tension band under considerable pressure which places the faceplate in compression where its strength is greatest. T-bands are simply the same tension band applied without a metal rim. The molded cover approach uses a contoured wraparound plastic or glass panel laminated to the faceplate.) Combinations of these four methods also may be used to achieve the desired protection to the viewer. Often the implosion protection is combined with mounting "ears," further simplifying assembly of the picture tube into the cabinet.[58–60] These methods continue to be used for color television picture tubes. Figure 5.41 illustrates several of these implosion protection techniques.

RCA introduced 114-degree tubes with their 20TP4 in 1967, four years prior to terminating black and white tube production in the United States.[61] Monochrome picture tube sizes using 114-degree deflection angles included 16-, 17-, 19-, 20-, 21- and 23-inch screens.

5.9 RECEIVER CHECK TUBES

As picture tubes increased in size during the 1950s and were mounted to the cabinet rather than the chassis, the difficulty of removing the tube to allow the operation and testing of the chassis on the service bench became a major headache for the service technician. Manufacturers and their service organizations often could have an appropriate picture tube permanently mounted on their benches since they were always testing or servicing a limited number of similar models, however the independent service technician was likely to encounter a wide variety of receiver makes and models. To solve this problem, Sylvania Electric introduced three small and easily carried tubes that could be used in virtually any chassis then in use. A further benefit was the ability to use them to determine in the customer's home if the receiver's picture tube was defective by using the substitution method.

The first tube announced by Sylvania in 1954 was the five-inch round 5AXP4 (Figure 5.42) with a 53-degree deflection angle. It was usable in receivers with up to 70-degree deflection although the tube would be somewhat overscanned. The standard duodecal base

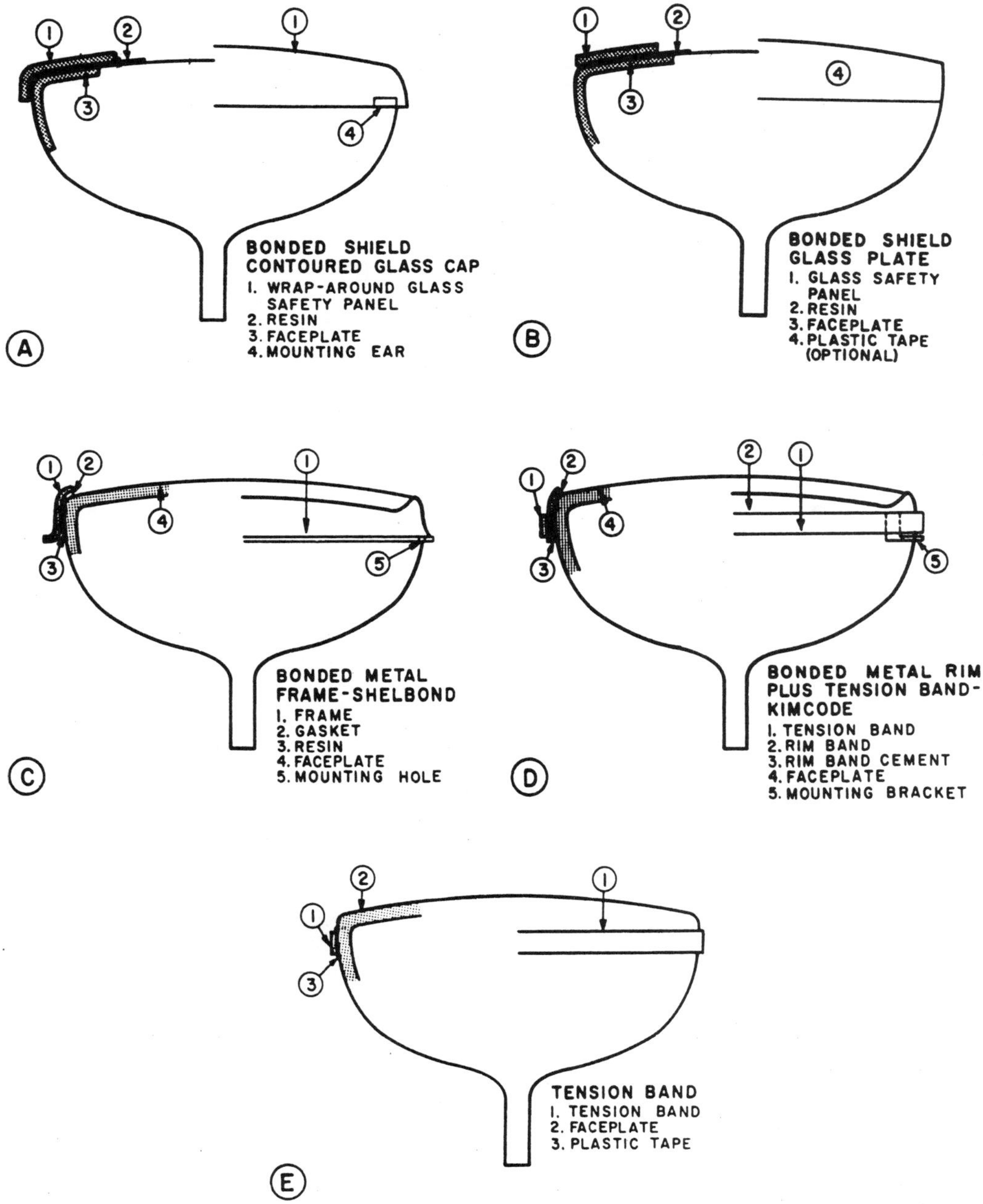

Figure 5.41 Implosion protection methods.[58] *(Courtesy of IEEE.)*

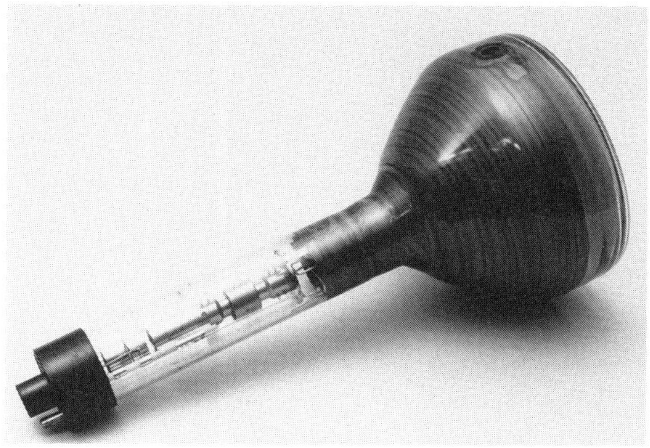

Figure 5.42 Sylvania 5AXP4 receiver test tube for substitution in place of normal picture tube during servicing, circa 1954. (Courtesy of Jerry Talbott.)

The last monochrome television receiver check tubes were the Sylvania 8YP4 (Figure 5.43) and Thomas Electronics 8JP4 in 1958. The 8YP4 had a "pancake" bulb with 110-degree deflection angle and a six-pin overcapped base for durability. An adaptor was supplied to adapt it to the newer and more fragile hard-pin base style coming into use at that time. A lucite overlay with printed outlines corresponding to the proper amount of scan at 52-, 70- and 90-degree deflection angles also was included with the tube.[62] A similar 110-degree receiver check tube, the 8JP4, was registered two months later by Thomas Electronics. The 8JP4's principal

and connections were used and a self-focusing gun allowed its use in receivers with either magnetic or electrostatic focusing. The bulb was basically similar to that of the 5FP7 radar CRT except that the anode connection was a small cavity connector compatible with that of most television receivers. Also, it was mounted close to the faceplate in contrast to the previous small ball cap connected midway along the flare. This permitted higher voltage operation, up to 18,000 volts for large-screen receivers. Only the anode connector, socket and deflection yoke from the receiver were required to be connected for operation. RCA also cataloged the 5AXP4 during the mid-1960s.

In 1956, Sylvania introduced the eight-inch rectangular type 8XP4 which used the same bulb as the previously described 8DP4 picture tube. It was intended for receivers having deflection angles up to 90 degrees and anode voltages up to 22,000 volts. Otherwise usage was similar to the 5AXP4. The 8XP4 was later available from RCA as well.

Figure 5.43 Sylvania 8YP4 receiver test tube for up to 110-degree deflection angles, circa 1958. (Courtesy of Jerry Talbott.)

difference was the hard-pin base that reduced the need for adaptors, but at the expense of durability.

5.10 MONITOR CRTS

One remaining television application for the cathode-ray tube is that of monitoring pictures and waveforms in the television station. Three distinct uses, each having specific requirements regarding display size and format, are common. These uses include camera viewfinders, studio and control room picture monitors and video waveform monitors.

Early television cameras (during the 1930s) used optical viewfinders to aid focusing and positioning.[63–65] Du Mont Laboratories pioneered the electronic viewfinder using a cathode-ray tube mounted on the camera side in or slightly before 1941.[66,67] From photographs published that year, the viewfinder appeared to have used a shortened, electrostatically deflected CRT measuring approximately five inches in diameter. This was in accordance to Du Mont's early preference for electrostatic deflection.

All post-war studio cameras incorporated electronic viewfinders with magnetically deflected CRTs. They began with the RCA five-inch round 5FP4-A (Figure 5.44) which was merely the venerable 5FP7 radar CRT with white phosphor and improved spot size. The 5FP4-A was housed in the top portion of the widely used RCA TK-30 camera.[68] General Electric later registered an aluminized version of the 5FP4-A in 1952 as the type 5QP4 (Figure 5.45).

Seven-inch round monitor CRTs began with the magnetically deflected 7AP4 in 1938 and 7CP4 in 1942. The 7AP4 used the five-pin radio tube base with no separate anode connector, and the 7CP4 used an eight-pin octal

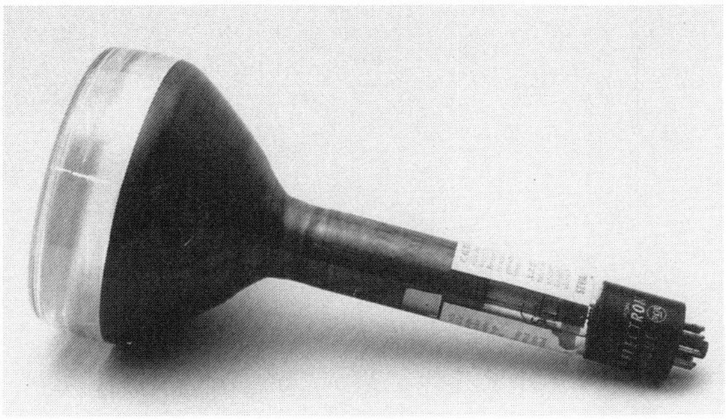

Figure 5.44 RCA 5FP4-A television camera viewfinder cathode-ray tube manufactured in 1955. (From the author's collection.)

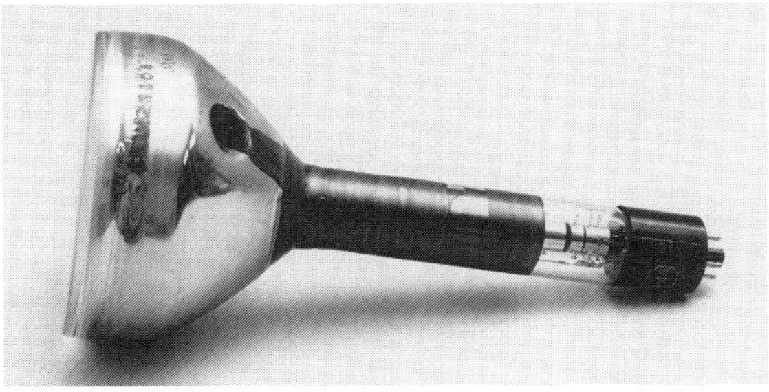

Figure 5.45 General Electric 5QP4-A aluminized viewfinder CRT manufactured in 1960. (Courtesy of C. E. "Sonny" Clutter.)

base with a small ball anode connector. Both used similar blown-glass bulbs with very rounded faceplates. These were predominantly used for portable monitors for remote pickups. They were followed by the RCA 7QP4 in 1950, the RCA 7TP4 (Figure 5.46) in 1951 and the General Electric 7ABP4 in 1956. All three used the bulb with a semi-flat faceplate that was previously used for the earlier 7DP4 picture tube as well as the now-standard duodecal base. The 7QP4 used magnetic deflection and focus while the 7TP4 was

Figure 5.46 RCA 7TP4 monitor cathode-ray tube manufactured in 1954. (Courtesy of C. E. "Sonny" Clutter.)

similar except for its high-voltage electrostatic focus and an aluminized screen. The 7ABP4 used low-voltage electrostatic focus and an aluminized screen. Several eight-inch rectangular CRTs followed in the period of 1958 to 1962. These included the Sylvania 8FP4, RCA 8HP4, Sylvania 8KP4, Du Mont 8MP4 and RCA 8NP4. All used the same 90-degree deflection angle bulb as the 8DP4 picture tube and had the added feature of aluminized screens. Low-voltage electrostatic focus was employed in all but the 8FP4, which used magnetic focus.

Larger control room and studio monitors usually used standard round-screen, magnetic-deflection picture tubes in television's early years. The first tube designed specifically for the purpose was the 10-inch round RCA 10SP4 registered in 1952. The 10SP4 featured higher resolution, an aluminized screen and high-voltage electrostatic focus. The 10SP4 was otherwise similar to the ordinary 10-inch picture tubes of the late 1940s. Larger-monitor CRTs appeared later with rectangular screens, including the General Electric 14UP4 in 1956 and the Westinghouse 21EWP4 in 1959. Standard or slightly modified rectangular picture tubes continued to be used, especially for industrial television monitors.

By the early 1960s, color television was rapidly replacing monochrome, however monochrome viewfinder CRTs made a comeback during the 1980s as home video recording became popular. Recent monochrome viewfinder CRTs from Japan (Figure 5.47) are a far cry from the early broadcast viewfinder tubes and are a marvel of miniaturization. Low power consumption from batteries, small size and light weight are important requirements in today's color camcorders and cameras. Screen sizes of 1/2 to 1-1/2 inches are universal with optical magnification to provide an easily viewed picture. All use flat faceplates with both round and rectangular tubes being common.

Waveform monitors allowing observation of video, synchronization and blanking signals are widely used for television production. Early picture monitors almost invariably had a small electrostatic waveform

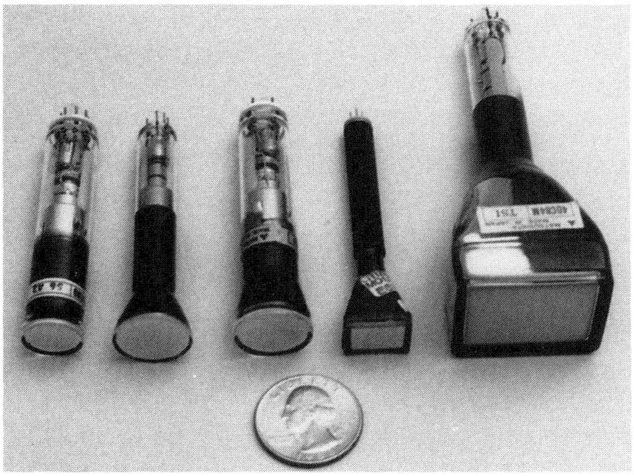

Figure 5.47 Several home television camera miniature viewfinder CRTs. (Courtesy of C. E. "Sonny" Clutter.)

monitor CRT mounted in the same cabinet to aid setting video, synchronizing signal levels and troubleshooting problems in the complex composite video waveform. Usually these used a 3BP1, 5CP1 or similar electrostatic deflection oscilloscope CRT. After all, the waveform monitor is merely a specialized oscilloscope designed to be scanned at either of the television horizontal and vertical deflection frequencies and to have a vertical bandwidth high enough to display the four megahertz video signal.

During the early 1950s, flat-faced CRTs such as the RCA 5ABP1 began to be used to reduce measurement parallax errors with the graticule scale. In 1955 Tektronix introduced their Model 525 rack-mounted waveform monitor, starting a trend toward separate waveform and picture monitors. The 525 used their round, flat-face T52 oscilloscope CRT (Figure 5.48), and the 527 waveform monitor announced in 1961 used their new rectangular flat-face T5270 CRT specifically de-

Figure 5.48 Tektronix 5CAP2/T52P2 electrostatic deflection CRT for television waveform monitors. (Courtesy of Tektronix, Inc.)

signed for it. From that point on, separate waveform monitors became a way of life.

Hewlett-Packard made the next CRT-related advance in waveform monitors with their Model 191A "television waveform oscilloscope" in 1966.[69] New vertical interval test signals (VITS) inserted in the video signal during the vertical blanking pulse were starting to be used to evaluate picture fidelity during signal transmission. The short duration of these signals made them difficult to view because of their low brightness. H-P designed a new CRT utilizing their domed scan-expansion mesh and a 20,000 volt acceleration potential to increase the brightness without sacrificing deflection sensitivity. At present television waveform monitors continue to use CRTs based on electrostatic oscilloscope tube technology.

Color television required another type of monitor for determining the phase and amplitude of chrominance signals using a polar coordinate display. Tektronix introduced the Model 526 vectorscope in 1959 for the color television broadcaster. Its Tektronix-made T526 CRT differed from conventional electrostatic oscilloscope CRTs only in having equal sensitivities in the vertical and horizontal deflection plates to facilitate polar coordinate presentations.

5.11 BACK TO SMALL SCREENS

By the 1970s, color television had almost made monochrome obsolete and the trend toward larger screens was reversed. Monochrome picture tubes were relegated to portable television receivers where cost, power consumption and weight constraints precluded the use of color tubes at the then-current state-of-the-art. Eventually only four- and five-inch receivers used monochrome tubes (Figure 5.49). All such tubes were produced in the Far East. Continuing the trend, today the monochrome tube is being replaced by color even for portable receivers. The only significant remaining television application for monochrome picture tubes is their use as a viewfinder in home video cameras and camcorders. We have seen the evolution of monochrome screen sizes increase by a factor of 10 from about three inches to 30 inches and back to less than one inch. Although not as extreme, only the automobile has experienced similar size evolution as the public's taste swings from economy to luxury and back.

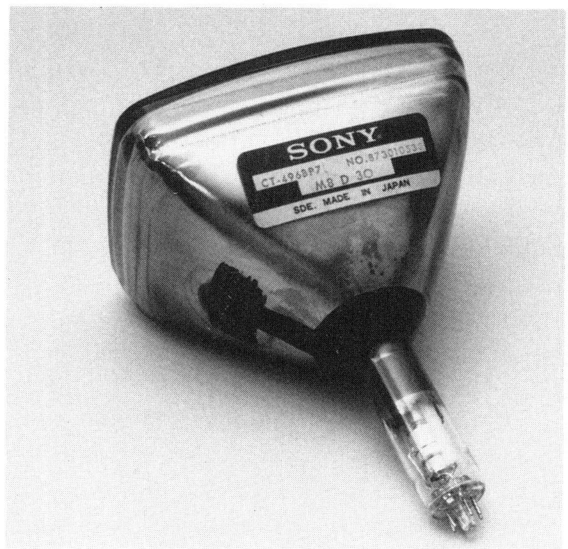

Figure 5.49 Sony five-inch CRT as used in portable television receivers. (Courtesy of James Richardson.)

6

Color Television Picture Tubes

6.1 Baird color systems

The idea of color television dates back to the mechanical television scanning era of about 1925. Oddly enough, one of the first proposed color systems was electronic as proposed in a patent application filed in 1925 by V. K. Zworykin.[1–3] This patent was based on a cathode-ray tube having a checkerboard of color filters in front of a white phosphor. The first demonstrated color television system appears to be that of John L. Baird on July 3, 1928 using a modified version of that period's mechanical scanning system.[1,2,4] The 1930s introduced a considerable amount of development of the monochrome cathode-ray tube for television. Baird demonstrated a two-color system (Figure 6.1) in 1939 using a projection CRT and rotating color filter consisting of red and blue-green filters[2,5] and a three-color stereoscopic version (Figure 6.2) in 1941.[6–8] Baird demonstrated two-color television in 1942 to 1943 using mirrors to combine two adjacent images through filters (Figure 6.3) in one case[9,10] and the Telechrome tube (Figure 6.4a,b) in 1944 having two or three electron guns on opposite sides of a mica screen in the other.[11–15]

6.2 CBS field-sequential system

In 1940, Peter Goldmark of CBS successfully combined field sequential mechanical scanning using a rotating three-color filter wheel (Figure 6.5) and a conventional monochrome cathode-ray tube.[1,2,15,16,127] This system produced some of the best quality pictures for several more years, although it was limited in screen size. It also was incompatible with the NTSC 525-line system approved for commercial black and white broadcasting by the Federal Communications Commission (FCC) the following year. One of the first uses of a rectangular screen cathode-ray tube in the United States is associated with the CBS system. A seven-inch, flatface tube is shown in the 1942 Proceedings of the IRE paper by Goldmark.[2]

In 1941 daily broadcasts using the Goldmark system began over WCBW in New York. Unfortunately, no practical all-electronic NTSC-compatible alternative was forthcoming during the 1940s and approval was granted by the FCC on October 10, 1950 for commercial color television broadcasting using the 405-line, 144 Hz field rate CBS field sequential system.[17] By that time, the shadow-

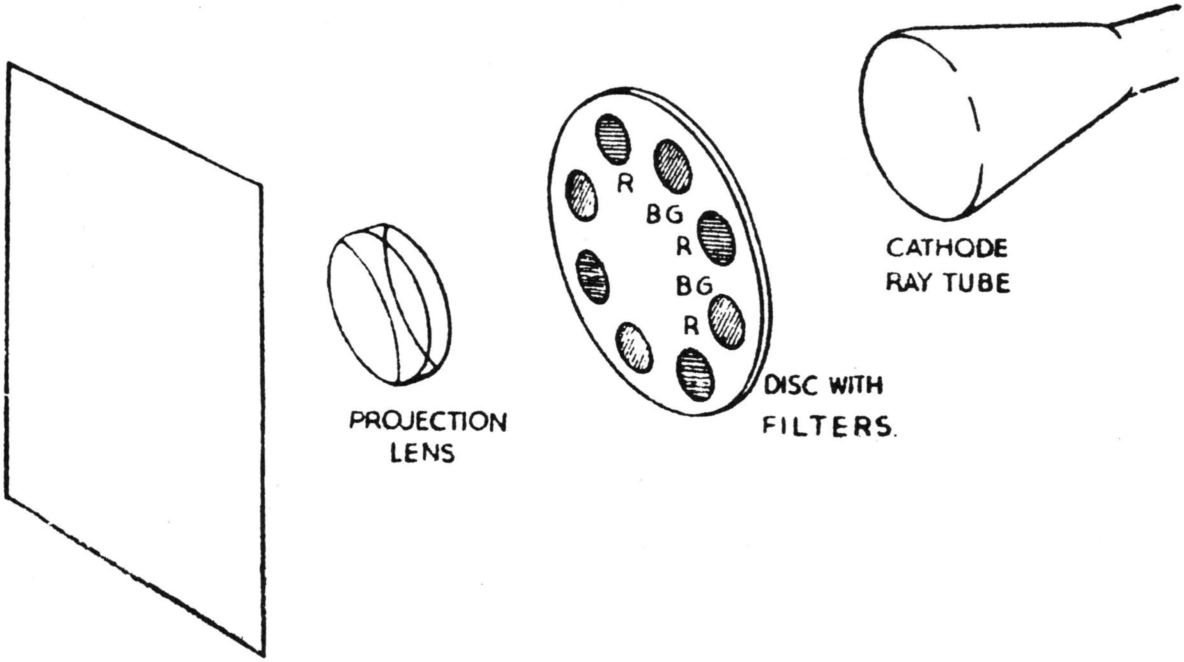

Figure 6.1 Baird two-color wheel of 1939.[5]

mask color tube and NTSC compatible dot-sequential system were being developed by RCA and the CBS victory was short-lived, but not without a considerably heated debate before its reversal.

Perhaps the most interesting event was the headline-making demonstration to the FCC (in 1949 or 1950) of a 20-inch color receiver modified for the CBS system by Dr. Thomas Goldsmith, Jr., director of research at Du Mont Laboratories. The four-foot diameter color wheel driven by a five horsepower motor and television receiver weighed a total of 700 pounds. The demonstration's intent was to show the impracticality of large-screen receivers using the CBS system.[18] A clipping from an unknown newspaper in the Du Mont collection at the Smithsonian Institution describes the debacle under the heading: "FCC Stops Du Mont Demonstration After 'Converter' Blows a Fuse." Despite the sideshow atmosphere, it appears that a major impression was made on the press if not the FCC commissioners.

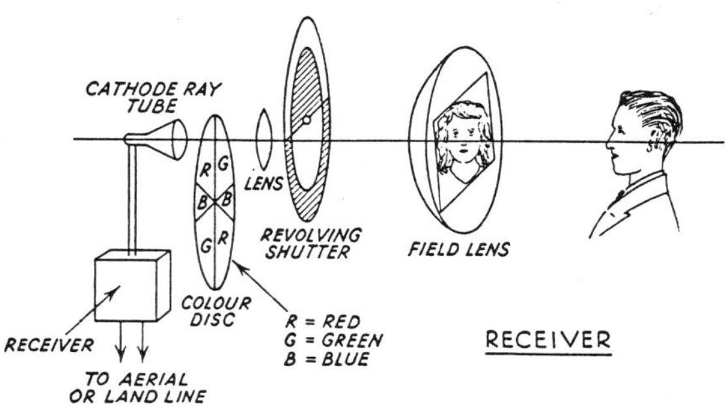

Figure 6.2 Baird three-color wheel of 1941.[6]

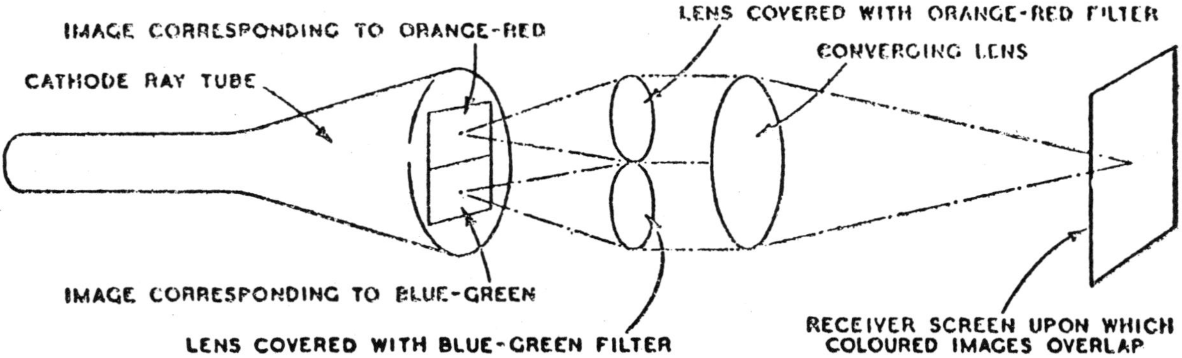

Figure 6.3 Baird two-color television receiver of 1943 which eliminated mechanical scanning.[10]

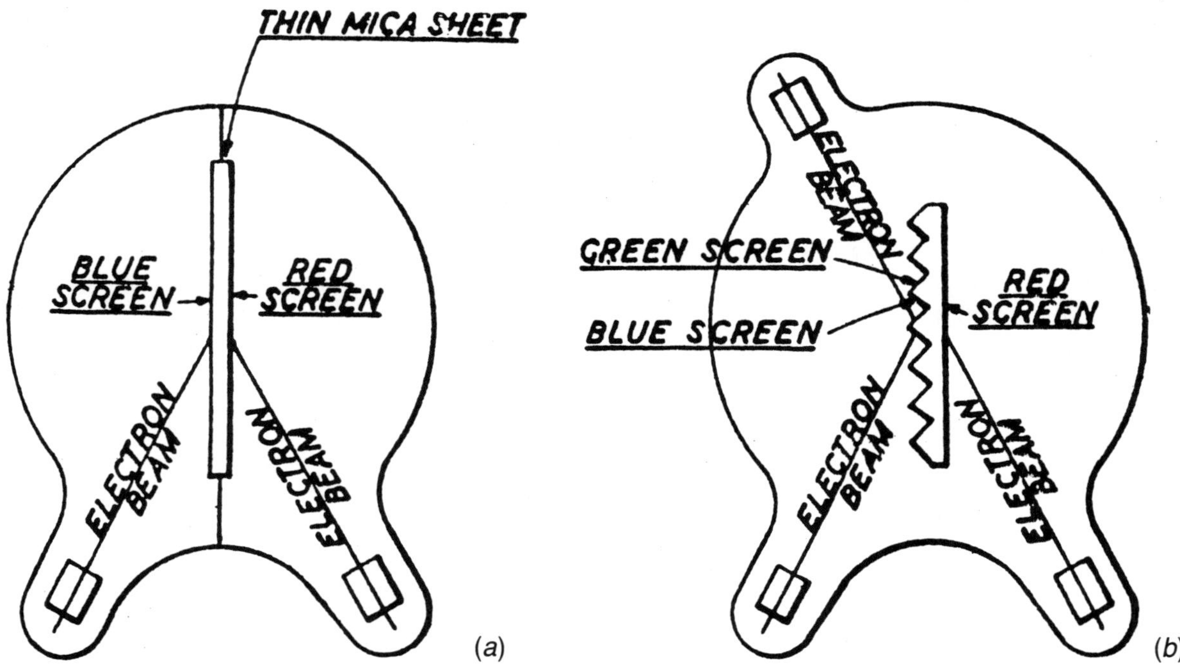

Figure 6.4 Baird Telechrome color television picture tube of 1944 for (a) two and (b) three primary colors, respectively.[12]

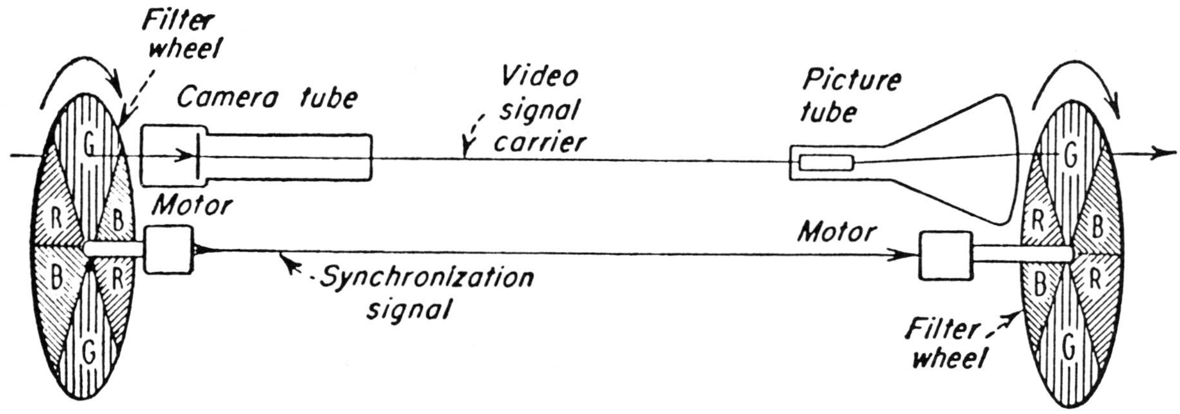

Figure 6.5 CBS field-sequential color system of 1940 with color wheels for three primary colors.[16]

Field-sequential color was revived by NASA during the 1960s. Slow scan pictures taken in space were scanned through three-color filters and combined on earth to make color photographs.

6.3 TRINOSCOPE

RCA pressed for an all-electronic approach to color television and in 1946 demonstrated a partially compatible system using three simultaneous channels for red, green and blue.[19] The green channel was similar to a standard monochrome signal and thus provided the compatibility. The CRT developed for the system was dubbed the Trinoscope and was made in two forms, both projection. In one version, the three separate CRTs with red, green and blue phosphors, deflection yokes and combining optics assembly were called a Trinoscope. The second version comprised a single CRT with three necks and electron guns and one common deflection yoke to produce three side-by-side pictures which were combined optically.[20] Later versions, described in 1949, combined the images using either two or three 10-inch picture tubes with dichroic mirrors.[21] The expense, large physical size and poor picture quality prevented these approaches from being a viable alternative to the CBS system.

6.4 GEER TUBE

C. W. Geer filed for a patent on a rather unique color cathode-ray tube in 1944.[22,127] The tube used a prismatic screen having red, green and blue phosphor deposited respectively on the three facets of each cubical-pyramid prism (Figure 6.6). Three electron guns were mounted at 90-degree angles to each other in order to bombard only one color phosphor each. Technicolor Corporation reportedly supported the development at Stanford Research Institute.[1] An article published in 1947 on experimental cathode-ray tube research at Du Mont Laboratories pictures a similar tube which they refer to as the Trichromoscope (Figure 6.7) with no mention of Geer.[23] General Electric licensed the device and a prototype color television receiver based on it now resides in the Smithsonian Institution archives. Needless to say, the tube was a bit cumbersome.

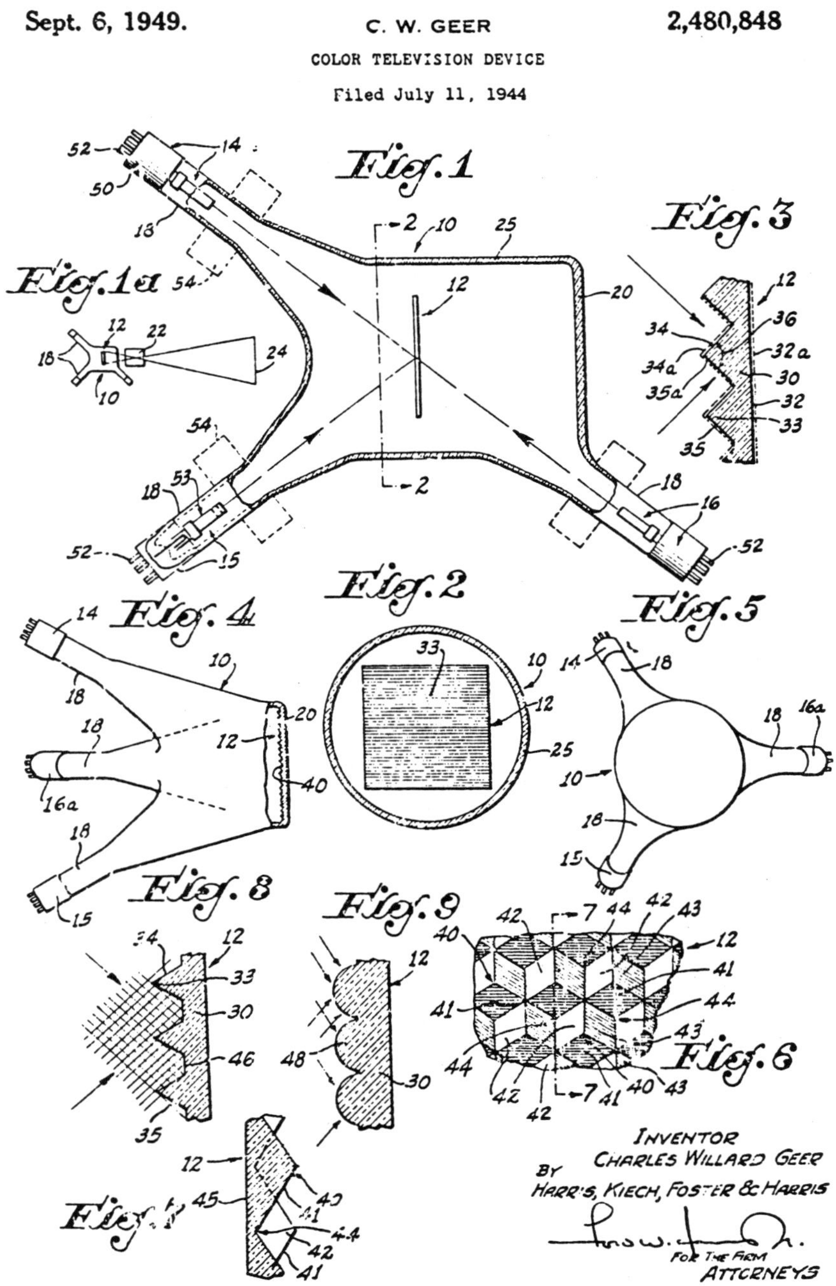

Figure 6.6 Geer tricolor picture tubes from patent filed in 1944.[22]

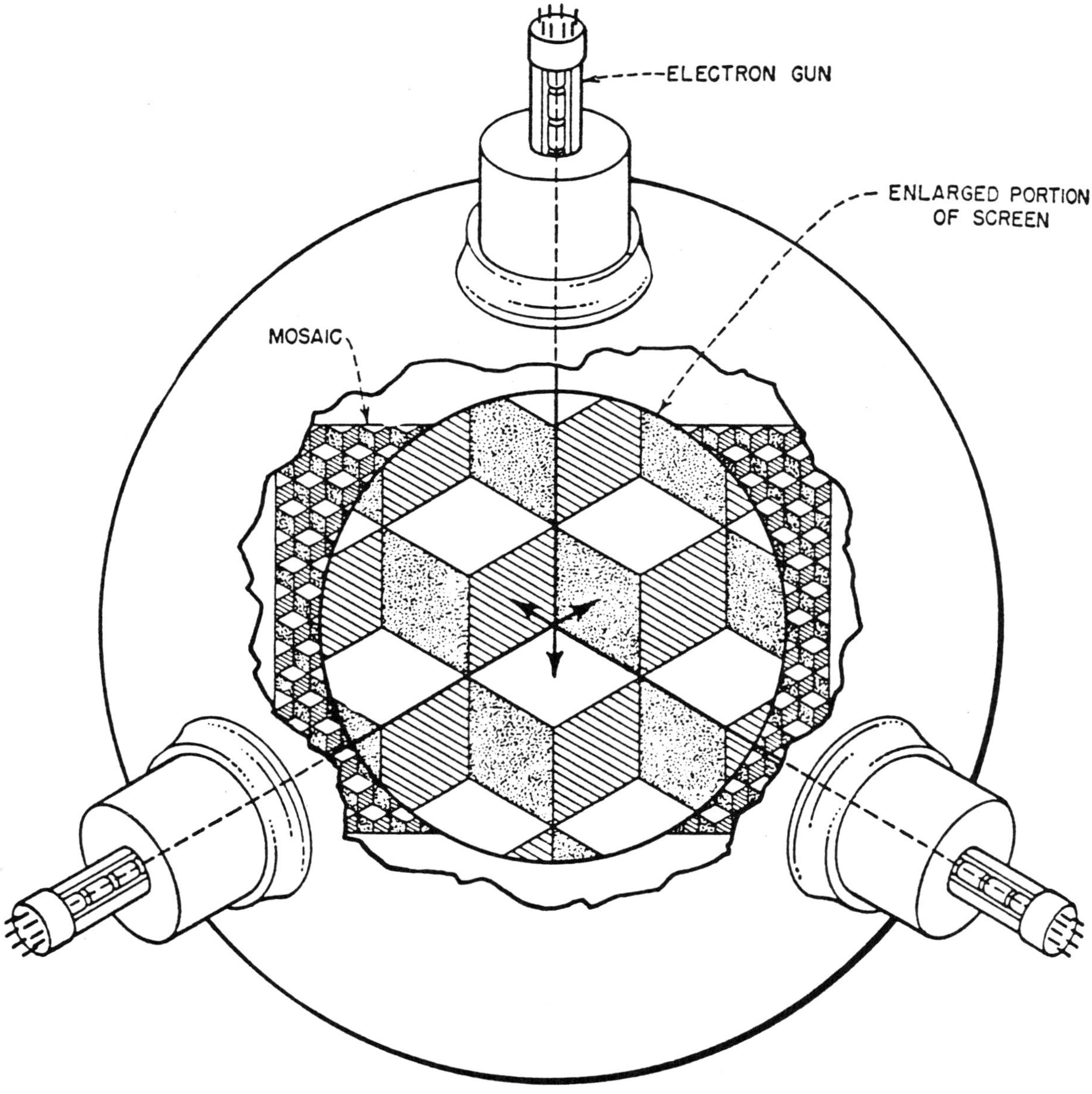

Figure 6.7 Du Mont Trichromoscope tricolor picture tube of 1947.[23]

6.5 SHADOW-MASK COLOR TUBE

Two things prevented the CBS system from becoming the permanent U.S. Standard: (1) a herculean effort (beginning in September 1949) by RCA to develop NTSC-compatible, all-electronic color and (2) development of suitable standards by the second NTSC committee meeting.[24] At the heart of the problem was the requirement for a color cathode-ray tube. RCA imposed no expense limits and all of the company's personnel were at the disposal of the research team led by Dr. Edward W. Herold.[25] RCA pursued several parallel paths to develop such a device and by 1950 had several different designs from which to choose. The *RCA Review* (September 1951) and the *Proceedings of the IRE* (October 1951) were devoted to color CRT developments and contain excellent coverage of these competing designs.[26,27] Articles in the two journals covered a one-gun shadow-mask tube,[28,29] a three-gun shadow-mask tube,[30–41] a grid-controlled tube[42,43] (Figure 6.8), a line-screen tube[44,45] and a reflection-type color tube.[46,47]

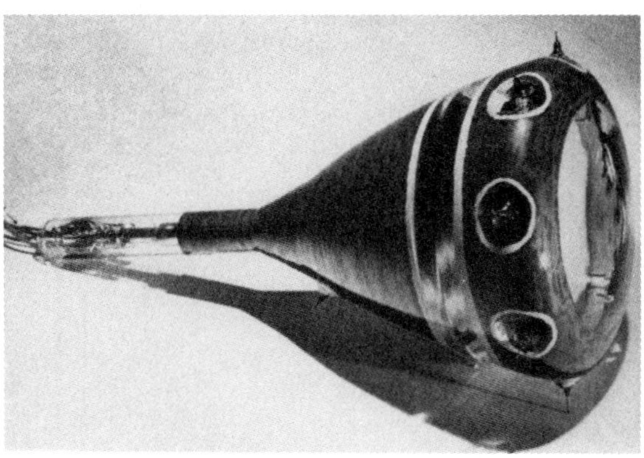

Figure 6.8 RCA experimental grid-controlled single-gun color kinescope of 1951.[42,43] (Courtesy of RCA/Thomson.)

The shadow-mask color tube quickly moved to the forefront of development activity due to its superior picture quality. Forty years later, the shadow-mask tube remains the heart of virtually all direct-view color television receivers. Interestingly enough, the shadow mask concept dates back to a German patent application filed in 1938 by Werner Flechsig.[25,127]

Less than three months after RCA initiated their crash program to develop a compatible electronic color television system, H. B. Law had constructed the first experimental shadow-mask picture tube using a photographic and lithographic process to produce the three-color phosphor dot pattern. A shadow mask was mounted a short distance behind the screen and allowed the electrons from the red, green and blue guns to strike only their corresponding phosphor dots (Figure 6.9). Law's first tube was nine inches in diameter and produced a 4 × 5-inch picture (Figure 6.10). An improved version of the three-gun, shadow-mask CRT was demonstrated by RCA to the FCC on March 23, 1950. Subsequently it was shown to the press and public on March 29. The tube demonstrated was about equivalent to a 12-inch black and white screen and had to be viewed in subdued lighting because of its low brightness. A metal cone envelope similar to the 16AP4 monochrome picture tube was used with a flat glass screen and shadow mask which were mounted internally behind the glass faceplate[25,33,39,48,49] (Figures 6.11, 6.12).

The next three years were devoted to the commercial introduction of color television. Many problems required solutions to convert the hand-built prototype CRTs into production devices.[50] The RCA 15GP22 (Figure 6.13) three-gun color tube solved many of these

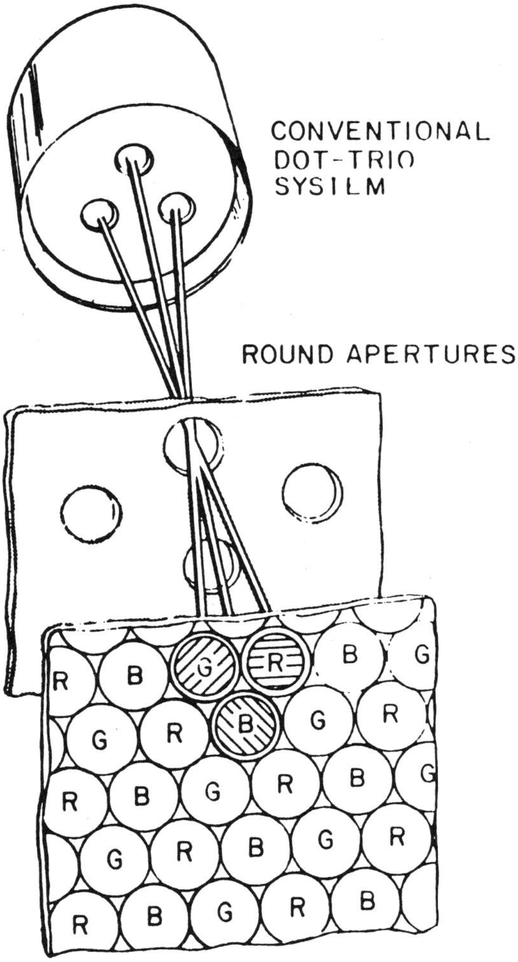

Figure 6.9 Shadow mask and phosphor dot pattern.[107] *(Courtesy of RCA/Thomson.)*

problems (1953) and was registered in February 1954 by RETMA. Unlike the prototypes, the 15GP22 used a new, round, 15-inch glass envelope with a metal flange between the faceplate skirt and funnel rim. The flange's two halves were attached to their respective glass parts with glass-to-metal seals. This flange provided the means for sealing the separate faceplate to the funnel by welding, after mounting the screen and shadow-mask, and also served as the high-voltage connection. The actual picture size was only 11-1/2 × 8-5/8 inches with a tube length of approximately 26 inches, which made television cabinets quite bulky. A 20-pin bidecal base provided connection to the three guns' many electrodes and a shallow deflection angle of 45 degrees (partly the cause of the great overall length) helped to maintain color purity and convergence over the entire screen

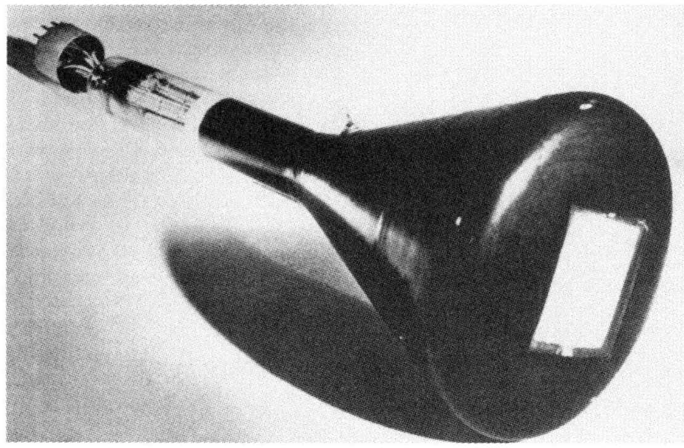

Figure 6.10 RCA shadow-mask kinescope prototype of 1949. (Courtesy of RCA/Thomson.)

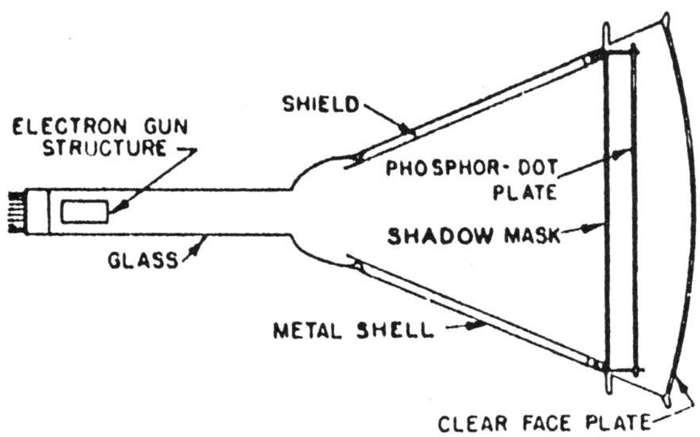

Figure 6.11 Cross section of 16-inch shadow-mask kinescope. (Courtesy of IEEE.)

Figure 6.12 RCA metal cone 16-inch shadow-mask kinescope of 1950. (Courtesy of RCA/Thomson.)

Figure 6.13 RCA 15GP22 shadow-mask glass kinescope of 1954. (Courtesy of C. E. "Sonny" Clutter.)

area. Electrostatic focus and convergence, an aluminized screen and a screen consisting of 585,000 individual red, green and blue phosphor dots were features of this landmark CRT. RCA produced the massive CT-100 console color television receiver using the 15GP22. It is reported that RCA sold these receivers at a substantial loss as part of their strategy to create a market for color. When properly adjusted, the CT-100 was capable of excellent, albeit relatively dim, pictures. Comparatively small production quantities of the 15GP22 were manufactured. Note that the P22 portion of the type number was used to denote the tricolor phosphor screen until 1981 when the new World-Wide Type Designation System was adopted by EIA.

RCA had chosen the difficult approach of having interchangeability of all shadow masks and screens. Shortly after the RCA 15GP22 was announced, a paper by Norm Fyler announced CBS-Hytron's development of a similar 15-inch tube with the phosphor dot pattern photographically deposited directly on the glass faceplate using its own individual curved shadow mask as the negative for the exposure.[51,127] Resultant advantages included mechanical simplicity, ease of manufacture, stability and improved picture quality. The CBS-Hytron tube was registered with RETMA as the 15HP22 (Figure 6.14) in the same

Figure 6.14 CBS-Hytron 15HP22 shadow mask glass kinescope picture tube. (Courtesy of C. E. "Sonny" Clutter.)

month as the RCA 15GP22. Except for screen structure the two tubes were mechanically and electrically quite similar. Even the developers of the RCA shadow-mask tube freely acknowledged the importance of this advancement.[48] All subsequent color picture tubes have been based on this process.

Larger screens soon followed. CBS-Hytron scaled up their 15HP22 to a 19-inch version, the Colortron 205, which was registered with RETMA as the type 19VP22 in July 1954. Du Mont registered a similar tube, the Chroma-Sync type 19TP22 (Figure 6.15) in November 1954. Both utilized a round glass envelope with a flange seal as in the previous 15-inch color tubes. Sixty-degree deflection was used to improve the ratio of screen size to length. The 19VP22 used the 14-pin diheptal base commonly used on oscillographic CRTs as the number of electrodes requiring electrical connection started decreasing. Neither tube achieved any significant degree of commercial success.

Figure 6.15 Du Mont 19TP22 Chroma-Sync color picture tube of 1954. (Courtesy of Jacob Ruiter.)

The sales appeal of the 21-inch screen combined with a number of technical advances led to the introduction of the 21AXP22 (Figure 6.16) by RCA in September 1954.[1,52] This design reverted to use of the metal envelope, which had been abandoned for black and white television in the early 1950s. The chrome-iron envelope was chosen for design flexibility and a weight-savings of 10 pounds compared to glass. A round screen continued to be used, as with almost all commercially produced color tubes, for the next 10 years. Greater dimensional stability of the envelope and internal parts, simpler construction and lower cost were the reasons given for use of the round envelope. For the first time in color

Figure 6.16 RCA 21AXP22-A metal-cone color picture tube of 1956. (Courtesy of James Richardson.)

tubes the 21AXP22 used 70-degree deflection, thus the overall tube length was held to 25-3/8 inches. This was only two inches longer than the 70-degree 21ZP4 monochrome picture tube then in common usage, and it produced a slightly larger picture.[53] The reduced diameter 14-pin (actually only 12 pins were used) neo-diheptal base was used and remained the standard color picture tube base for the next 10 years.

The 21AXP22 was the first color CRT produced in volume and proved the soundness of the shadow-mask color tube as well as the NTSC compatible color system. Several improvements were made in 1956 to the *lighthouse* photographic exposure technique used to produce the tricolor phosphor dot pattern on the curved screen. An important part of these improvements was the identification of beam misregistration (beam landing errors) due to change in effective center of deflection as the beam was deflected farther from the screen's center. A radial corrector lens was added to the lighthouse exposure systems to correct the location of the phosphor dots.[54] The improved version was designated the 21AXP22-A and had considerably improved white-field uniformity and color purity over the entire screen.

Focus mask designs were first explored experimentally around 1955 with several developments being reported at the IRE conventions of 1955 and 1956. These designs were aimed at improving the efficiency of the shadow-mask color tube by allowing use of larger shadow mask openings to transmit more beam current, subsequently focusing the beam between the mask and phosphor screen by connecting each to a different voltage. Thus, each shadow-mask aperture formed a small electron lens for focusing the beam on the phosphor dots. Generally the concept is attributed to a French color cathode-ray tube screen patent application filed by Walter Flechsig in 1938, but it took 18 years to put it into practice.[55]

R. C. Hergenrother of Raytheon first reported on the design considerations pertaining to the application of mask focusing to the color shadow-mask picture tube in 1955.[56] Rauland demonstrated mask focusing shortly thereafter in experimental 19-inch round glass tubes as well as 24-inch rectangular metal cone tubes; both types being of two-piece construction with flange seals, as was typical of the period.[57] Also described in the same paper was the addition of an auxiliary fine stainless steel mesh in close proximity to the rear of the shadow mask and connected to the final accelerating voltage. This served to collect secondary electrons released from the shadow mask under electron bombardment and prevent them from reaching the phosphor screen, resulting in contrast improvement and a wider color gamut. In 1956 Norm Fyler of CBS-Hytron described a developmental *Unipotential Mask Focusing* (UMF) version of their previously announced 19-inch 19VP22 Colortron as well as a 22-inch rectangular glass picture tube of similar flange-sealed construction.[58] The same year, C. G. Lob of General Electric used a wire grill and phosphor stripe pattern for similar purposes, also in a 22-inch rectangular envelope.[55] RCA also reported on development work for the focus grill color tube at the same time.[59] No commercial color picture tubes resulted from this research.

RCA permanently returned to the all-glass envelope in 1957[60] with their 21-inch 21CYP22 (Figure 6.17). This tube was another major advancement because of the frit

Figure 6.17 21FBP22-A glass color picture tube similar in appearance to the 21CYP22. (Courtesy of James Richardson.)

or solder glass seal method used to seal the faceplate to the funnel in place of the previously used welded metal flanges, which were first sealed to the separate parts with a glass-to-metal seal. The glass frit, developed by Corning Glass Works, is applied like toothpaste to one of the glass edges and the other part is held in contact with it under a constant pressure. A high temperature bake converts the glass-containing frit to a ceramic in an irreversible process known as *devitrification*. The frit sealing technique has been used for color picture tubes ever since. Other features of the 21CYP22 included an improved shadow mask with tapered holes for higher contrast by reducing scattered electrons and removing a limiting aperture from the electron guns for increased beam current, hence greater brightness. Electrically, it was quite similar to the 21AXP22 and was used as a higher performance replacement.

The 21CYP22 spawned several improved versions. During the late 1950s and early 1960s the emphasis was on increasing the brightness of the color pictures. The screen efficiency of the 21CYP22 was only about 4 percent of that of a similar size monochrome picture tube due to shadow mask losses and the low efficiency of the red and blue phosphors, which accounted for two-thirds of the screen area. The red phosphor, zinc phosphate: manganese, was particularly troublesome. Not only was its efficiency poor but the color was more orange than red. The 21FBP22 was introduced in 1961 utilizing an all-sulfide phosphor screen patented by Austin Hardy of RCA.[61] A 50 percent gain in brightness was achieved. Furthermore, all three electron guns operated at about the same beam current to produce white. Previously, short life of the red gun cathode was a serious problem since it had to be driven harder than the green and blue cathodes for proper color balance. It was very common to encounter early color tubes with only two out of three guns working properly with the red gun almost always being the "sick" one. Improved brazed cathode assemblies also increased stability of color balance and contributed to longer life. An otherwise similar tube, the 21FJP22 by RCA (1961), was the first color tube to use a tempered-glass implosion shield laminated to the faceplate. Separate glass implosion windows were prone to dust accumulation between the tube and window because of the high-voltage charging of the faceplate glass and frequent disassembly and cleaning were required. Multiple reflections of ambient light also was a problem from the separate glass surfaces.

One of the most important developments during the early color years was the so-called rare-earth phosphor screen in 1964. Actually only the red phosphor was a rare-earth material, usually based on yttrium with europium activator. The 21FBP22-A used a yttrium vanadate phosphor originally developed by Sylvania and although it did not produce a quantum leap in screen brightness, it did serve to focus phosphor research on the possibilities of rare-earth emitters which would soon become a permanent solution to the brightness problem.[62]

6.6 CHROMATRON (LAWRENCE TUBE)

During the 1950s several other color tubes were under development as alternatives to the shadow-mask tube. In 1951 Dr. Ernest O. Lawrence (1901–1958) (inventor of the cyclotron and 1939 Nobel physics prize winner), of the University of California, demonstrated a color CRT, which became known as the Chromatron or PDF Chromatron.[63,64] About that time development of the shadow-mask color tube was beginning to show promise. The major feature of the Chromatron was its use of a single electron gun with 400 sets of vertical three-color phosphor stripes on the screen. A control voltage applied to a wire grid placed 0.4 inch behind and aligned with the phosphor stripes was used to deflect the beam (which normally was aimed at the green stripes) slightly to the left or right to strike either the red or blue stripes (Figure 6.18). Only 3,000 to 4,000 volts accelerating voltage was applied between the gun and wire grid for ease of deflection, while an additional 9,000 volts was applied between the grid and phosphor to increase the brightness. The voltage and grid-to-phosphor spacing were chosen to provide focusing of the beam at the

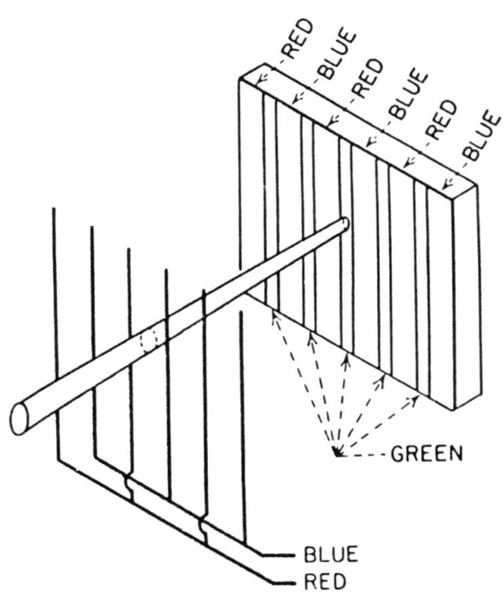

Figure 6.18 Chromatron screen structure. Voltage applied to grid wires deflects beam to land on proper phosphor color.[64] (Courtesy of Electronics magazine.)

phosphor screen, hence the term *PDF* (Post-Deflection Focusing). (This was similar to the focus-mask concept described earlier.) Chromatrons having three guns aligned horizontally to independently strike phosphor stripes of a particular color also were constructed. Their advantage was that no deflection voltage had to be applied to the grid wires to cause the single beam to strike the different phosphors. This was a forerunner of the in-line CRT.

Chromatic Television Laboratories in New York worked toward commercializing the Chromatron. Tubes were fabricated using 22-inch round metal envelopes although the actual screen size was considerably less than the tube diameter.[65] Chromatic Labs licensed Crosley Corp. and Thomas Electronics in the mid-1950s to manufacture the Chromatron. Neither had manufactured the tube by 1956. Crosley no longer had a picture tube plant and Thomas had chosen the shadow-mask tube. Du Mont Laboratories signed a license agreement in 1956 and intended to inaugurate tube and receiver production within one year.[66] It was a corporate marriage of convenience—Paramount Pictures owned 25 percent interest in Du Mont Labs and 50 percent of Chromatic Television Labs. Circa 1960, later than the company's original schedule, a lab employing 35 to 50 people was set up to develop the Chromatron. No product emerged and the lab (located on 34th Street in New York City) was closed in 1965. Sony was licensed to use the Chromatron patent in 1962 and Du Mont worked with them to transfer the technology. Sony introduced a 17-inch Chromatron television receiver for the Japanese market in 1964. This problem-plagued receiver was discontinued after 13,000 had been sold. Sony launched an intensive effort to recover a belated position in the color television market, which eventually resulted in their development of the Trinitron.[67] The Chromatron's grid structure may have influenced the design of Sony's immensely successful Trinitron. A great similarity exists between the vertical wire grid structure in each device.

Litton Industries, also under license from CTL, registered a five-inch round Chromatron, the 5CGP29, with EIA in 1958. Only two long-persistence phosphor colors were used, P2 and P25. That particular phosphor stripe screen also was registered simultaneously by Litton as P29. The intended application was for radar moving-target identification, IFF and a number of other uses where color coding of certain radar and navigation information would aid recognition or identification.

6.7 BEAM-INDEX (APPLE TUBE)

The Apple beam-index tube was announced by Philco in 1956 following several years of development.[68–71] No mention is made in the published literature of the name's origin, but one suspects that it was an internal project code name. The Apple tube was a simplified approach to color television using a single electron gun and no shadow-mask structure.

The Apple tube consisted of an aluminized tricolor phosphor stripe screen with one magnesium oxide (MgO) indexing stripe for each trio of color phosphor stripes (Figure 6.19). The magnesium oxide stripe which was deposited on the gun side of the aluminizing had a high secondary emission yield, i.e., when struck by an electron beam, many secondary electrons were released. The aluminum layer was connected to a 27,000 volt acceleration potential while the conductive dag

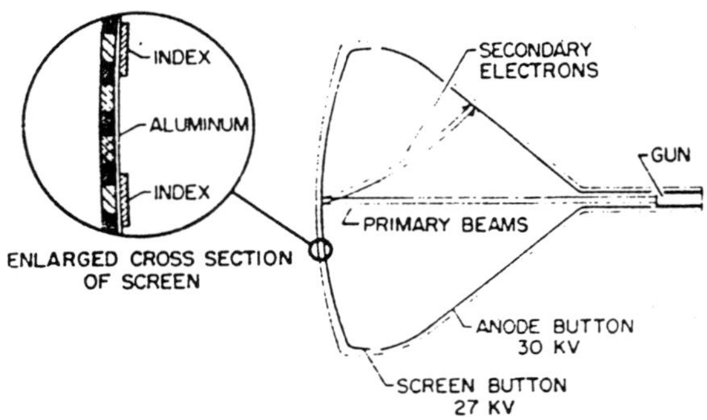

Figure 6.19 Apple tube cross section.[68] *(Courtesy of IEEE.)*

coating on the interior of the funnel was at 30,000 volts, thus forming a collector. The more positive voltage of the collector attracted the negative secondary electrons which were received in bursts as the electron beam traversed the MgO index stripes. The electron gun produced two beams, the pilot beam and the writing beam, with individual control grids. This signal from the collector was used to determine the low current pilot beam's position relative to the color stripes so that the color information displayed by the higher current writing beam was aligned with the appropriate color stripe. A convergence electrode ensured that the two beams tracked together across the screen.

Philco manufactured preproduction quantities of a 21-inch rectangular glass, 72-degree deflection tube for evaluation and life-testing. While the results were technically satisfactory, the timing was wrong as the shadow-mask tube already was becoming well entrenched, thus Philco disbanded their efforts.[1]

Sony described a 30-inch, 114-degree, beam-index picture tube in 1981.[72] By that time the previous electronics complexity required for beam indexing had been largely solved by advances in integrated circuits. Beam-indexing has been revived recently as an approach to cockpit and computer color displays by Sony, Thompson-CSF, Thomas Electronics and others.

6.8 BANANA TUBE

Although the source of the term *Apple tube* may not be readily apparent, the term *banana tube* is self-explanatory (Figure 6.20). Mullard Research Laboratories in England began the development of a simpler color cathode-ray tube in 1955, which resulted in the banana tube being reported in 1960. The banana-shaped tube contained only one lengthwise strip of phosphor for each of the color primaries. The beam from a single electron gun in one end of the tube was deflected to the phosphor stripes near the side of the tube with the horizontal sweep (Figure 6.21). A slight additional up and down deflection allowed the beam to address the desired color phosphor stripe. Vertical deflection was accomplished with a rotating drum surrounding the entire length of the phosphor stripes and containing three cylindrical, lengthwise lenses.[73–78] The banana tube was never commercialized but is presented as one of the many tried and discarded approaches in the color cathode-ray tube's evolution.

6.9 ZEBRA TUBE

The Zebra tube was another single-gun, beam-index, color picture tube and was described in early 1962. Sylvania-Thorn Color Television Laboratories in England developed it using short-persistence ultraviolet emitting phosphor stripes on the rear of the screen. A photomultiplier tube detected the emission as

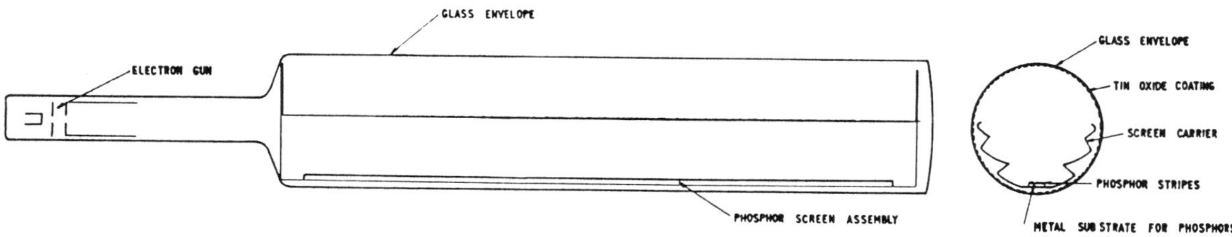

Figure 6.20 Banana tube cross section.[74] (Courtesy of Institution of Electrical Engineers.)

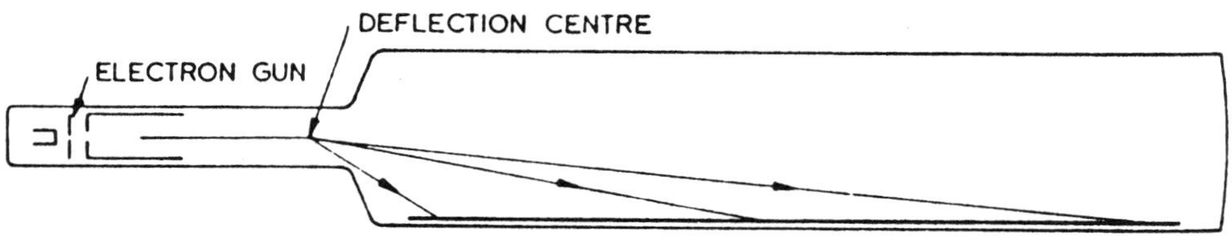

Figure 6.21 Banana tube showing electron beam trajectory.[74] (Courtesy of Institution of Electrical Engineers.)

the electron beam traversed the stripes, thus providing position information relative to the tricolor phosphor stripes and allowing modulation by the proper color information. A total of 1,100 phosphor stripes were used. The Zebra tube did not become a commercial product at that time.[79]

6.10 RECTANGULAR SHADOW-MASK COLOR TUBES

The first commercially available rectangular color tube was the 22-inch, all-glass 22EP22 (Figure 6.22) developed by Westinghouse Electric Company in their Elmira, New York, plant (1956). A 24-hour-long pump cycle combined with a high-temperature bake (400 degrees) resulted in a very good vacuum and long life. The 22EP22 was used in a television receiver manufactured by Westinghouse, but economics doomed it after only one year of limited production. The long pump

Figure 6.22 Westinghouse 22EP22 rectangular glass color picture tube of 1956. (Courtesy of James Richardson.)

cycle limited production and Westinghouse was unable to compete with RCA who was selling receivers at a loss in order to develop the market.

It remained for RCA to produce the 25-inch rectangular tube in 1964 before the rectangular screen format spread through the color television market as it had the black and white market 15 years earlier. Rectangular color tubes made their successful commercial debut with RCA's 25-inch, 90-degree deflection 25AP22.[80] Rectangular tubes had long been the norm for black and white television, but the less-critical round tube had served the color television industry through its comparatively long incubation period and by the early 1960s the processes for mass-production were established and well understood. The 25AP22 had a living room-sized screen while being over four inches shorter than the round 21-inch series of tubes. Several new features were introduced in this tube, including a flatter screen, an aluminum foil electron shield between the edge of the shadow mask and the bulb to prevent scattered electrons from reaching the screen and degrading contrast, and a more compact electron gun. The latter allowed the use of a smaller diameter neck (1-7/16-inch (36 mm) versus the previous two-inch diameter) which reduced the deflection yoke power and size requirements, and improved convergence by placing the electron beams closer together. Also new to color picture tubes was the *hard-pin stem* which eliminated the need for a separate base by employing short, stiff, feed-through wires on the stem as pins to mate with a socket—a trend initiated in the late 1950s for monochrome picture tubes.

With the 25AP22 and the soon-to-follow 25BP22, 19EXP22 and 19EYP22 by RCA in 1965,[81] the floodgates opened for new rectangular color picture tube designs by all of the major U.S. manufacturers. Up to the late 1960s all tubes utilized the early *dot triad* screen with the three electron guns arranged in a triangular fashion about the neck axis in what was termed a *delta gun* configuration. (Sections 6.12 and 6.13 discuss the further major evolution of the shadow mask color picture tube in the Trinitron and the slot mask/in-line tubes.)

In the late 1960s picture tube faceplates became somewhat squarer and tube designations changed as a result of a Federal Trade Commission ruling requiring actual viewable screen dimensions to be advertised rather than overall glass dimensions as previously used.[82] The letter "V" was added after the screen size in EIA tube registrations to indicate viewable inches, e.g., the Rauland 19VMJP22. At that time tube type numbers had become quite lengthy.

During the period of 1967 to 1970 one major advance occurred stemming from Sylvania's earlier rare-earth phosphor work. RCA developed yttrium oxysulfide phosphor and Sylvania developed yttrium oxide, both europium-doped, resulting in substantial brightness gains. Further screen processing developments produced still greater brightness gains.[62] Note that despite the considerable chemical and spectral differences of the newer phosphors, the RETMA/EIA designation still remained P22 with only the descriptor "sulfide/silicate/phosphate," "all-sulfide," etc., differentiating the registered variations.[83]

Another important enhancement was the black matrix screen. Under ambient light the white phosphor body color is detrimental to picture contrast as it reflects most of the ambient light back at the viewer. A gray glass faceplate helps somewhat as it absorbs room light twice in its path to the phosphor and back, however, it also absorbs the light emitted by the image on the phosphor once on its path to the viewer. The black matrix, or black surround (Figure 6.23), developed by Rauland for their Chroma-Color picture tube (1969) uses a black coating on the faceplate except for the areas where the phosphor dots are deposited.[48,84,127] Because about 50 percent of the screen area is black, the reflected ambient light is cut in half with no picture brightness sacrificed by absorption of light emitted by the phosphor. Variable phosphor dot sizes also were utilized to aid the balance of brightness between the three color primaries.

A later refinement in 1970 further improved contrast by depositing a pattern of small three-color filters on the faceplate which pass only the color emitted by the individual phosphor dots. A metal resinate or luster process was employed by which colored inorganic material decomposed to leave a residue of the appropriate color at the location where each phosphor dot was to be subsequently deposited. *Negative Guard Bands* consisting of overlapped areas of the color filters formed the black surround between phosphor dots to provide absorption of ambient light.[85]

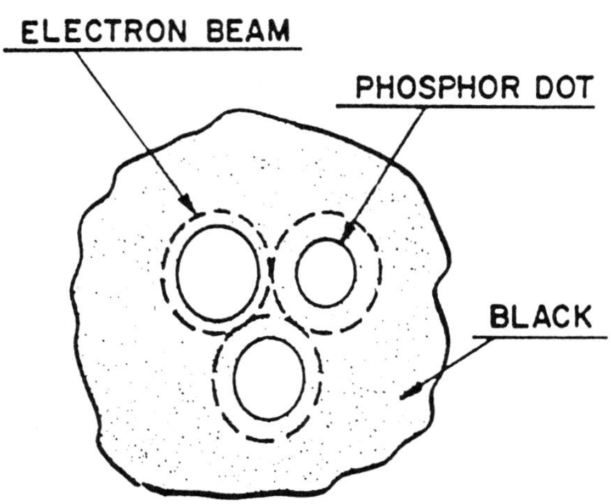

Figure 6.23 Black matrix screen configuration. (Courtesy of IEEE.)

Wider deflection angles were achieved by RCA in an 18-inch color picture tube in 1970[86] and a 25-inch tube in 1971,[87] both having 110-degree deflection. The 25-inch tube, the RCA A67-150X was intended for European sales only. New deflection yoke designs played an important role in this development which allowed still shallower cabinet designs. Reduced deflection power was required by use of a 29 mm diameter neck instead of the 36 mm size used for 90-degree deflection and the tube length was shortened by four inches for the 18-inch tube.

Focus-mask research resurfaced in Japan in 1976 with a delta-gun, dot triad, 110-degree deflection tube (Figure 6.24) being described by Hitachi.[88] Philips Research Laboratories[89–91] in the Netherlands and RCA Laboratories[92,93] in the United States reported on dipole and quadrupole focus-mask research in the period of 1980 to 1984. The Philips work was with magnetic mask focus-

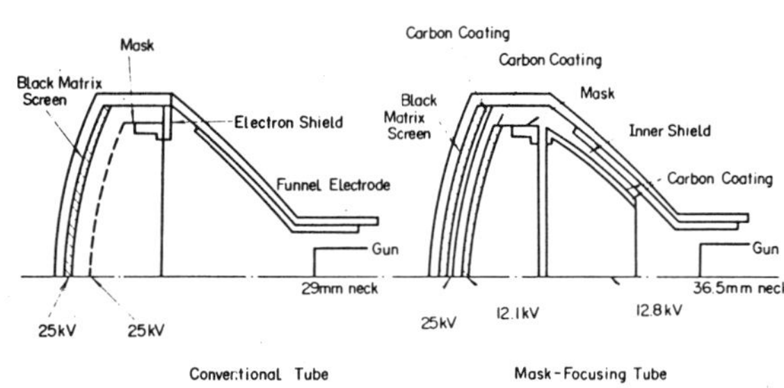

Figure 6.24 Conventional and focus-mask color picture tube construction. (Courtesy of IEEE.)

ing whereas RCA used electrostatic fields. Again, no commercial products resulted due to cost, resolution, purity and stability factors.[77,94]

The aforementioned were most of the final developments for the delta-gun shadow-mask color picture tube. Beyond 1972, most development work concentrated on the Trinitron and the in-line-gun color tube which will be discussed later. (An excellent text on color picture tube design up to 1974 is the Morrell book, *Color Television Picture Tubes*.)[77]

6.11 THE X-RAY SCARE

All electron tubes produce X-radiation internally in the process of electrons being accelerated toward and striking an anode or other electrode which is positive with respect to the cathode. To paraphrase an old saw, "It ain't the acceleration, it's the sudden stop." Fortunately, at accelerating voltages below approximately 15,000 volts, virtually all X-rays are absorbed by almost any type and thickness of glass, metal or ceramic envelope used to construct the tube. It is only when accelerating voltages reach 20,000 to 30,000 volts, as used to obtain adequate brightness from the relatively inefficient shadow-mask color picture tube, that X-radiation becomes a significant design consideration. X-radiation increases linearly with beam current or current density, but increases approximately with the 20th power of the acceleration voltage,[95] therefore the acceleration voltage must be controlled to prevent X-ray emission.

In early 1967, the U.S. Public Health Service discovered that some color television receivers exceeded the 0.5 mR/hr (milli-Röntgen per hour) X-ray limit recommended by the National Committee on Radiation Protection (NCRP). Although no evidence of viewer injury was evident, the finding understandably caused considerable concern among the viewing public, particularly those with small children who tended to watch television while lying on the floor directly in front of the receiver.[96] This problem led to much investigation by the Electronic Industries Association, the receiver manufacturers and, naturally, the government. (Actually, television program content may have been more harmful than the X-rays!)

Three sources of potential excessive X-radiation were identified: the picture tube, the high-voltage shunt regulator tube (which was the source of most X-radiation) and the high-voltage rectifier tube.[97] The problem could be compounded by failure or misadjustment of the high-voltage regulator circuit which can cause excessively high voltage on any or all of the three vacuum tubes.

Several solutions were implemented both by industry's voluntary changes and government's regulatory changes. NCRP soon became the Bureau of Radiological Health (BRH) and was assigned the role of collecting annual reports from all manufacturers and importers of television receivers in the U.S. under the 1969 Federal Performance Standard for Television Receivers. All receivers manufactured or imported after June 1, 1971 were required to meet a 0.5 mR/hr limit with any possible adjustment of user and service controls and a single component failure which maximizes radiation. The reports contain test results from sample production lots and identification of models, dates, components critical to safety, etc. Provision for recall of receivers in the event a problem is found after shipment also was included.

The manufacturers quickly modified receiver designs, switched to lead glass bulbs for high-voltage shunt regulator and rectifier tubes and eventually eliminated the shunt regulator tube altogether. The color picture tube's faceplate glass composition was altered to include strontium, an excellent absorber of X-rays.[98,99] Lead glass, although highly suitable for necks and funnels, could not be used for the faceplate due to its susceptibility to X-ray and electron browning and the resultant loss of brightness with time. Extensive measurements to characterize the X-ray attenuating properties of various picture tube envelopes were taken and published by EIA to illustrate the maximum accelerating voltage and current that they could safely handle with minimum glass thicknesses. Conservatism abounded in all of the aforementioned measurements, changes and regulations. Today's picture tubes and receivers are very "clean" regarding X-radiation.

6.12 TRINITRON

One of the most significant advances in color picture tubes was the advent of the Sony Trinitron in 1968. During the 1970s the Trinitron led the way toward the displacement of U.S. manufacturers from being major domestic suppliers of color television receivers and television receivers. Zenith is the sole U.S. manufacturer left today and their position appears tenuous.

In 1961, Sony was producing small-screen, transistorized monochrome television receivers, but had no definite plans to enter the color market. It became apparent in the early 1960s that the color market was immense and that Sony might miss out on it. After their abortive attempt at developing the Chromatron (Section 6.6) Sony launched a development program reminiscent of RCA's color effort. Teams of engineers led by Masaru Ibuka explored several alternative approaches in parallel with the aim of choosing the best. Sony wanted to offer something unique rather than merely being a "me too" producer. General Electric approached Sony in 1966 with a color tube having the three electron guns mounted in-line rather than in a triangle configuration, as in the conventional delta-gun color picture tubes. Although there were things to be learned from the in-line design, the feeling at Sony was that it was alien to the Sony spirit to become a mere licensee to General Electric.[67]

Meanwhile, one of Sony's engineers, Senri Miyaoka, produced a blurred picture using a single-gun CRT with three cathodes. This information was included in his daily progress report late on a Saturday afternoon. He was immediately called to a conference with Ibuka and other company leaders. He was unsure of the importance of his experiment and whether it could be refined to produce a satisfactory picture. Because he had a weekly practice commitment as a cellist with a community orchestra, he was eager to leave, so when pressed to give a definite answer as to whether it would be successful, he answered "Yes." From that moment on, Sony's full resources were devoted to the new electron gun. A good picture was produced in February 1967.

Attention now turned to the screen's structure. After investigations of the conventional shadow mask, the Chromatron-style grid and other approaches, the aperture grill was suggested by A. Ohkoshi in early summer of 1967. Superior pictures were produced but many manufacturing problems remained to be solved; including elimination of distortion of the aperture grill with heating by the electron beams, and a source of glass CRT bulbs with cylindrical faceplates. After working through the night, the first Trinitron was demonstrated on October 16, 1967. A press release in April 1968 announced the Trinitron publicly with production promised in the impossibly short time of six months. Using the Manhattan Project as a model, a crash program was initiated to set up a plant in Osaki. Many new processes were required but 10,000 television receivers were produced by Christmas. A detailed account of the birth of the Trinitron is recalled in *The Sony Vision* by Lyons.[67]

Several unique features distinguish the Trinitron.[100] They include the use of a single electron gun to produce three electron beams, a single large aperture focus lens for improved brightness and focus uniformity, an aperture-grill, and a resurrection of the cylindrical faceplate originally used for monochrome picture tubes in the early 1950s. All three beams are aimed to pass through the center of the common focus lens (Figure 6.25). The three beams, which diverge after passing through the focus lens, are then reconverged with a set of electrostatic convergence electrodes to bring them together at the screen. The aperture grill, which is flat in the vertical axis, consists of an etched metal sheet with a series of vertical strips. A heavy frame keeps the aperture grill under vertical tension, thus reducing deformation at high beam current due to localized heating. This allows greater brightness without color purity loss. Other stated advantages included higher transparency of the aperture grill with greater brightness than with the dot-triad shadow mask, less sensitivity to terrestrial magnetism because of vertical stripe orientation and freedom from moiré patterns caused by beating of the horizontal scanning lines and the dot array pattern. Convergence adjustment is considerably simpler than with the conventional shadow-mask color tube and is very stable. The usual assortment of convergence and other magnetic adjustments associated with the shadow-mask tube was greatly reduced. Receiver manufacturing and servicing time was correspondingly low. All this was not without some shortcomings—higher tube cost and weight and greater length were the more se-

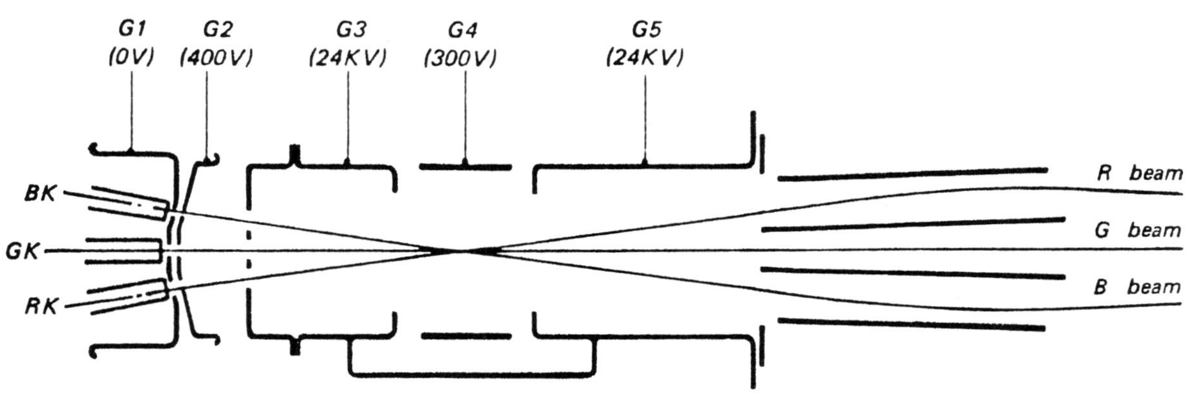

Figure 6.25 Sony Trinitron electron gun cross section showing all three beams passing through common focus element.

rious consequences although progress has been made on these fronts recently.

The first Trinitrons were 12-inch rectangular glass tubes such as the 330AB22 (Figure 6.26). Note the type designation under the Electronic Industries Association of Japan (EIAJ) system where the first digits indicate the screen diagonal in millimeters, the A is an arbitrary sequential assignment and B22 indicates the phosphor with the same digits used in the U.S. for tricolor phosphor screens (P22). The Trinitron has been scaled to at least 3.7-, 5-, 8-, 9-, 13-, 15-, 19-, 20-, 25-, 27-, 30- and 32-inch sizes for television and has been widely used in high-resolution versions for computer displays and ground-based aviation applications, such as flight arrival and departure displays, and air traffic control.

Wide deflection angle Trinitrons arrived in 1973 with 114-degree tubes of 17- and 19-inch screen size.[101] A 114-degree, 25-inch Trinitron followed in 1974[102] and a 30-inch with the same deflection angle was announced in 1977.[103]

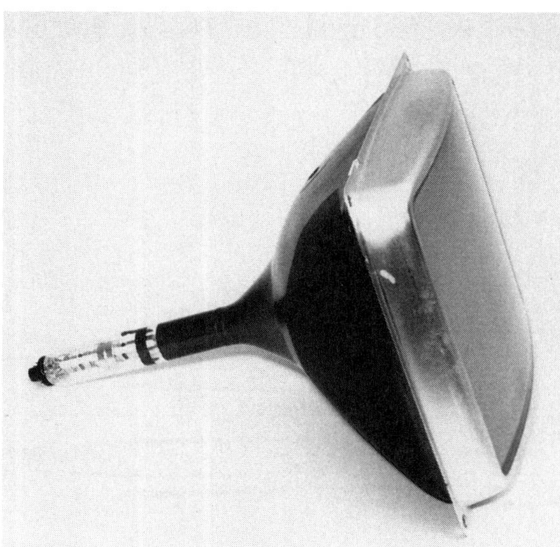

Figure 6.26 Sony 330AB22 12-inch Trinitron picture tube. (From the author's collection.)

High-resolution versions possessing narrower, closer-spaced phosphor stripes and shadow-mask slots have been produced since 1980 when 12V and 19V tubes were announced.[104] All the high-resolution tubes have used 90-degree or less deflection angles. The principal applications have been television studio monitors and computer graphics displays.

Sony described a 45-inch flat-square Trinitron in 1988. It has a 110-degree deflection with an overall length of approximately 28 inches. The reported brightness was only about 32 foot-lamberts, less than that of similar sized projection receivers although the viewing angle is considerably wider.[105]

6.13 SLOT-MASK/IN-LINE GUN

The in-line gun concept dates back to the Lawrence Chromatron in 1951[64,65] and was later attributed to General Electric about 1966.[67] It did not arrive commercially until 1972 when RCA announced the Precision In-Line System.[106] As with the Sony Trinitron, the three electron beams are arranged in-line horizontally and vertical phosphor stripes are applied to the screen (Figure 6.27). Three completely separate electron guns are used rather than the Trinitron's shared elements. The shadow mask consists of slots with connecting webs for strength so that a spherically shaped mask and faceplate may be used instead of the cylindrical shape of the Trinitron. Employing the spherical shape retains low manufacturing cost using existing equipment. The in-line design simplifies convergence adjustment, i.e., the adjustment of all

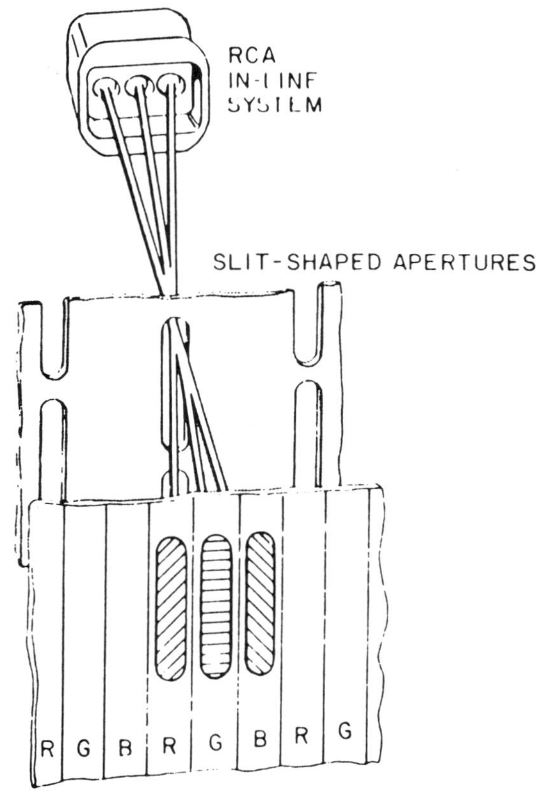

Figure 6.27 In-line screen structure. (Courtesy of RCA/Thomson.)

three beams to produce overlapping colors. At time of manufacture a special factory-installed, self-converging yoke on the tube's neck further simplifies convergence adjustment and eliminates the need for dynamic convergence correction circuits.[107–109] Since the tube and yoke are supplied as an assembly, replacement in the field is nearly as simple as that of a monochrome picture tube.

Another advance was made in 1975 by RCA with their *filtered phosphors* which contained pigmented red and blue phosphors.[110] These phosphors absorbed light of colors other than those emitted by themselves. The darkened phosphor body color absorbed much of the ambient light previously reflected by the conventional white-appearing phosphor and significantly improved picture contrast.[1]

Many electron gun refinements have occurred in recent years, primarily offshore, with the goals of better focus and focus uniformity. These are identified with a bewildering array of abbreviations such as the Toshiba QPF (Quadra Potential Focus) gun with a mere 22.5 mm neck diameter,[111] the RCA XL Hi Pi gun in their COTY-29 (Combined Optimum Tube and Yoke—29 mm),[112] the Matsushita DAF (Dynamic Astigmatism and Focus) gun, the Mitsubishi DBS (Dynamic Beam Shaping) gun, the Philips CFF (Conical Field Focus) gun, etc. The importance of improved electron-optics will continue to grow with the increased use of high-resolution computer displays and high-definition television (HDTV).

Finally, one of the most impressive CRTs ever developed is the Matsushita 43-inch direct-view CRT (Figure 6.28) described in 1986

Figure 6.28 Matsushita 43-inch in-line color picture tube. (Courtesy of Matsushita Research Center.)

papers.[113,114] A deflection angle of 110 degrees was used to hold the overall tube length to about 24 inches. The tube weight alone was 187 pounds. At the other extreme is the Matsushita 1.5-inch "Pocketable"[115] described in 1983 literature. The tiny in-line tube had a neck size of 13.5 mm diameter, 30-degree deflection angle and 0.25 mm shadow-mask pitch with a phosphor dot pattern.

The recent trend in color picture tubes has been to wide usage of the flat-square or full-square (FS) faceplate shape. These screens are nearly flat on the sides and have squarish corners. In addition, the faceplates are noticeably flatter. These are aesthetic improvements for more modern styling and required special treatment of corner defocus and glass strength problems.

6.14 PROJECTION

As direct-view picture tubes increase in screen size, glass strength, implosion safety, size and weight become major factors. Beyond about one meter screen diagonals, projection systems become an attractive alternative to direct-view picture tubes.

Several problems plague projection systems. Most are associated with the large amount of light required from a small CRT in order to produce a comfortably bright picture when that light is spread over the considerably larger projection screen. High beam current, high-voltage and efficient, nonsaturating phosphors are used to produce an extremely intense small image on the projection CRT. High beam current and high voltage make X-ray shielding, heat dissipation from the screen, high voltage arcing and breakdown, and glass and/or phosphor browning important design considerations in both tube and system design. Color projection systems have the added problem of obtaining exact registration of the red, green and blue images which are produced by three separate CRTs (each producing one of the primary colors). Projection systems do have one advantage other than size over the shadow-mask color tube. Since no shadow mask and phosphor dot or stripe pattern is required to produce the separate colors, no color structure (such as moiré patterns) is visible in the picture.

Aside from the pre-NTSC color systems in the 1940s, described at the beginning of this chapter, projection color television for home use was not again seriously considered until 1972. At that time Henry Kloss, founder of the Advent Corporation introduced a three-CRT Schmidt projection television receiver with about a 78-inch screen diagonal.[1,116–121,126] The development of the Kodak Ektalite screen having a gain of ten-times brightness by concentrating the light on-axis toward the viewer and this began to make projection television a viable alternative to the directly viewed cathode-ray tube. The trade-off is a somewhat restricted viewing angle which limits viewing to positions directly in front of the screen. Kloss initially used unique CRTs with built-in Schmidt mirrors to provide efficient coupling of the light from the phosphor to the optical system (Figure 6.29). Systems using mirrors are referred to as reflective projection systems.

Refractive projection systems became practical with the development of high-quality, fast, plastic aspheric lenses which permit lower cost optical systems than can be achieved with reflective optics.[121,122] Sony made good use of refractive plastic optics with their liquid-cooled 5-1/2-inch rectangular projection CRTs announced in 1981.[123] A coffee table-type television receiver pro-

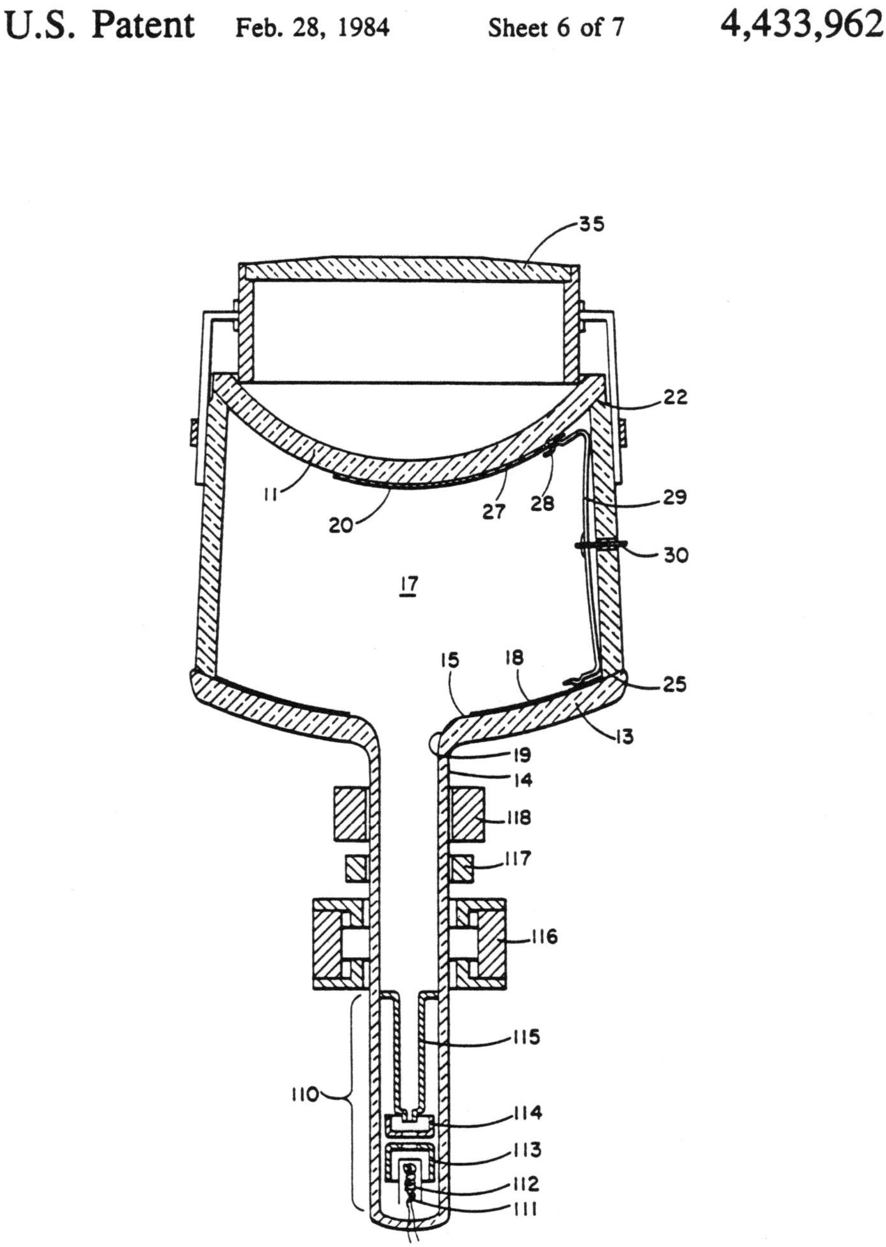

Figure 6.29 Kloss projection CRT cross section.[119]

duced 120 foot-lamberts of brightness on a 50-inch screen and 60 foot-lamberts on a 72-inch screen. The liquid-cooling took place using ethylene glycol between an externally mounted faceplate and the screen panel (Figure 6.30). Cooling of the screen allows high beam power without the severe decrease in phosphor efficiency that occurs at high temperature, as well as providing improved screen life.

Zenith Radio described a self-converging, three-tube projection system in 1982.[124] Three rectangular CRTs with liquid-cooled faceplates were arranged in an in-line configuration. The unique feature was the use of tilted faceplates on the two outer CRTs to correct for the trapezoidal optical distortion that results from their longer "throw" distance to the screen.

The popularity of projection television in the home has increased dramatically in recent years and a number of projection CRTs and receivers are presently available, mostly from Japan. A number of the previously mentioned problems have been solved or at least significantly reduced to the point where projection television is a viable alternative to the direct-view CRT.

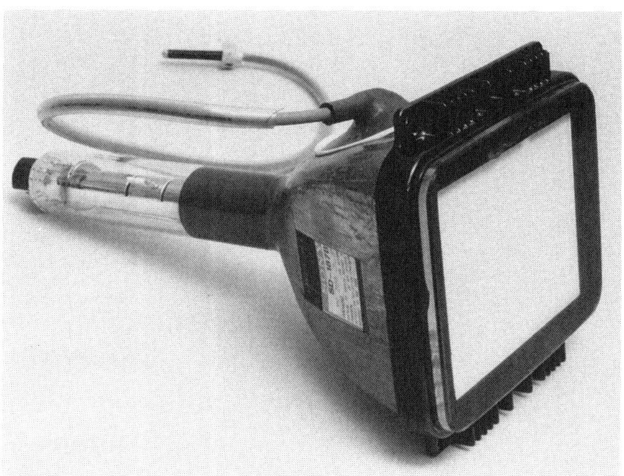

Figure 6.30 Sony SD-187 liquid-cooled projection CRT.

6.15 HIGH-DEFINITION TELEVISION

On the near horizon lies High Definition Television (HDTV). A battle for technical dominance and market share already is shaping up with competing systems undergoing development in Europe, Japan and the United States. Three aspects of HDTV will affect the CRTs yet to be designed: (1) the desire for a large picture size, (2) the need for high-resolution and (3) picture aspect ratio (width to height) of 16:9 to more closely resemble a wide screen motion picture screen. New glass envelopes will be required and projection may become a strong contender. The need for higher resolution will again make adequate brightness an issue since beam current must be sacrificed for smaller spot size. Prototype CRTs have been demonstrated, such as the 28-inch Sony type SD-168 and 40-inch Panasonic type T101-01,[125] but commercialization of HDTV awaits the adoption of transmission standards. Naturally, compatibility with existing NTSC signals is desirable, but there is some thought that the immensely successful NTSC system has been stretched to its limits and that it may be time to develop a new system without the compromises required by the NTSC format. The next few years hold the promise of many new technical advances in the field of HDTV with further CRT evolution initially being a significant contributory factor. Great effort is being directed toward development of flat panel displays for HDTV but CRTs will probably be used for at least the first-generation HDTV receivers.

7

Data Display Cathode-Ray Tubes

7.1 Early Computers

Until the end of World War II, all computing machines were just that—machines. The term persists to this day despite the fact that very few mechanical components are used. Electronic computing began in the early post-war years utilizing many of the wartime radar developments and were used by the engineers responsible for them. Electronic computing was far faster than mechanical means and made it easier to scale-up computing capacity. Early computers such as the IBM 701, the UNIVAC, Remington Rand 409-2 and a host of one-of-a-kind models (the SWAC, SEAC, DYSEAC, EDVAC, MSAC, NAREC, ENIAC and ILLIAC) used mechanical input/output (I/O) devices rather than cathode-ray tube displays as are used universally today. The I/O devices included paper tape readers and punches, card readers and punches, typewriters and page printers similar to teletype machines.[1] Any CRTs found in the early computers usually were for servicing or monitoring applications rather than display of I/O data.

Cathode-ray tubes did find one important role in a number of the early computers. This was in the form of digital memory devices using charge-storage on the phosphor screen. In 1948 F. C. Williams of Ferranti Ltd. first described the charge-storage method[2–4] which employed standard low-voltage electrostatic oscilloscope CRTs. These were referred to as Williams tube storage or electrostatic storage. The phosphor (usually P1) was left with a charge where addressed by the electron beam during the *write* mode. A close-fitting metal cover over the screen then capacitively picked up a signal from the various charge patterns as the beam scanned the screen in the *read* mode. Relatively low-cost oscilloscope CRTs, such as the RCA 3KP1, 5CP1 and 5UP1, were found quite suitable for the purpose.[5,6] One early computer[7] used a bank of 37 CRTs to store 256 words of 37 bits each, just over one KB of memory by today's standards! Later, specially constructed CRTs optimized for charge-storage with internal readout electrodes such as the IBM-79 were commercially produced[8,9] but by then, magnetic-based computer memories were becoming firmly established. The RCA 6499 Radechon[10] and the RCA 6571 are examples of special charge-storage CRTs developed specifically for computer memory during the early 1950s. Flying-spot scanning with CRTs also was briefly tried by Bell Labs for reading photographically recorded data in the range of one to 100 MB storage capacity (see Chapter 9).[11–13] The result (Figure 7.1) was a three-foot long 10-inch diameter CRT somewhat reminiscent of the Davisson tube described in Chapter 5.

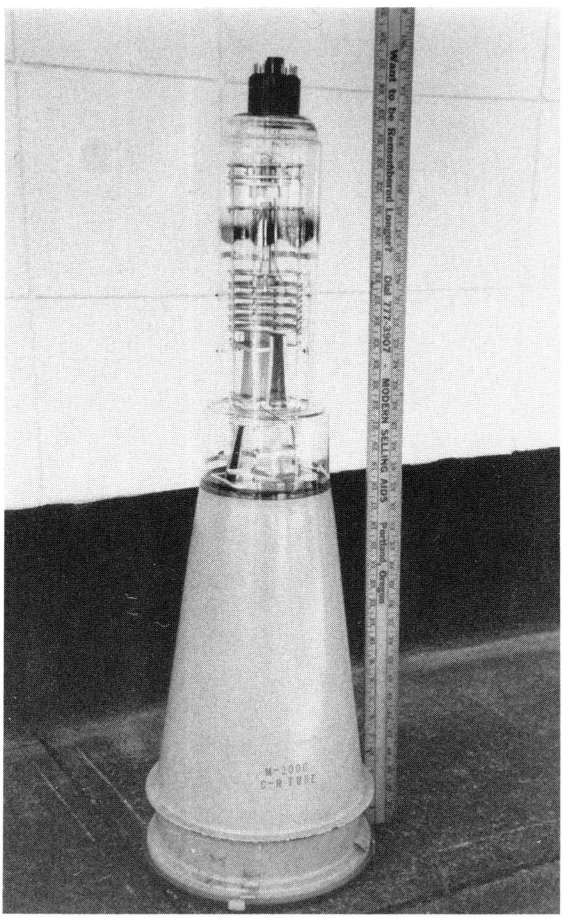

Figure 7.1 Ten-inch flying-spot film store CRT used by IBM. (Courtesy of Larry Lockwood.)

7.2 CHARACTRON

The Charactron cathode-ray tube, developed first by Consolidated-Vultee Aircraft (Convair) in 1952 and later refined by Stromberg-Carlson, was designed to produce character displays on its screen for computer displays.[14–16] A matrix mask containing an array of 64 apertures in the shape of alphanumeric characters along with deflection electrodes to address the desired character by the electron beam are the fundamental differences between the Charactron and conventional CRTs. The mask openings shaped the electron beam which projected the character image on the phosphor screen (see Figure 3.21). This technique is known as *shaped beam* or *extruded beam*. Versions were constructed with both electrostatic and magnetic deflection.

Improvements were later made to enhance the usefulness of the Charactron. In 1971, several versions were constructed possessing greater brightness and resolution.[17] These included a 21-inch tube using a standard TV glass bulb with P4 phosphor, a seven-inch microfilm printer tube, a five-inch projection Charactron and other assorted screen sizes with P31 phosphor. The dual-purpose tubes were intended to display both alphanumerics and graphics. Reduced manufacturing costs were anticipated through external electromagnetic addressing of the characters instead of using electrostatic deflection electrodes in the gun. Additional brightness gain and higher resolution was reported in 1976.[18] This time, a 15-inch rectangular bulb with 8-1/2-inch usable screen area allowed 30 lines of 132 characters each to be displayed using P31 phosphor producing a brightness of 100 foot-lamberts.

A related device was the Matricon described by Rank Precision Industries in 1972.[19] Instead of a character matrix addressed by shifting the beam through the appropriate character stencil, the Matricon used a matrix grid assembly in a 7 × 5 pattern to form 35 individually controllable parallel beams. The individual beams could be turned on by the matrix grid to form characters and those characters could be positioned to any location on the screen since all 35 parallel beams would be deflected as a group.

7.3 TYPOTRON

When using the Charactron, the display on the screen had to be continually updated or refreshed in order to produce continuous information on the screen. Refreshing of a display required a portion of the computer's memory to be devoted to this task. In the early years of computer development, memory was very expensive, therefore the storage tube became attractive to perform both image display and retention of the screen data for prolonged viewing. The Typotron (Figure 7.2) developed by Hughes Aircraft was announced circa 1956. It combined the character generation method of the Charactron with a storage-tube screen structure. The result was a display that required no further use of computer memory once the information was "written" on the storage target until one desired to change the displayed information. In all other respects operation was similar to that of a conventional storage tube.[20]

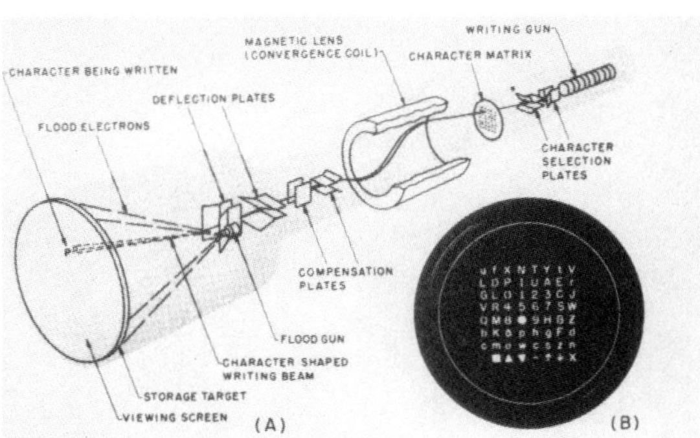

Figure 7.2 Hughes Aircraft Typotron operation.[20]

7.4 DIRECT-VIEW STORAGE TUBES

Electronic character generation circuits, made practical by solid-state electronics in the 1960s, eliminated the need to generate the characters by shaping the electron beam. It was merely necessary to make a series of short, rapid electron beam movements to form a character with larger beam movements directing the beam to the proper location on the screen. This process was referred to as *stroke writing, directed beam writing, vector addressing* or simply *X-Y addressing*. Storage was still desirable to avoid tying up computer memory for display functions.

Small five-inch displays such as the Tektronix Model 601 (T6010 CRT) storage X-Y monitor and the non-storage 602 (T6020 CRT) (Figure 7.3) began to find some application for computer displays, but the trend toward larger screen sizes was evident in the emerging alphanumeric display market. Tektronix was committed to the task of scaling up their T5640 five-inch, bi-stable, direct-view storage tube (DVST) developed by Anderson (see Chapter 4) to an 11-inch tube with the screen size equivalent to a standard written page. The task was not easy as it was difficult enough at that time to produce consistently uniform five-inch storage targets, let alone one having more than four times the area.

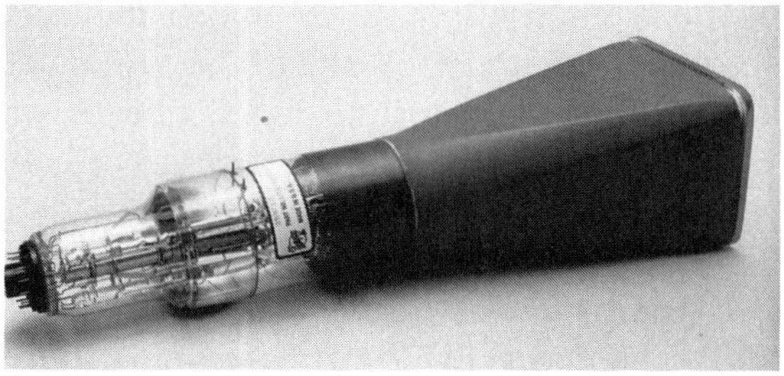

Figure 7.3 Tektronix T6020 electrostatic deflection data display CRT.

More than \$1 million went into the project and the result was the T6110 CRT (Figure 7.4) for the highly successful Model 611 alphanumerics/graphics display in 1968 and a subsequent series of computer terminals.[21–23] The T6110 was a rectangular tube using a 1/2-inch thick, flat faceplate that was frit-sealed to a Tektronix-made ceramic envelope resembling a rectangular flower pot. The resolution was excellent and fully supported page-sized text due to use of a high-resolution magnetically deflected electron gun. Only two drawbacks were present. The T6110's brightness was low, about 10 foot-lamberts, and there was a tendency to age differentially. The aging problem was due to the fact that computers write most often in the upper left corner of the screen, thus causing that area to decrease in light output more rapidly than the rest of the screen.

Despite these shortcomings, the T6110 and the later T4014 (Figure 7.5) and T4016 (Figure 7.6) storage tubes used in several Tektronix-manufactured computer terminals were a mainstay of the CRT manufacturing operation throughout the 1970s. The latter departed from the usual Tektronix ceramic envelope and used standard 19- and 25-inch rectangular color picture tube faceplates and funnels, respectively.[24–26] The screens were spherically shaped as opposed to the flat faceplate of the T6110.

Write-through capability was added to the T4014 in 1978.[27] Write-through was an operating condition whereby beam currents be-

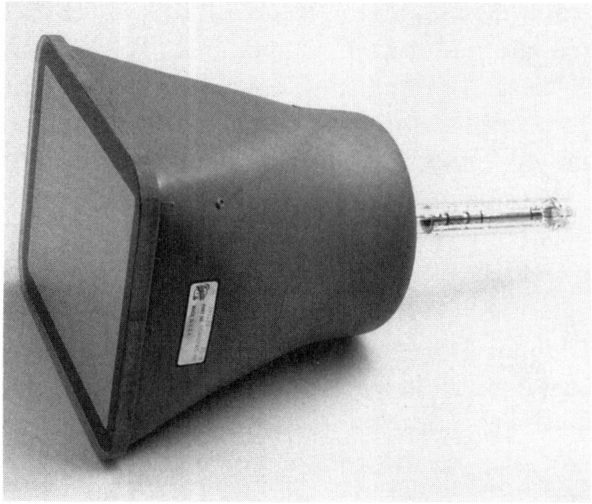

Figure 7.4 Tektronix T6110 page-size ceramic storage tube.

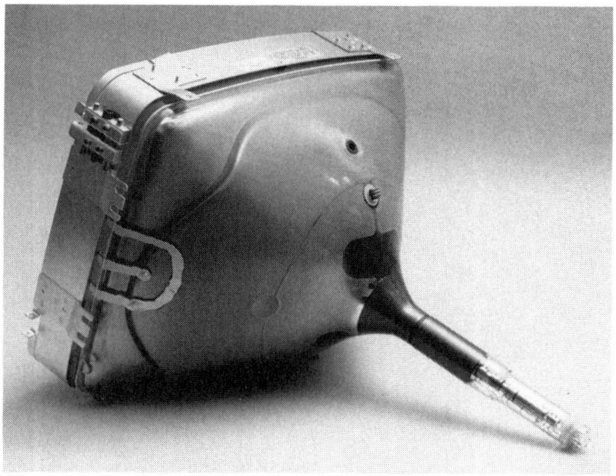

Figure 7.5 Tektronix T4014 19-inch storage tube.

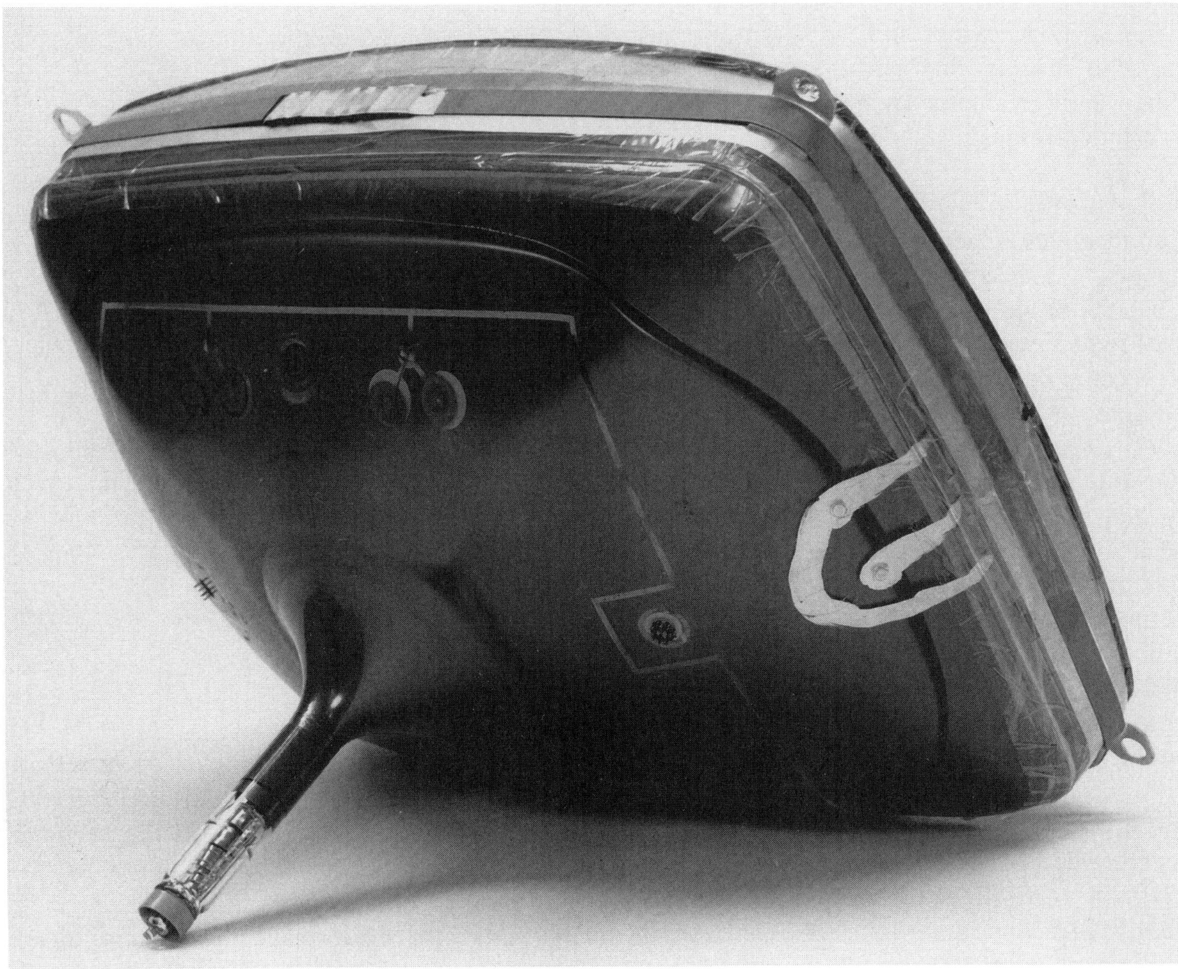

Figure 7.6 Tektronix T4016 25-inch storage tube. (Courtesy of Tektronix Inc.)

low the storage threshold could be used to superimpose a cursor or other nonstored information over the stored information. Color write-through was a further improvement in 1981 that presented the overwritten information in a different color than the stored information (which is green).[28] Two phosphors, red and green, made up the storage target. The green phosphor, which was on the surface, was excited by the low-voltage flood-gun electrons and also provided storage of the writing beam. The higher voltage writing beam electrons penetrated deeper to the red phosphor. Some excitation of the green phosphor also occurred producing a yellow-orange color.

The bi-stable DVST was unique to Tektronix and bridged the gap between early mechanical I/O devices and the lower cost raster-scan display that became cost-effective when semiconductor memory cost dropped dramatically during the 1980s.

7.5 MONOCHROME RASTER DISPLAYS

Semiconductor memory underwent rapid increases in data storage capacity and large-scale offshore manufacturing brought substantial cost reductions during the 1980s. As a result it became feasible to devote a portion of a computer's memory to providing a continuously refreshed display. Raster scanning similar to that used for television was the logical format with characters typically reproduced as a 7 × 9 pattern of dots with the information to be displayed *bit mapped* within the computer memory.

Raster scanning greatly eased the requirements on the CRT. Conventional monochrome and color television CRTs, which could be inexpensively produced in large volume, were readily adapted to computer displays. Simultaneously, the personal computer (PC) further aided demand for mass-production of CRTs as they expanded into nearly all business and many home endeavors. Rectangular CRTs with 70- to 110-degree deflection angles and screen sizes of from nine to 19 inches are common (Figure 7.7). Monochrome CRTs for PCs are now available with several phosphor screens to suit user preferences,[29] the most common being highly efficient green (P39 or P31), amber, television white (P4) and paper-white (less blue than P4). Of course, resolution is an important consideration in the CRT design. Greater care must be paid corner-focus than with television picture tubes in order to ensure readability over the entire screen area. Many CRTs for computer displays are now produced in the Far East, a fact of life dictated by economics. Complete monochrome displays are available for less than $100, thus placing severe cost constraints on the CRT. Despite their low cost, the performance and reliability of these mass-produced CRTs is excellent.

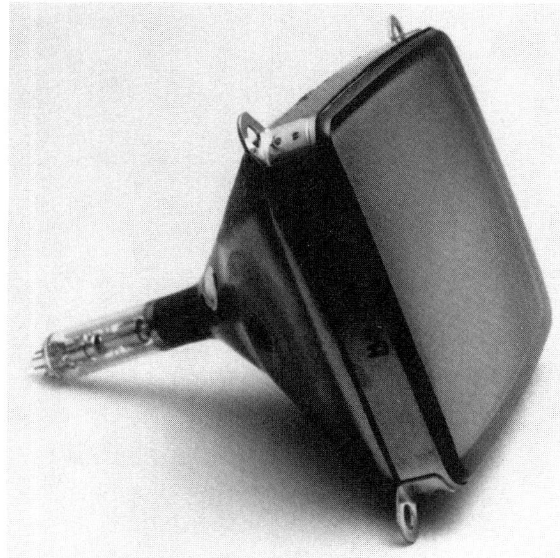

Figure 7.7 Typical low-cost monochrome data display CRT.

The conventional television screen shape is well-suited to most computer applications. They may be used with the long-axis oriented horizontally (landscape format) which allows two pages of text to be displayed side-by-side for larger CRTs or oriented vertically (page or portrait format) to display a single page as normally printed or X-ray images. Some square-format screen CRTs such as the 16-inch Clinton Electronics CE857M16 have been produced for specialized data display applications.

7.6 HIGH-RESOLUTION MONOCHROME CRTS

Medical imaging, photoreconnaissance, page layout and phototypesetting became important applications during the 1980s. All depend on computer-stored images and the ability to view fine detail is crucial. Limitations on color CRT resolution imposed by the shadow-mask structure dictate the use of monochrome CRTs for the highest resolution displays.

An important trend in monochrome displays is toward higher resolution, which fulfills a need that cannot be met with color shadow-mask CRTs. Great progress was made in resolution of both monochrome and color CRTs and displays during the 1980s, although in some cases the improved specifications were achieved in the advertising department rather than the engineering department. Advertised figures for resolution must be viewed with a degree of caution. Addressability (the ability to position the electron beam to a certain number of points on the screen) and resolution (the ability to actually see that number of distinct individual pixels or discrete picture elements) often are used interchangeably even though they are not the same thing. Naturally the best of the two quantities all too often appear in product announcements. Ideally the two specifications should be close to the same. RAR, the Resolution/Addressability Ratio proposed by Murch, Virgin and Beaton in 1985,[30–33,75] goes a long way toward sorting these factors out.

Generally speaking, high-resolution CRTs are very similar to conventional monochrome television picture tubes except for details of the electron gun and the use of shallower deflection angles (typically 90 degrees) to reduce defocusing between the center and corners of the screen. The guns are optimized for small spot size and usually include a beam-limiting aperture to trim the electron beam diameter. Of course, this reduces the total beam current and thus brightness is the primary trade-off for resolution. The driving electronic circuitry for high-resolution CRTs have stringent demands placed upon them to permit full utilization of the CRT performance characteristics. Greatly increased horizontal scan rates, video amplifier bandwidth, power supply regulation, high voltage and deflection amplifier stability, and additional electrical and magnetic shielding are all required resulting in substantially higher cost.

Nineteen- and 20-inch screen diagonals are best-suited to high-resolution displays at the viewing distances normally employed. The Clinton Electronics 20-inch type CE801M20 introduced in 1983 is a popular example of such a tube.[34] The nominal center spot-size of approximately 0.005 inch theoretically allows about 3,000 × 2,400 pixels to be displayed on its 15.9 × 11.9-inch screen quality area if one ignores effects of increased spot size in the corners. Other manufacturers of similar tubes include Thomas Electronics, Thompson CSF (now Hughes), Matsushita and Toshiba.

The CRT with the highest overall resolution for imaging displays probably are the 19-inch, 90-degree deflection Tektronix T4201 and T4202 (Figure 7.8) developed in 1983.[35–37] Several unique features distinguish these tubes from the usual data display CRT. These include the use of added dynamic astigmatism electrodes, a low-capacitance grid

Figure 7.8 Tektronix T4202 very high resolution 19-inch CRT for electronic imaging applications.

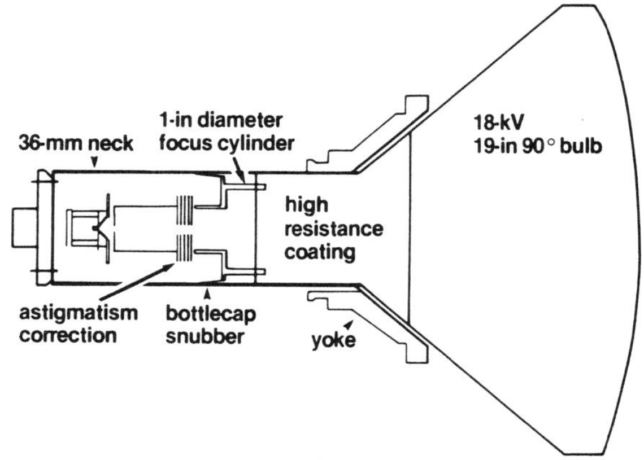

Figure 7.9 Tektronix T4201 very high resolution CRT cross section.[36] (Courtesy of Society for Information Display.)

assembly to allow 200 MHz minimum video amplifier bandwidth, large diameter focus lens to minimize spherical aberration and a particle trap to prevent arcs due to any residual particles remaining from processing (Figure 7.9). The dynamic astigmatism arrangement goes well beyond the capability of dynamic focus as used with most high-resolution CRTs. Dynamic focus often is used to adjust the focus voltage constantly as the electron beam is scanned over the screen. Since the distance from the gun to the screen corners is greater than the distance to the screen center, a different focus voltage is required to focus the beam at any specific point on the screen. Parabola-shaped correction voltage waveforms from the deflection amplifiers usually are used to control the focus, depending on the beam position. Because the beam in the corners also is astigmatic when adjusted for optimum focus, i.e., elongated in the direction of the corner, the corner spot size is typically about 1.5 times of that in the center. The added astigmatism electrodes (Figure 7.10) of the T4201 provide the ability to dynamically correct for astigmatism in the corners in a similar manner to focus correction.

Focus and astigmatism correction wave-

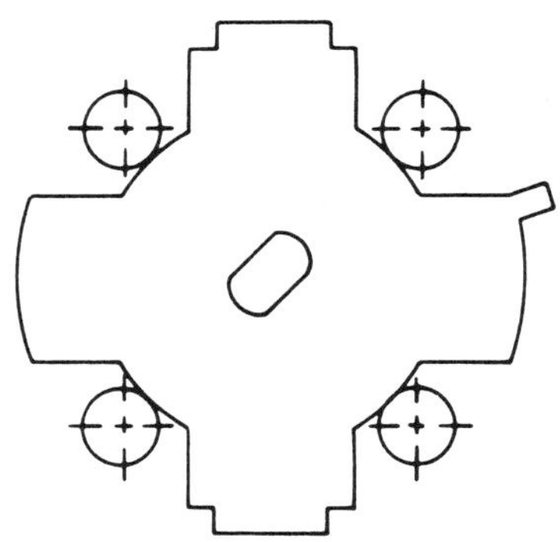

Figure 7.10 Astigmatism electrode used for dynamic control of astigmatism for improved corner focus.[36] (Courtesy of Society for Information Display.)

forms are individually programmed in read only memory (ROM) chips for the individual CRT. The result is a corner spot size virtually identical to the center, usually 0.005 to 0.006 inch, and the ability to conservatively display 2,048 × 1,536 addressable and fully resolvable pixels. Cost is the obvious trade-off for such extreme performance.

The most recent trends are toward the flat-square (FS) faceplate (Figure 7.11) and the wider 110-degree deflection angles that have become standard for television. The corner focus and astigmatism are the limitations imposed and at present restrict these features to medium-resolution CRTs (Figure 7.12).

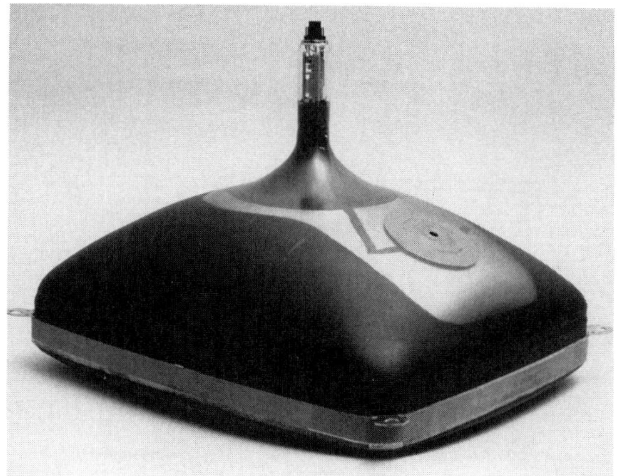

Figure 7.12 Clinton CE945M20 20-inch 110-degree deflection data display CRT.

7.7 MULTIBEAM MONOCHROME CRTS

A GTE Sylvania concept for multibeam CRTs for realtime display of slow-scan information from 1973[38] and one by Litton Electron Tubes for multichannel data recording from 1981[39] was also applied to monochrome data displays as a means to present greater amounts of data without increasing the bandwidth of the signals applied to the video amplifiers.[40] Some work was performed by IBM, but the only commercial product, as short-lived as it was, was a 19-inch monitor having a CRT with eight electron beams (Figure 7.13 and Figure 7.14) developed by now-defunct Azuray with a custom-made Tektronix CRT

Figure 7.11 Flat-square screen shape. (Courtesy of Matsushita Research Center.)

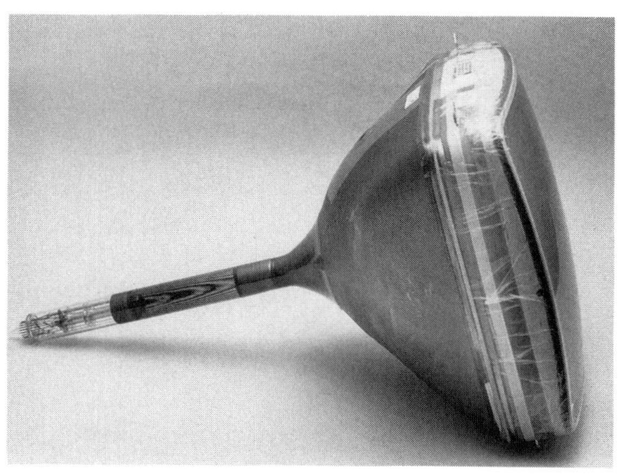

Figure 7.13 Tektronix multibeam 19-inch CRT.

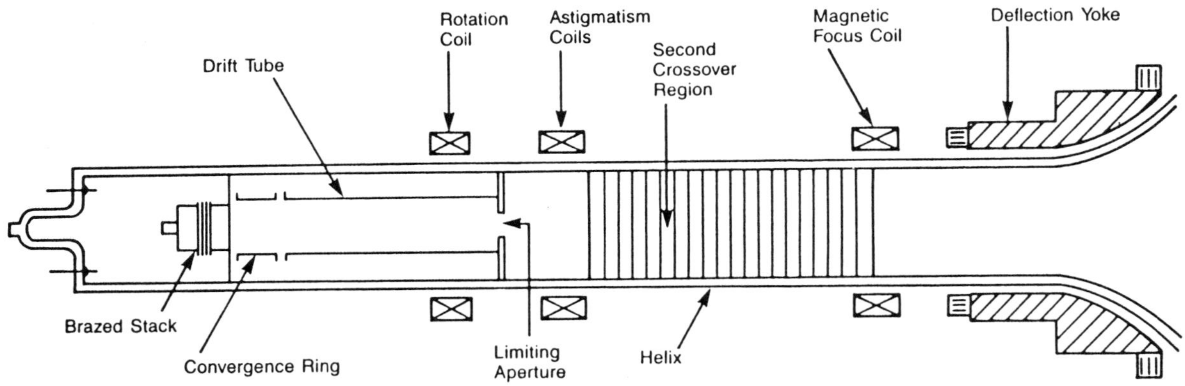

Figure 7.14 Electron gun for multibeam CRT.[41] (Courtesy of Society for Information Display.)

called the Gatling-Gun tube.[41,42] The name's derivation arose from the placement of the eight beams at the grid-cathode assembly (Figure 7.15). The difficulty of manufacturing the tube, cost and the number of operating adjustments made the display impractical.

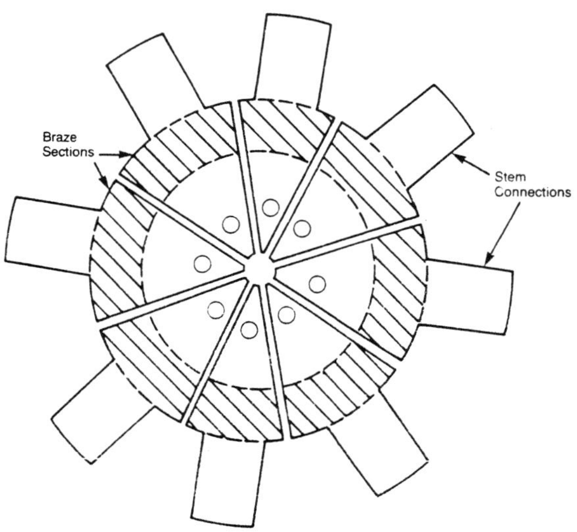

Figure 7.15 Multibeam grid assembly.[41] (Courtesy of Society for Information Display.)

The multibeam data display CRT differs from previous multigun CRTs such as the shadow-mask color picture tube and oscilloscope tubes which use multiple guns. Instead the multibeam data display CRT uses a single large area uniform cathode with a grid that simultaneously forms and modulates several beams. These are subsequently focused and deflected as a group. By aligning the beams in a vertical orientation, several raster scan lines are displayed with a single horizontal sweep of the screen. Since the lines are presented simultaneously, rather than one after the other, slower scanning and lower video amplifier bandwidth are required. The video bandwidth and horizontal scan rate requirements are reduced inversely proportional to the number of beams.

7.8 CURRENTRON

Like the voltage-sensitive Penetron color CRT described in Chapter 3 (Section 3.14), current sensitive phosphors also may be used to produce limited color gamut displays with a single electron gun. Work sponsored by NASA aimed at the development of a single-gun color CRT was reported by that agency and General Telephone Laboratories during the period of 1968 to 1971.[43–45] Sony later described various nine- and 15-inch CRTs, dubbed the Currentron, in the years 1982 through 1985 for use in word processing and color graphics applications.[46–48]

Two phosphors having different colors and intensity linearity characteristics are mixed and screened together using conventional screening techniques. Red and green phosphors are used with one being sublinear with current, i.e., the efficiency decreases with increasing electron beam current, while the other is superlinear and has increasing efficiency with current density. Only the beam current needs to be changed to shift color from red to yellowish-green, which is far simpler than the high-voltage and deflection amplifier gain changes that must be made for a similar change in the Penetron color. Primary advantages of the Currentron are good resolution and low cost. Despite these benefits, current sensitive color displays have not found wide application. The advantages of a large color gamut and gray scale combined with steadily improving resolution have made the shadow-mask type color CRT the norm for color graphics and data processing applications.

7.9 COLOR SHADOW-MASK CRTS

Color displays are most useful for graphics display applications such as computer-aided design (CAD), although a great many personal computers are purchased with color displays. Color aids rapid identification of complex details in drawings and is helpful for word-processing, spreadsheet and other office applications.

Shadow-mask type color television picture tubes were not as easily adapted to computer displays as were monochrome tubes. The color shadow-mask tube has inherently poorer resolution because of the screen structure itself and also the need for higher beam current to overcome beam current wasted by interception of electrons by the shadow mask. The higher beam current necessary for adequate brightness is achieved at the expense of larger spot size. Nevertheless, the shadow-mask color tube (Figure 7.16) and the related aperture grill Trinitron have been the only serious contenders for color data and graphics displays.

High-resolution color CRTs as developed by Matsushita began to appear in about 1972.[49] IBM began use of the color display in their Model 3279 terminal in 1979.[50] Early tubes were of the delta-gun type with phosphor dot triads. In-line guns for data displays were described in 1982 by RCA.[51] Phosphor dot triads, however, were retained. The in-line gun resulted in lower cost circuitry and more stable operation.

Conventional television picture tubes use shadow-mask pitches (the spacing at which

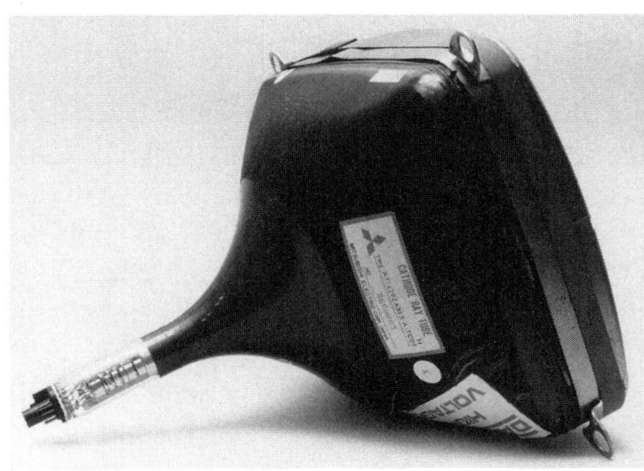

Figure 7.16 Mitsubishi in-line color data display CRT.

one complete set of red/green/blue dot or line patterns repeats) of about 0.31 mm. Through the use of smaller dots or stripes, the spacing may be reduced for higher resolution at the expense of brightness. Many color tubes for data displays now use mask pitches of approximately 0.2 mm as described by Hitachi in 1983[52,53] allowing resolution of up to 1,280 × 1,024 pixels on a 19- to 20-inch screen. Although often reported, development of shadow-mask tubes having pitches of 0.15 mm, has not yet resulted in a commercial product, due to low efficiency and manufacturing difficulties. Electron guns possessing smaller spot-sizes and deflection yokes that produce more uniform spot-size over the entire screen have been developed by the Japanese in parallel with the finer-pitch screens. Flat-square or full-square (FS) screen shapes began to appear in computer display CRTs in 1986, not long after being introduced in the color television market.[53] These screens have straighter sides, squarer corners and are flatter. The ability to better utilize the entire screen with rectangular rasters, reduced optical distortion of the display, the pleasing aesthetics and the ready availability of low-cost television glass envelopes are all major factors in the change to FS screens.

One unique but relatively short-lived approach to improving the quality of color displays was the Auto-Convergence CRT (T4115) developed through a joint effort by Philips ECG and Tektronix for the latter's Model 4115 color computer terminal (1983).[55–59] A 19-inch color display with resolution of 1,280 × 1,024 pixels was achieved, which was about the limit for that time. A pattern of chevrons (Figure 7.17) was sprayed through a stencil onto the rear of the shadow mask using ultra-short persistence P47 phosphor. A photomultiplier

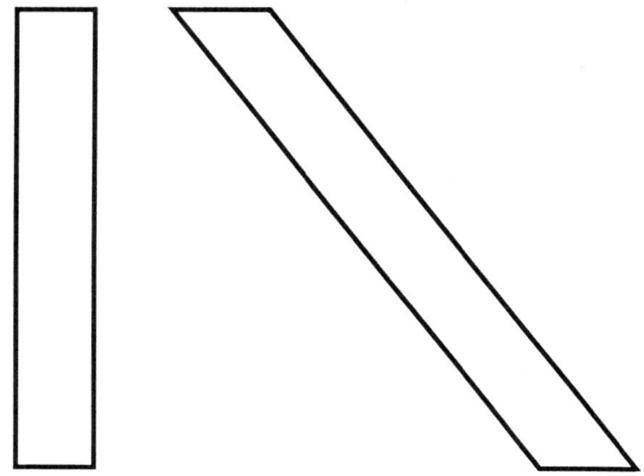

Figure 7.17 Chevron index pattern.

tube (PMT) mounted to a side of the funnel where no conductive coating was present detected light emitted by the chevrons as the beams were swept across the screen. The fast pulses provided positional information used to make all three beams land at the same point at the same time. The chevron's vertical portion provided horizontal position information while the diagonal portion defined the vertical position. Misconvergence from the earth's magnetic field or local magnetic fields were thus corrected automatically, regardless of where the display was placed physically.

The Sony Trinitron in higher resolution versions such as the 12-inch SD-121F and 19-inch SD-112F announced in 1980[60] has been particularly well-suited to computer displays. Both utilize 90-degree deflection. Many other sizes are currently available for data display, electronic imaging and picture monitoring applications including 4-, 5-, 7-, 9- and 15-inch versions. A 20 × 20 inch Trinitron developed in 1986 for air traffic control displays also has been used in CAD/CAM and computer graphics applications.[61] An improved electron gun with a complex prefocus lens was announced in 1989 for the 20 × 20-inch tube. The new gun has a spot size measuring about one-half of that of earlier guns, thus allowing display resolution of up to 2,000 × 2,000 pixels.[62] See Chapter 6 for information on the Trinitron's early development.

Zenith demonstrated a flat tension mask (FTM) color tube (Figure 7.18) for data displays in 1986.[63–67] This tube was an outgrowth of an experiment by K. Palac of Zenith in 1973 and the development of avionics color CRTs by R. C. Robinder et al at Tektronix in 1983.[68] Palac frit-sealed a shadow-mask foil between the faceplate panel and a tube's funnel. Upon cooling, the mask was held taut due to different rates of thermal contraction of the mask and glass. Despite the promising results, the idea was shelved due to lack of an application.

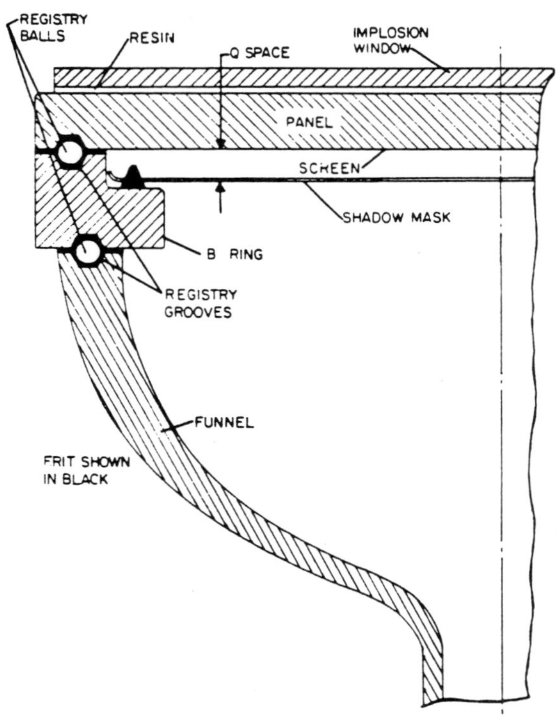

Figure 7.18 Prototype of Zenith flat tension mask construction.[65] Production versions used steel frame for tension mask rather than B ring. (Courtesy of Society for Information Display.)

The idea was revived, however, as the need for brighter, higher-resolution color displays increased. The flat tension mask can be operated at much higher beam current than conventional shadow masks which are curved to the glass faceplate's contour. Curved masks are sensitive to heating by the electron beam which causes *doming* of the mask with attendant loss of color purity because of thermal expansion. The Zenith tube currently is manufactured in a 14V-inch diagonal size with a flat glass faceplate. Zenith has been investigating the possibility of scaling up the design with the future HDTV market as incentive. Recently, Hitachi developed an 18V-inch diagonal FTM CRT for display terminals and workstations.[69]

Many of the evolutionary improvements to the color television picture tube have benefited the color data display CRT as well. These improvements include convergence, brightness, contrast, physical size and cost improvements. Without the tremendous production quantities of tubes required by the consumer market, it would be impossible to justify the development and manufacturing costs for the smaller data display market.

7.10 BEAM-INDEX COLOR CRTS

Despite several attempts, the beam-index tube has not yet found a home in computer displays. With some of the recent advances reported by Sony and others it still may happen. The beam-index color CRT has been under development since the early 1950s as a simpler, lower-cost and more rugged alternative to the color shadow-mask CRT, first for television and later, military and computer displays. It appears that vehicular and avionics applications may be the first market for beam-indexing. (See Chapters 6 and 8 for more detailed information on the beam-index tube.)

7.11 LIQUID-CRYSTAL COLOR SHUTTER

Not strictly a cathode-ray tube development, the liquid-crystal color shutter developed and marketed by Tektronix provides a significant advance in CRT performance. Liquid-crystal technology is more often viewed as a competing technology to the CRT with the advances now being made in flat panel displays. The liquid-crystal color shutter approach to a color display combines the unique features of each to produce a high-resolution color display using a conventional single-gun monochrome CRT. The color shutter is a field sequential system closely akin to the CBS color system of 1940 with one major difference: The cumbersome mechanical filter wheel is replaced by an electro-optical means of color selection.

The color shutter consists of a liquid-crystal Π cell halfwave retarder sandwiched between two color polarizers and a linear polarizer. The color polarizers permit white light to pass through in one axis and only the particular color for which it is designed in the axis oriented 90 degrees. One filter passes red light and in the latter orientation the other passes green light. These filters are positioned 90 degrees apart so that while one passes its color, the other allows white light to pass, thus each filter permits light of the color selected by the other to pass. Light from the CRT screen is polarized by the linear polarizer and is rotated by 90 degrees when the Π cell is energized. This causes the light from the screen to be polarized such that only light of the color selected by the color polarizer oriented in that plane is seen by the viewer. When the Π cell is not energized, the light is polarized 90 degrees from that plane, causing the other color polarizer to determine the color of light transmitted to the viewer. By energizing the Π cell during alternate raster frames, one frame will present red information and the next, green.[70–72]

The CRT's resolution is essentially unimpaired and color misregistration is nonexistent with the liquid-crystal color shutter. These advantages, however, are not without some performance compromises. The brightness is about 15 to 25 percent of that of the CRT because of absorption and polarization by the various optical elements. And, of course, the color gamut is somewhat limited to colors that can be produced by mixtures of red and green, as with the Penetron CRT.

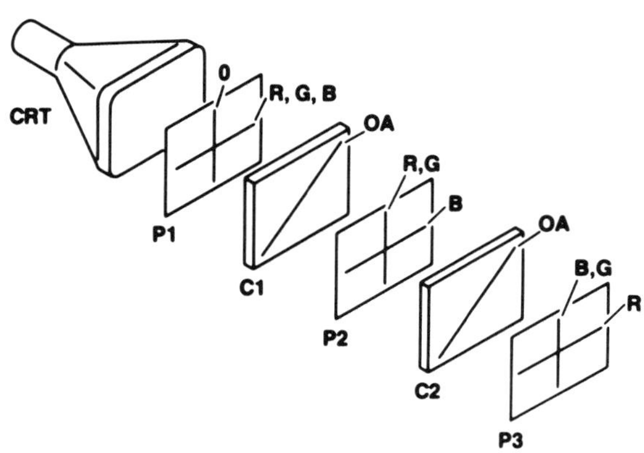

Figure 7.19 Full-color liquid-crystal color shutter display.[73] (Courtesy of Society for Information Display.)

Color switching of the Π cell is easier than the high voltage switching of the Penetron, however. The liquid-crystal color-shutter does impose a resolution limit although it is not spatially limited as is the color shadow-mask CRT. The requirement to divide the video signal into the two color components for alternate frames reduces the amount of picture detail that may be displayed by a factor of two for a given video amplifier bandwidth and choice of scan rates.

Full-color displays using two liquid-crystal color shutters (Figure 7.19) were disclosed by Tektronix as early as 1984.[73] Recent improvements in shutter and CRT efficiency have shown promising results and excellent quality, full-color pictures similar in color gamut to shadow-mask CRTs have been demonstrated. Resolution is significantly greater for the color shutter method.

Similar techniques have been employed for stereoscopic monochrome displays using liquid-crystal shutters, polarizers and polarized viewing glasses.[74] This approach has the advantage of using passive viewing glasses with no electrical connection as required by some other methods.

8

Avionics and Vehicular Cathode-Ray Tubes

8.1 World War II Airborne CRTs

Only occasional use of cathode-ray tubes in airborne applications was made prior to World War II. A cathode-ray aircraft radio compass was invented by Edward J. Hefele, chief radio engineer of the Airplane and Marine Direction Finder Corporation and was submitted to the U.S. Coast Guard for flight testing in 1936.[1] A three-inch electrostatically deflected CRT, a Du Mont ruggedized type 34-XH high-deflection sensitivity tube, was used as the cockpit direction indicator.[2]

The development of radar in the late 1930s led to its adoption during the war for airborne use as operating frequencies increased and antenna size became reasonable for aircraft mounting. Equipment ranged from the relatively simple AN/APS-6 intercept radar for night-fighter aircraft, such as the Grumman F6F Hellcat,[4] to complex search radar systems for bombers and patrol aircraft, such as the Boeing B17 Flying Fortress and the Consolidated PBY Catalina, respectively. Smaller airborne indicators usually used three-inch electrostatically deflected 3FP7 CRTs (Figure 8.1) by RCA and National Union or the RCA magnetically deflected 3HP7 (Figure 8.2). Larger indicators for airborne search radar used the type 5FP7 manufactured in large quantities by RCA, National Union, Sylvania, Norelco, Du Mont, General Electric and Research Enterprises, Ltd.[3] of Toronto, Ontario. The display formats were quite varied and included PPI, B, G and O scan.[4–7] (See Chapter 1 for illustrations of radar indicator formats.)

Cathode-ray tubes found several other airborne applications during the war years. These included radar countermeasures, airborne interception (AI), air-to-surface vessel (ASV), Loran (Long-Range Navigation) and engine performance analyzers. Examples of typical CRTs used were the electrostatically deflected 3BP1, 3CP1, 3DP1, 5BP1 and 5CP1. All but the radial deflection 3CP1 and 3DP1 were widely used oscilloscope CRTs of the period. The 3DP1 was similar to the 3BP1, except for an added radial deflection electrode through the tube face's center (Figure 8.3). (See Chapter 3, Section 3.3 for additional information on the 3CP1 and 3DP1.) In England, the six-inch VCR97, 3-1/4- inch VCR138 and six-inch VCR517 were commonly used electrostatic CRTs for airborne

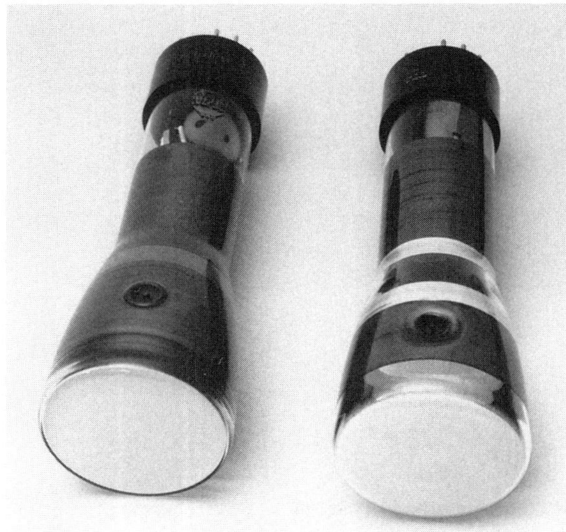

Figure 8.1 RCA 3FP7 electrostatic-deflection CRTs manufactured in 1973 and 1944, respectively. (From the author's collection.)

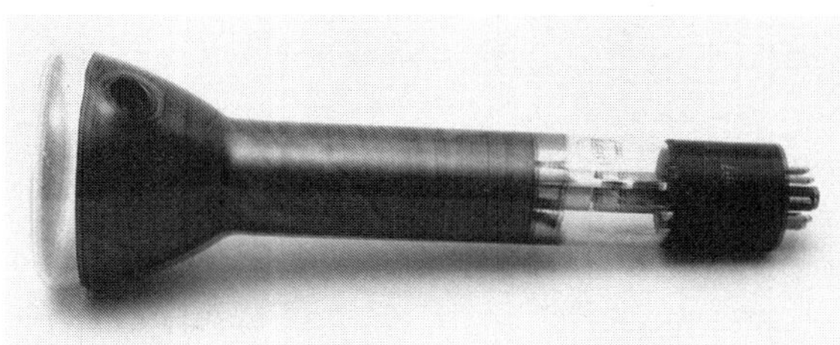

Figure 8.2 RCA 3HP7 magnetic-deflection radar CRT, circa 1944. (From the author's collection.)

radar indicator applications.[8,9] (Chapter 4, Section 4.3 contains further information on the electrostatic oscilloscope type CRTs.)

8.2 POSTWAR MONOCHROME CRTS

The years immediately following World War II brought little change in CRTs for airborne use. Magnetic deflection and focus cathode-ray tubes, which had become standard for airborne radar indicators in the late war years, continued as the predominating choice for new equipment design. With a few refinements the ubiquitous 5FP7 remained one of the most widely used tubes. The nomenclature became 5FP7-A with RCA's development of an improved electron gun having a beam-limiting aperture permitting sharper focus (Figure 8.4). P14 phosphor, possessing color and long-persistence characteristics especially suited for airborne radar using sector scanning, was registered by General Electric in 1945 and immediately introduced in their type 5FP14 CRT.

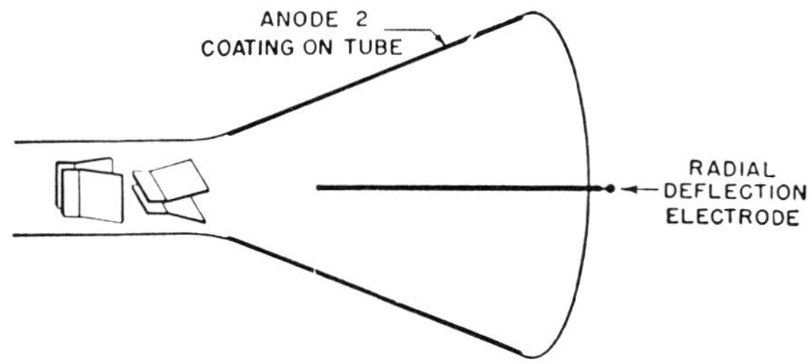

Figure 8.3 3DP1 radial-deflection CRT construction.

Electrostatic focus for lighter weight and conservation of strategic materials (cobalt and copper) along with aluminized screens for brighter displays followed similar developments in television by about four years. One example is the Du Mont 5AHP7 announced in 1953, and its aluminized counterpart the 5AHP7-A in 1954, which were essentially electrostatically focused 5FP7s. During the post-war years weather radar for safety and smoother flights became standard with the airlines. The five-inch round CRT was almost universally used for these displays during that period.

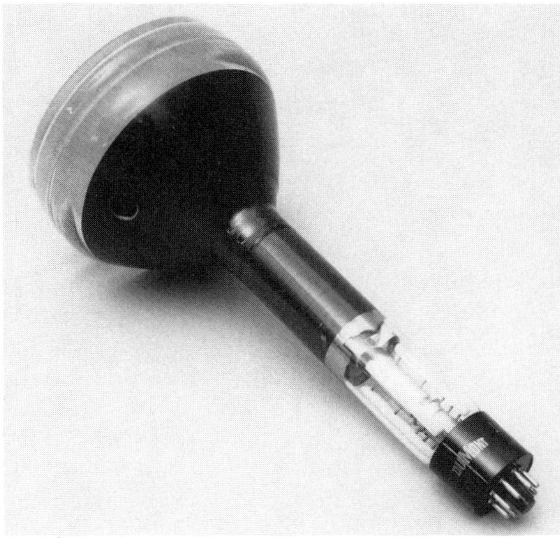

Figure 8.4 Du Mont 5FP7-A radar CRT manufactured in 1957. (From the author's collection.)

Around 1955, tubes such as the five-inch Du Mont B1125, which later was registered as the 5BCP7 (Figure 8.5) were introduced to aid size and weight reduction in airborne radar indicators. Length was reduced to only seven inches through use of a 70-degree deflection angle instead of the usual 50 degrees. A smaller 7/8 inch diameter neck offset the higher deflection power called for by the 70-degree deflection angle. A glass hard-pin stem similar to standard-nine pin receiving tube bases reduced socket weight and size. Both deflection and focus were magnetic. Heater power was lowered to two watts from the previous four watts to reduce power and equipment cooling demands. The 5BCP- was available initially with either P1, P4, P7 or P11 phosphors for other display applications. A similar seven-inch version, the Du Mont B1142, and three-, five-, seven- and 10-inch electrostatically focused versions, the B1173, B1174, B1175 and B1191, followed shortly thereafter. Use of electrostatic focus particularly saved weight by eliminating the focus coil. These four tubes had the added feature of aluminized screens to improve brightness.

Another development from 1956 or 1957 was the sector-scan CRT with off-center neck such as the Du Mont three-inch K1517. This tube also was electrostatically focused and aluminized, and used *potted* base and anode connections to prevent arcing at high altitude while providing greater reliability than afforded by conventional sockets. Other sector scan CRTs included the five-inch diameter Du Mont 5EAP7 and 5EBP7, and Sylvania SC3179 and SC3180 of the early 1960s. All utilized offset necks.

Seven-inch and larger CRTs usually associated with ground-based and shipboard radar were used for airborne early-warning radar in larger aircraft, including the Lockheed

Figure 8.5 Thomas Electronics 5BCP7 short radar CRT. (Courtesy of Jake Brain.)

EC121 Super Constellation which permitted installation of full-sized display consoles.

CRT characteristics required for the demanding environment of military aircraft include shock and vibration resistance, reliability, high brightness and contrast for viewing in sunlit bubble canopies, good resolution, freedom from arc-over at high altitude, minimum volume and weight for a given screen size, and low power consumption.[10] To a lesser degree the same requirements apply to commercial aircraft display CRTs.

As CRT technology was refined for the severe environmental conditions of tactical aircraft and manned spacecraft, tubes began to evolve during the 1960s as integral assemblies consisting of CRT, laminated contrast filter, deflection yoke, mu-metal shield and mounting hardware potted together (Figure 8.6). Connections became permanently attached "flying leads" instead of conventional bases and anode buttons to help reduce arc-over of high voltage at high altitudes and provide greater reliability.

About 1960 the Grumman A-6A and Northrop F-5 were the first U.S. tactical aircraft to use instrument panel-mounted CRT displays which were fully ruggedized for the severe environmental conditions encountered in modern combat aircraft. Their success led to similar CRT displays on the Boeing B-52, Ling-Temco-Vought A-7, McDonnell-Douglas F-15, General Dynamics F-16 and the Lockheed AC-130 gunship during the late 1960s and early 1970s.[11–13]

The ultimate in ruggedized monochrome cathode-ray tubes was required for the Space Shuttle orbiter during the early 1980s. Initially these shuttles included eight CRTs. Three orbital display units (ODUs) using eight-inch rectangular, magnetically deflected CRTs

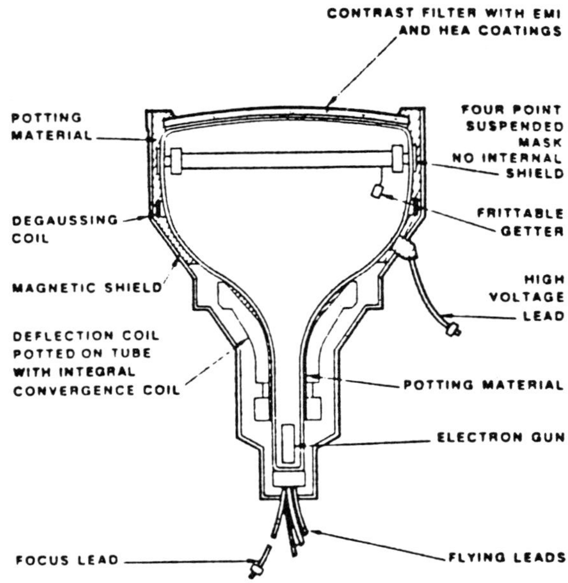

Figure 8.6 Color shadow-mask CRT assembly.[35] (Courtesy of Society for Information Display.)

(8M172P43M) by Thomas Electronics interfaced to onboard computers for systems status information. The ODUs were the primary displays. Improved ruggedness was achieved by sealing the end of the electron gun that was closest to the screen to the neck tubing. Low power consumption, narrow deflection angle and high-efficiency P43 phosphor contributed to its performance. Other Space Shuttle CRTs included a three-inch heads-up display CRT, the Thomas Electronics 3M199P1M, for use during landing; a line-scan fiber-optics tube, the 10M256P46MFO also by Thomas, for producing hard-copy printouts of information transmitted by ground stations or derived from onboard sensors; and the eight-inch Thomas 8M202P4M for television monitoring of the cargo bay operations.[14] Later Space Shuttle developments included the use of color CRTs and helmet-mounted CRTs, the latter for extravehicular activities.

8.3 DIRECT-VIEW STORAGE TUBES

The direct-view storage tube (see Section 4.17) offered two distinct advantages for airborne radar indicators. They retained the image between scans without fading and they produced high brightness in an environment that can exceed 10,000 footcandles. A. V. Haeff of the Naval Research Laboratory described a *memory tube* in 1947. One of the tube's proposed applications was radar indicators.[15] Much of the early development of the storage tube occurred during the early 1950s at RCA laboratories with M. Knoll as the key figure associated with it.[16–19] Development of storage tubes for military cockpit radar displays began during the early 1960s. U.S. manufacturers of storage tubes suitable for airborne applications included RCA, Capehart-Farnsworth, Du Mont, IT&T, Westinghouse and Hughes Aircraft.[20] The latter was especially successful with their Tonotron (Figure 8.7; see Figure 3.23) announced in 1956 which was produced in many variations with those intended for aircraft use potted in a mu-metal magnetic shield as an assembly. The Hughes H1038 and H1084 were ruggedized five-inch Tonotrons tested to five g vibration and 15 g shock.[21]

The screen of a mesh type storage tube may be made of almost any conventional cathode-ray tube phosphor although yellow-green P20 was used in many storage tubes because of its high visual efficiency and medium to medium-short persistence.[22]

8.4 HEADS-UP DISPLAY CRTS

Heads-up displays (HUDs) usually are used in tactical aircraft to permit radar, flight attitude or landing approach information to be displayed directly in the pilot's line of vision as he looks through his windshield. Thus, electronic information is superimposed on outside visual cues which prevents the pilot from "having his head in the cockpit," which is the situation with panel-mounted CRT displays.

HUD displays consist of ruggedized, high-brightness, cathode-ray tubes with an optical system to project a virtual image on a dichroic mirror between the windshield and pilot. Current manufacturers of HUD CRTs include Rank Brimar, Thomas Electronics, Thomson CSF, Telefunken AEG, ISTC (formerly Westinghouse Electric) and others. Screen sizes are under five inches in diameter. In many respects, HUD tubes are similar to projection tubes. High accelerating voltage is used to obtain adequate brightness in bright ambient lighting. Phosphors must have high visual efficiency and good aging properties. Precision physical dimensions are maintained from one tube to another to permit replacement in the field without recollimation of the optics.

HUD displays for gunsighting first appeared in England during World War II under the descriptor *windscreen projection* and used a 1-1/2-inch diameter CRT, the VCR522, comprising electrostatic deflection and focus in a compact design.[9] Subsequent mention of HUD-type displays included the 1945 patent application by Phillip Wheeler for naval gunsighting[23] and another by E. E. Flint in 1946 for an aircraft instrument-landing approach system.[24]

During the 1950s, the Army/Navy Instrumentation Program (ANIP, later JANAIR) was the driving force in the United States that resulted in increased interest in CRTs for cock-

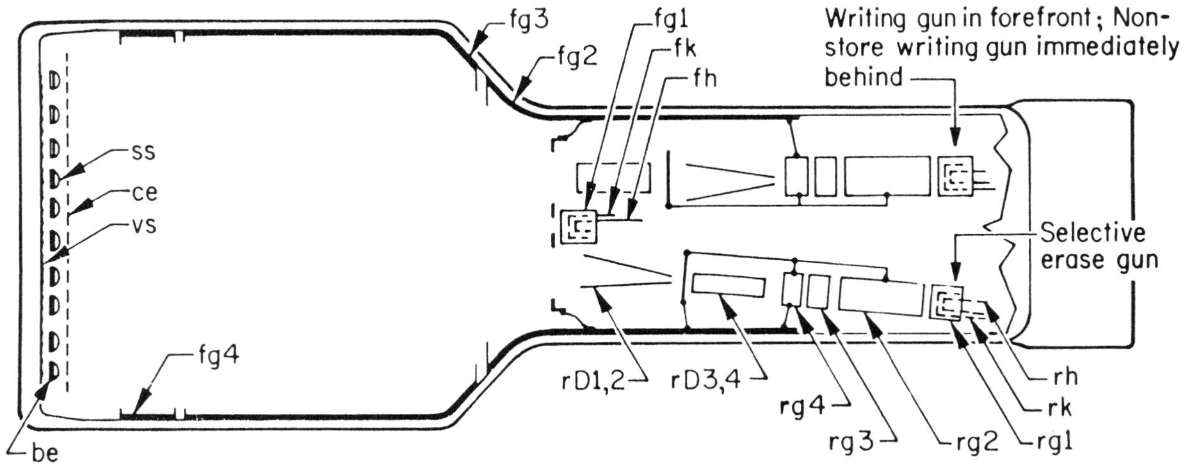

Figure 8.7 Hughes multi-mode Tonotron drawing.[21] (Courtesy of Society for Information Display.)

pit displays to reduce pilot workload and handle the increased information display requirements. Prototype systems were developed for both helicopters and conventional fixed-wing aircraft. The Kaiser-Aiken thin tube (see Chapter 10) was flown in 1956 at Edwards Air Force Base. This device was mounted in the windshield and the pilot looked through it rather than the customary projection arrangement. Helicopter HUD systems, as well as one presenting terrain clearance cues for the North American A-5J Vigilante carrier-based attack bomber, were developed by Autonetics during this period. In England, the Rank-Cintel Company developed CRT HUD flight display systems for the Royal Aircraft Establishment (RAE) at Farnborough. Rank received a contract in 1959 for producing several hundred HUDs for the Buccaneer aircraft.[13]

The Navy Integrated Light Attack Avionics System (ILAAS) program started in 1965 was an important factor in promoting the use of HUDs for the short-lived General Dynamics/Grumman F-111B and the successful Ling-Temco-Vought A-7. Later, the F-111D used dual HUDs with reflective rather than refractive optics. An unusual scoop-shaped CRT envelope was used with a P31 screen applied to the end opposite the electron gun (Figure 8.8). A window in the top of the scoop allowed light from the gun side of the screen to be projected, thus achieving higher efficiency.[25] During the late 1960s, Kaiser Electronics supplied combined HUD/panel CRT display systems for the F-14A Tomcat. The 1960s closed with full acceptance of HUD systems in the military and several of them in operational status. All HUD systems except the F-111D CRT used P1 (green) phosphor at that time.[13]

HUD systems experienced explosive growth during the 1970s in the U.S. and abroad and today have become commonplace for tactical aircraft. Commercial aircraft use of HUDs has lagged behind considerably.

Phosphors for heads-up displays have progressed from the old faithful P1 (zinc or-

Figure 8.8 Thomas Electronics heads-up display CRT. (Courtesy of Thomas Electronics.)

thosilicate) used until the early 1970s and beyond, through P43 (gadolinium oxysulfide), to the recent P53 (yttrium aluminum garnet), the latter two being rare-earth phosphors.[26,27] P43, registered with EIA in 1971 by Thomas Electronics, was very successful for HDDs (heads-down displays or instrument panel-mounted) as well as HUDs due to its very narrow spectral output bandwidth which allows a narrow-band contrast enhancement filter to be used. Such filters appear nearly black and when antireflective coatings, such as magnesium fluoride, are applied to the outer side to reduce specular or mirrorlike reflections from the glass surface, very high image contrast is achieved, even in the direct sunlight encountered in bubble canopies. P53 was registered by Ferranti Ltd. in 1980 and is particularly suited to high brightness applications since it continues to produce bright images at high beam current without saturation. Other phosphors, although more efficient at low beam current, saturate more readily at higher current where their efficiency falls off rapidly. Aging characteristics of a phosphor are important at the brightness levels at which HUD displays are operated and P43 has proven to possess the best characteristics for this use.

8.5 SHADOW-MASK COLOR CRTS

Somehow, commercial aircraft largely avoided the monochrome CRT display except for radar indicators. When CRT displays arrived on the commercial scene, color CRTs were already well advanced and soon became the norm on the modern flight deck. Today, the glass cockpit is rapidly evolving with its influence even being felt in General Aviation aircraft. We can expect to see substantial inroads by flat panel displays due to their obvious space advantages in a market that is not strongly cost conscious.

Early color displays were of the Penetron or beam penetration phosphor type due to their inherent ruggedness. One CRT of this type, which was optimized for radar display in a cockpit environment, was developed in 1981 by Phil Krzykowski and Elliot Schlam of the U.S. Army (ERADCOM) at Fort Monmouth, New Jersey. Red and green transparent phosphors and a black backing layer were deposited on a high temperature sapphire faceplate, which in turn was frit-sealed to the CRT envelope. Ambient light passed readily through the transparent phosphor layers and was absorbed by the black backing layer. Very high contrast and ruggedness with limited color were the result. A three-inch diameter round screen and magnetic deflection were employed.[28]

The limited color gamut of the Penetron type CRT led to considerable development effort toward the cockpit use of color shadow-mask CRTs during the 1980s. Mitsubishi described the development of a prototype vibration-resistant, color shadow-mask CRT assembly in 1981. A lower mass shadow-mask with stiffeners on the flange was designed to have a higher resonant frequency, and potting materials of differing hardness were used in locations requiring different thickness between the tube envelope and magnetic shield to control vibration dampening characteristics.[29]

The color shadow-mask cathode-ray tube made its debut in commercial aircraft during 1982 with the Boeing 757/767 program. Toshiba announced several new 0.3 mm pitch shadow-mask tube assemblies that year for use as primary indicators for the Electronic Flight Information System (EFIS) for both commercial and General Aviation.[30] Displays were used for both flight attitude and engine performance monitoring. The result was a clean, uncluttered instrument panel. All recent commercial aircraft use color CRT displays with use trending downward to the General Aviation market for business jets and, ultimately, private aircraft.

CRTs for EFIS displays are quite similar to the color television CRT with the exception of the shadow-mask pitch and some strengthening, although not to the degree required for tactical aircraft (Figure 8.9). Two problems exist with airborne use of shadow-mask tubes. One is the change in color purity that can occur as the aircraft traverses areas of the world with the changing strength of the earth's magnetic field, particularly in the southern hemisphere where the polarity is opposite to that of the northern hemisphere (where the equipment is usually installed or

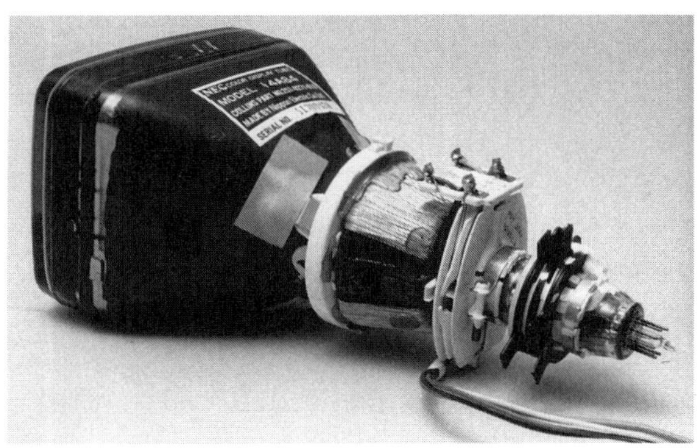

Figure 8.9 NEC shadow-mask color CRT for commercial avionics display. (From the author's collection.)

adjusted). The other problem is caused by the engines' low-frequency vibration of the shadow mask which may color break-up and purity loss.

The first use of a color shadow-mask CRT in a tactical aircraft was a 5 × 5-inch CRT by Matsushita for the McDonnell-Douglas F15 in 1985. The CRT was supplied as a bare tube to Honeywell who packaged it in a display. Stroke writing was employed rather than the more efficient raster scanning techniques currently in use.[31]

The conventional shadow-mask CRT suffers from an additional important limitation for tactical aircraft. Brightness is inadequate for the high ambient light encountered in the bubble canopy often used for fighter aircraft. The shadow mask, which is curved to match the glass faceplate's contour, is subject to *doming* or distortion of its shape at high beam current due to heating and resultant thermal expansion. This causes a loss of color purity and limits the brightness obtainable from the CRT.

To overcome brightness limitations of the shadow-mask CRT, Tektronix developed the taut-mask structure for tactical avionics applications in 1983.[31,32] A series of tubes was produced in five-inch-square (Figure 8.10) and six-inch-square screen sizes with both delta (Figure 8.11) and in-line guns. The ceramic envelope, long used by Tektronix for oscilloscope CRTs, was applied to these tubes and allowed design flexibility not available with glass envelopes. A flat, tightly stretched shadow mask was mounted behind a flat glass faceplate. The flat shadow mask permitted beam currents fourfold to tenfold that of ordinary shadow-mask color tubes to be used without deformation and loss of purity. A shadow-mask dot pitch of 0.20 mm provided good resolution. These tubes evolved into fully integrated assemblies which incorporate deflection yoke, purity magnets, magnetic shielding, shock mounting, potted electrical connections and laminated contrast enhancement filter in one compact package. Zenith makes a larger, taut-mask, flat-screen color CRT for data display applications as described in Chapter 7.

AEG Corporation in Germany introduced their high-resolution full-color M18E851 avionics CRT in 1986. This 5 × 5-inch shadow-mask CRT used a curved mask, but relied on a special Invar mask material that controlled movement with heating to achieve the ability to resist doming at high beam current densities.[33,34]

Larger ruggedized shadow-mask CRT assemblies patterned after the small avionics color tubes were developed by Thomson-CSF of France in 1988.[35] Intended applications include aircraft, armored vehicle, mobile system and shipboard raster displays. Nineteen-inch screens with 0.21 mm shadow-mask pitch offer 1,280 × 1,024 pixel displays while withstanding extremes of shock, vibration and temperature.

8.6 BEAM-INDEX COLOR

Beam-index color CRTs date back to the Philco Apple tube of the early 1950s (see Chapter 6). The beam-index tube has appeared in many variations by several manufacturers since then, yet is only just beginning to be considered commercially. Philco, RCA, Westinghouse, Zenith, Sony, Hitachi, Philips, EMI, Sylvania-Thorn, Thompson-CSF, AEG and Thomas Electronics have all explored the beam-index principle at one time or another. Over 250 U.S. patents have been issued on beam indexing.[36] Peter Barten

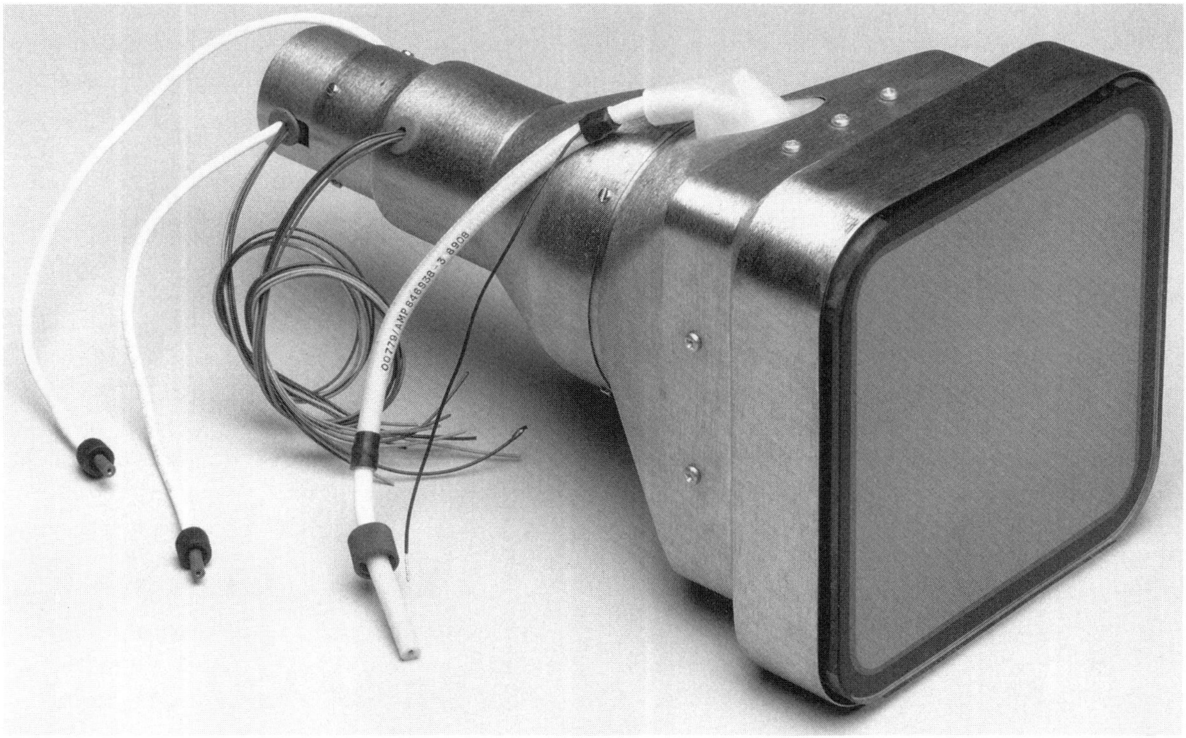

Figure 8.10 Tektronix T8650 five-inch square, flat tension mask color CRT assembly for tactical aircraft. (Courtesy of Tektronix Inc.)

summed it up with the statement: "The beam-index tube is the most attractive alternative to the shadow-mask tube. This is so nowadays; it has been so in the past and it will be so in the future. On no other type of color tube apart from the shadow-mask tube, has so much effort been spent by so many companies over a period of so many years."[37]

The single electron gun and lack of a shadow mask result in a rugged device without convergence adjustments or the power loss associated with absorption of electrons by the shadow mask. On the negative side is the requirement of small spot width to allow addressing of only one phosphor stripe at a time for good color purity. The shadow-mask tube does not suffer from this constraint where typically at least two complete sets of phosphor dot triads or stripes are covered by the electron beam. Elliptically shaped beams are now used to limit spot width to that of one phosphor stripe while allowing more total beam current for higher brightness by virtue of the increased spot height which ideally is exactly the same dimension as the spacing between the raster scan lines.[38,39]

Where the Apple tube used secondary emission effects to sense the beam location for indexing, most attempts since then have relied on ultraviolet index stripes with a pho-

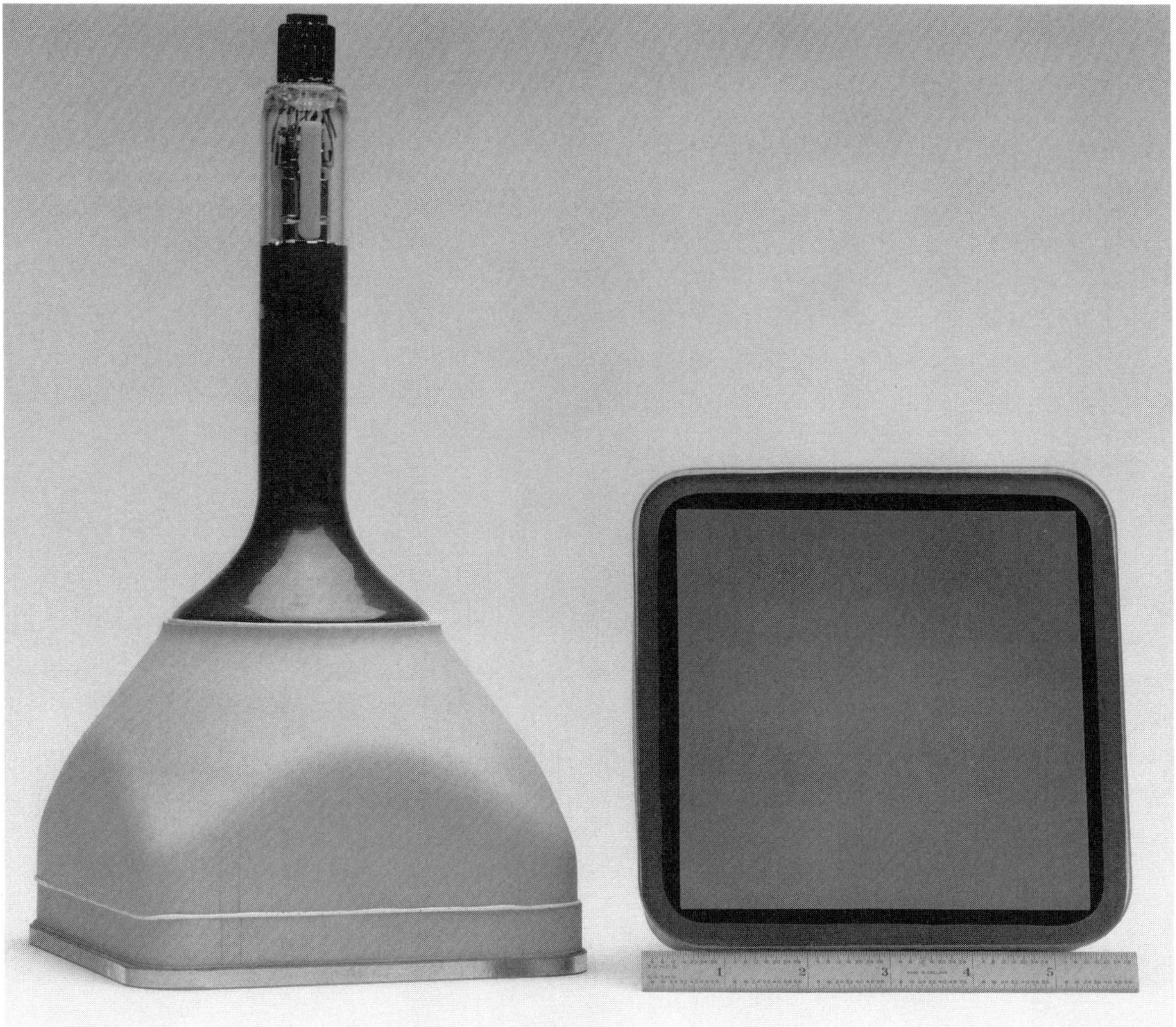

Figure 8.11 Tektronix ceramic T8650 color CRT before assembly in magnetic shield. (Courtesy of Tektronix Inc.)

todetector to sense the beam's position (Figure 8.12). In one case, direct pickup of the electron beam itself was used to sense position in a device known as the Turner tube (1976).[40,41] The conventional aluminum layer was photo-etched to form an interdigitated toothcomb pattern with individual connections brought out to sense which segment the beam was striking, hence its location relative to the phosphor stripes (Figure 8.13).

Vehicular and/or cockpit displays may be the first serious application of beam-index

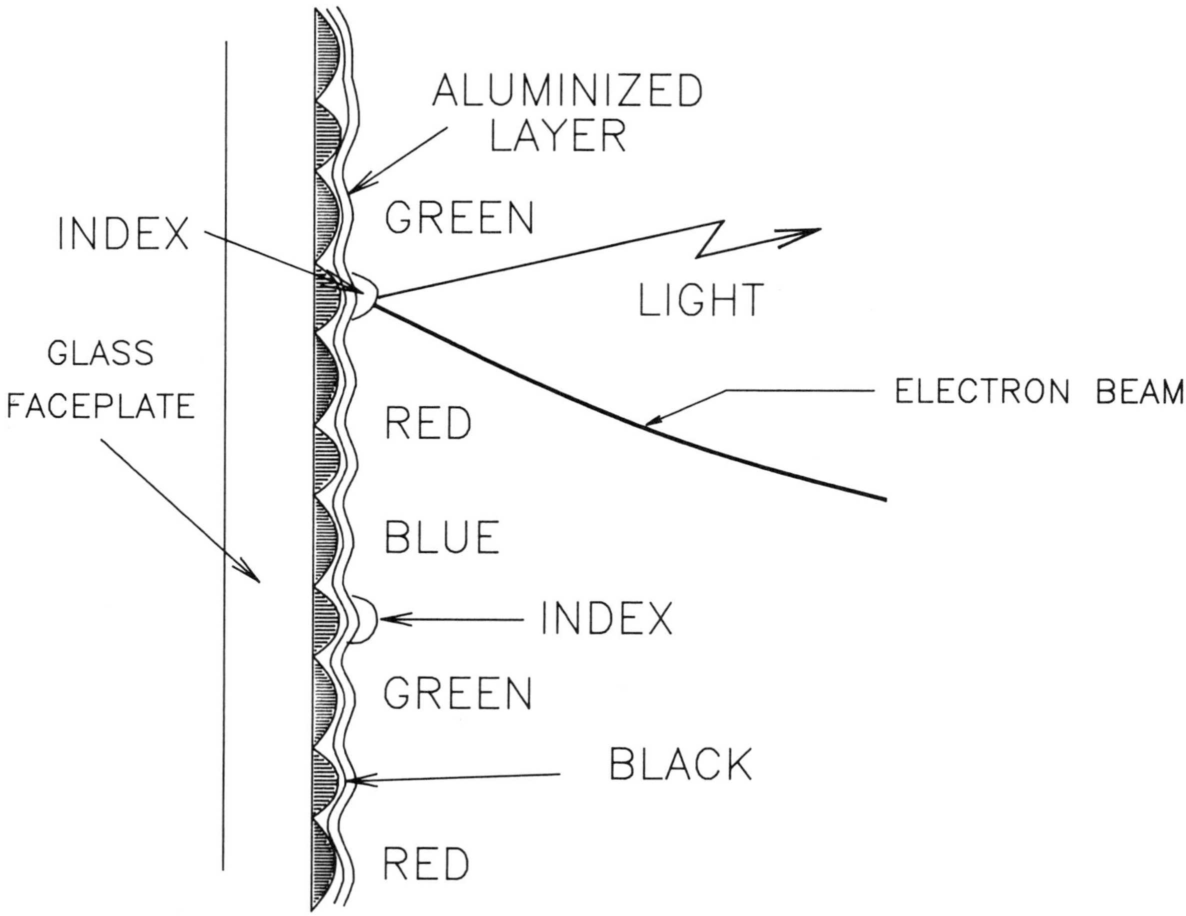

Figure 8.12 Beam-index screen structure. (Courtesy of Jake Brain.)

CRTs due to their inherent ruggedness and simplicity. The years 1987 and 1988 ushered in the commercial introduction of several new beam-index CRTs aimed at this market. These included 6.3-inch-square and five-inch diagonal tubes from Sony,[42,43] four tubes ranging from 1.5-inch diagonal to seven-inch square from Thomson,[44] an eight-inch diagonal tube from Thomas Electronics and one (size not listed) by AEG.[45] With all of this development work, the long-awaited beam-index tube may finally become a serious contender to the shadow-mask CRT for some applications.

8.7 HELMET-MOUNTED CRTS

Miniature CRTs, that is, tubes of approximately one-inch diameter, have been used for several vehicular applications including helmet-mounted displays for pilots and drivers

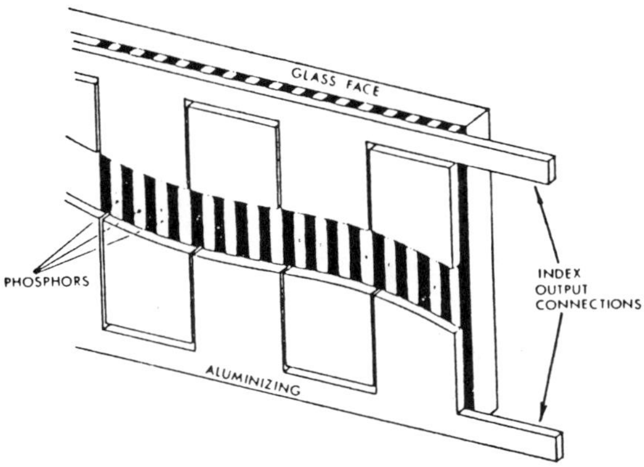

Figure 8.13 Turner tube beam-indexing structure.[40] (Courtesy of Society for Information Display.)

Figure 8.14 National Union 1DP1 miniature CRT manufactured in 1963. (From the author's collection.)

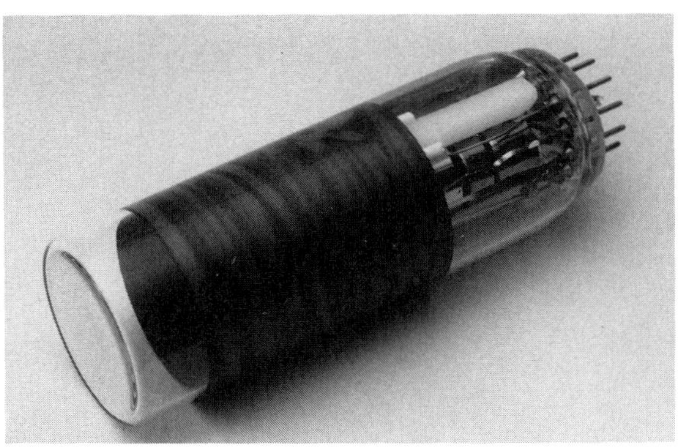

Figure 8.15 RCA 1EP2 miniature CRT, circa late 1950s. (From the author's collection.)

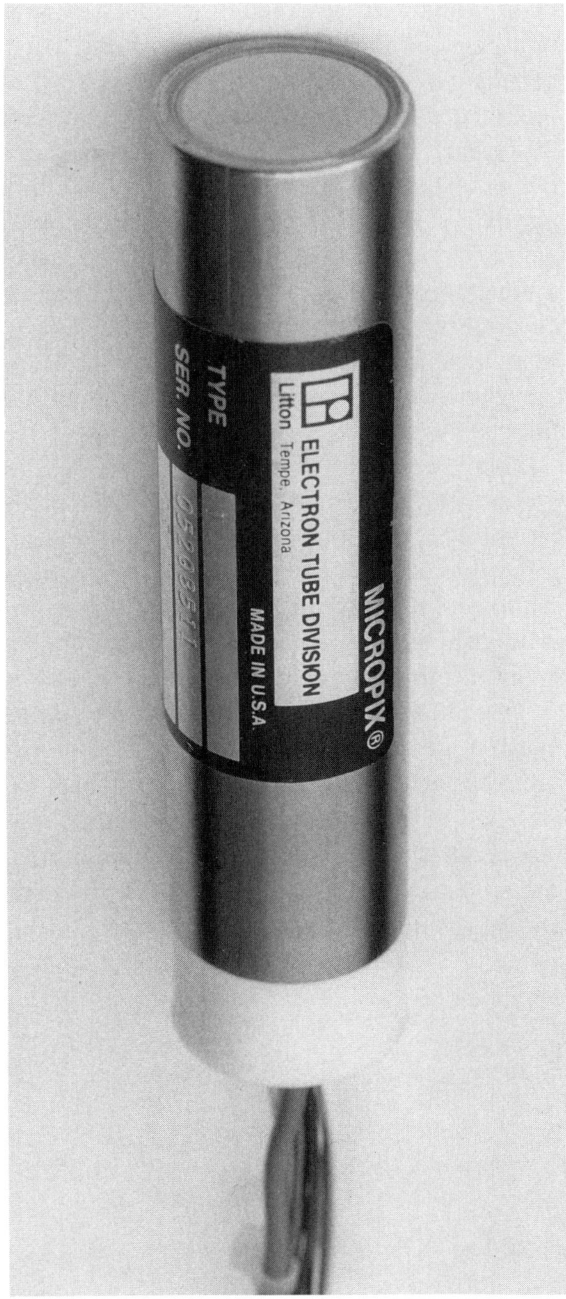

Figure 8.16 Litton L4272 Micropix™ miniature CRT for helmet-mounted displays. (Courtesy of Litton Electron Tube Division.)

of armored vehicles including tanks as well as a number of psychophysical research experiments. Development work of helmet-mounted display CRTs dates back more than 30 years.[46] The one-inch electrostatically deflected National Union 1DP- (Figure 8.14) and RCA 1EP- (Figure 8.15) were forerunners of the genre in the mid-1950s. One early avionics use of miniature CRTs was for recording timing information on aerial film.

Work began in the late 1960s by the U.S. Air Force for helmet-mounted displays for pilots.[47]

The small CRT is coupled to an optical system mounted in the helmet and forms a virtual image some distance in front of the pilot that allows him to view the display without diverting his attention from other tasks. The display may be from a low-light-level television camera.[48]

The current miniature cathode-ray tubes usually are magnetically deflected and use high-voltage electrostatic focus.[49,50] They are supplied in packaged assemblies that include magnetic shielding; a small diameter, concentric deflection yoke and potted leads; and are only slightly larger in diameter than the screen. Deflection angles are about 40 degrees to minimize power consumption and pattern distortion. Spot sizes on the order of 0.001 inch permit as much information to be displayed as on many larger CRTs. Length is generally less than six inches and screen diameters are usually about one inch although the Hughes type 1401 is only 1/2 inch in diameter and approximately three inches long. A typical miniature CRT assembly is illustrated in Figure 8.16. Rank Brimar, Litton Electron Devices, Thomas Electronics and Hughes Aircraft are among today's suppliers of helmet-display CRTs.

Common phosphor types include P1 (silicate green), P43 (rare-earth green) and P53 (rare-earth green).[51] Fine-grain phosphor screens are important to provide high resolution for the small screen sizes where the phosphor crystal size becomes an appreciable factor.[52] Recently, thin film and single crystal phosphor screens have been investigated by AT&T.[53,54] Despite their low efficiency due to trapping of light, they offer several important advantages in resolution and good thermal conductivity which permits use of higher beam currents without burning or aging.

9

Photo-Recording Cathode-Ray Tubes

9.1 Flying-Spot Scanner CRTs

Flying-spot scanning is the scanning of an object or photographic transparency, usually in a raster format, with a narrow beam of light and detecting either the reflected or transmitted light by photoelectric cells to convert the picture to an electrical signal capable of reproduction on a suitable display system. Initially the technique was proposed in a 1910 Swedish patent application by A. Ekstrom using a bright light source and oscillating mirrors to scan film or transparencies. An arc lamp and another oscillating mirror arrangement reconstructed the image on a screen.[1,2]

During the 1930s, the cathode-ray tube was adapted to flying-spot scanning by using an optical system to project the rapidly moving spot onto a slide.[3] Naturally, the CRT's inertialess electron beam did not suffer from the scanning speed limitations of a mechanical system. The CRT had its own limitations which included the difficulty of obtaining adequate video amplifier bandwidth, good resolution and sufficiently short persistence of the CRT phosphor screen. Flying-spot scanning also was applied during the 1930s to television pickup of scenes with limited success.

Electronic flying-spot scanning with a CRT uses synchronized raster scan of the CRT and the receiver, usually to NTSC standards. The light from the flying-spot CRT screen is imaged through a lens onto the slide to be displayed. The light transmitted through the slide is detected by a photomultiplier tube and the resulting electrical signal is fed to a video amplifier before going to the television transmitter (Figure 9.1).

Flying-spot scanners were highly suitable for televising still photographic images and often were used in the early post–World War II era for test patterns and station identification slides as well as the "please stand by" slides often seen during television's infancy. Resolution and grey scale rendition from flying-spot scanners exceeded that of any other pickup device of the period.[4]

Virtually all flying-spot scanner CRTs are magnetically deflected with relatively shallow deflection angles of 40 to 50 degrees to minimize corner defocus and raster distortion. Short decay phosphors, high accelerating voltage and low beam current for small

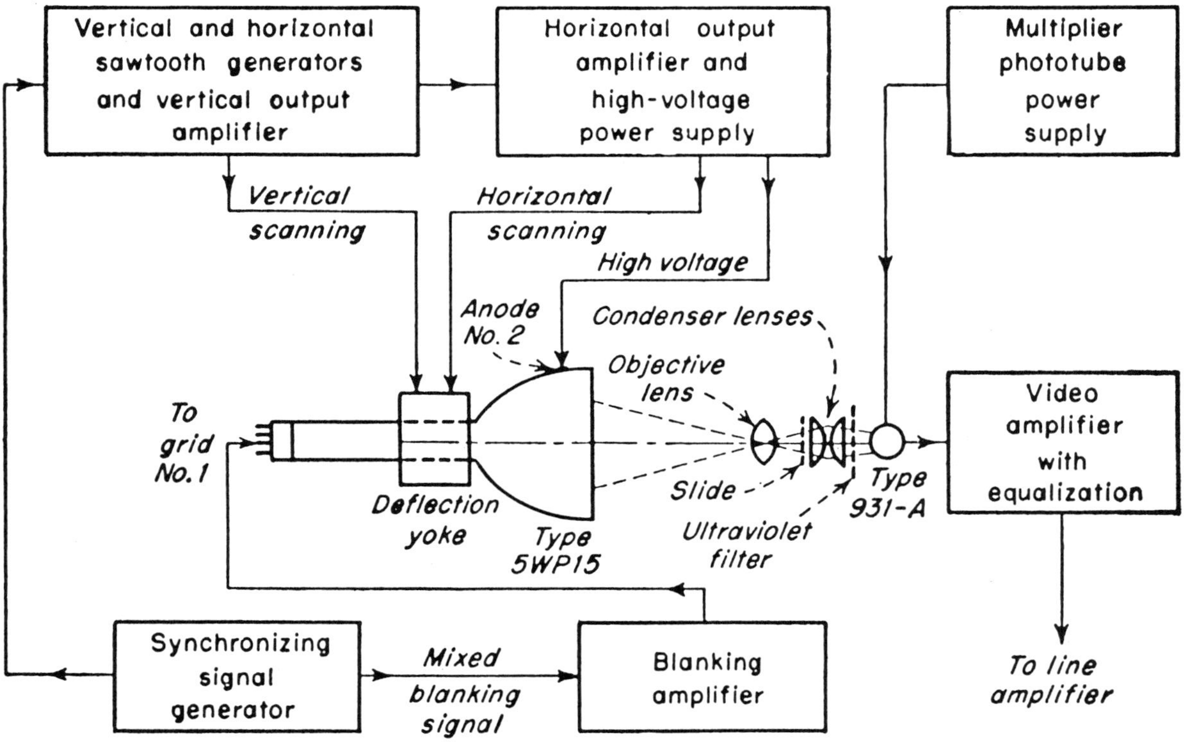

Figure 9.1 Flying-spot scanner block diagram. (Courtesy of RCA/Thomson.)

spot size also characterize these devices. Screen sizes range from three to 10 inches with round glass bulbs used almost exclusively. Flat or nearly flat faceplates of optical quality glass are used to reduce optical distortion away from the screen's central area with the short focal length lenses that are used to maximize the transfer of light from the screen. Manufacturers of flying-spot cathode-ray tubes have included Thomas Electronics, Westinghouse, Sylvania, Litton, RCA, Du Mont, Brimar Ltd. and Rank.

RCA produced the first cathode-ray tube in the U.S. designed specifically for flying-spot scanning, the type 5WP15 (Figure 9.2). P15 phosphor (zinc oxide) was developed by RCA for the 5WP15. The P15 emission appears as a pale green to the eye and has short persistence, however, it also has considerable energy in the violet and near ultraviolet portion of the spectrum with very short persistence of less than 50 nanoseconds to 10 percent of initial light output. It is this light that is utilized because of the persistence characteristic which prevents horizontal smearing of the picture with the fast horizontal scanning of the NTSC format raster. Additionally, it is spectrally well-matched to the photomultiplier tube.[5]

The 5WP15 was derived from the 5TP4 projection television picture tube described in Chapter 5. The 5WP15 differed from the

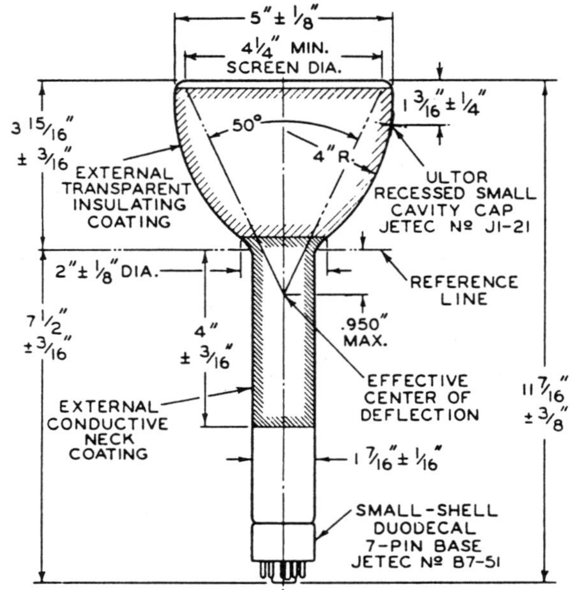

Figure 9.2 RCA 5WP15 flying-spot scanner CRT. (Courtesy of RCA/Thomson.)

5TP4 in phosphor and use of a flat faceplate. It had a five-inch aluminized screen, 50-degree magnetic deflection, high-voltage electrostatic focus and a duodecal 12-pin base similar to television picture tubes then in use. It operated at an accelerating voltage of 27,000 volts which produced a very small spot size capable of resolving 700 lines.

An improved flying-spot scanner CRT was registered in 1951 as the RCA type 5ZP16. A shallower 40-degree deflection angle and a higher focus voltage produced higher resolution exceeding 1,000 lines. Another new phosphor, P16, was used which had its emission concentrated in the violet and near ultraviolet spectrum with persistence only slightly longer than P15. The 5ZP16 was also manufactured by Westinghouse and Sylvania. Du Mont and Sylvania later cataloged it with P15 and P24 phosphors.

Ten-inch tubes also were used as flying-spot scanners. These included the 10VP15 registered in 1953 by Rauland and an unidentified tube used and presumably manufactured by Du Mont for their Model TA-150A scanner. The 10VP15 had a maximum anode voltage of 12,000 volts and does not appear to have had particularly high-resolution. It was second-sourced by Sylvania Electric. Other than the use of P15 phosphor, it is similar to the 10FP4 television picture tube discussed in Chapter 5.

Color television placed additional demands on the flying-spot CRT. Three photomultiplier tubes with red, green and blue color filters, respectively, detected signals for the three color primaries. Violet and near ultraviolet phosphors could no longer be used to produce pictures from color slides. In 1953, RCA again developed a new phosphor, P24, which was a broad spectrum zinc oxide material not unlike P15. The difference was its enhanced red output that provided more light for the red sensing photomultiplier tube, which was handicapped by low red sensitivity. The blue and green sensitivity of P15 were already adequate. RCA introduced the five-inch 5AUP24 the same year. Except for the phosphor, it was quite similar to the 5ZP16. Signal-to-noise ratios versus adequate resolution was a problem with television color flying-spot scanners and the vidicon camera tube replaced it in the 1950s.[6]

Telecine made extensive use of flying-spot scanning for conversion of motion picture film to video or videotape. Initially, most Telecine tubes were produced in England by Rank with Litton later manufacturing similar tubes in the United States. CRTs for Telecine scanning were seven-inch diameter with triode guns, magnetic focus and a blend of phosphor similar to what was later registered as P48, a mix of short persistence phosphors P46 and P47. Electronic frame storage later made easier the task of converting 24-frame per second motion pictures to interlaced 30-frame per second NTSC video standards.

Other flying-spot scanner CRTs developed for transparencies as well as 16 and 35 mm motion picture film during the 1950s included the seven-inch high-resolution Du Mont K1080P15 and the improved versions, the 7BCP15 and 7BCP24 registered in early 1960. These were magnetically focused and although they did not have flat faceplates, they did use optical quality glass of uniform curvature and thickness to prevent image quality degradation. A five-inch tube, the Sylvania Electric 5BNP16, was registered in early 1956. This relatively inexpensive electrostatically focused tube was used under the B&K designation 5BKPV1 for generating test patterns from slides for television service shops in the B&K 1075 Television Analyst. Low-voltage electrostatic focus and approximately 11,000 volt acceleration potential were adequate for this application, which was less demanding than that required for broadcast television.

By 1960, CRTs with spot sizes as small as 0.001 inch on a five-inch diameter screen were becoming available, e.g., the Du Mont K1725P16 and Litton Industries L4119 and L4146 (Figure 9.3). These were intended for higher resolution video signal generator and photo-recording applications than required for broadcast television. The quality of associated components such as deflection yokes and focus coils and the precision of their mechanical positioning became critical with the 0.001-inch spot size. Isolation from mechanical vibration, magnetic shielding and elec-

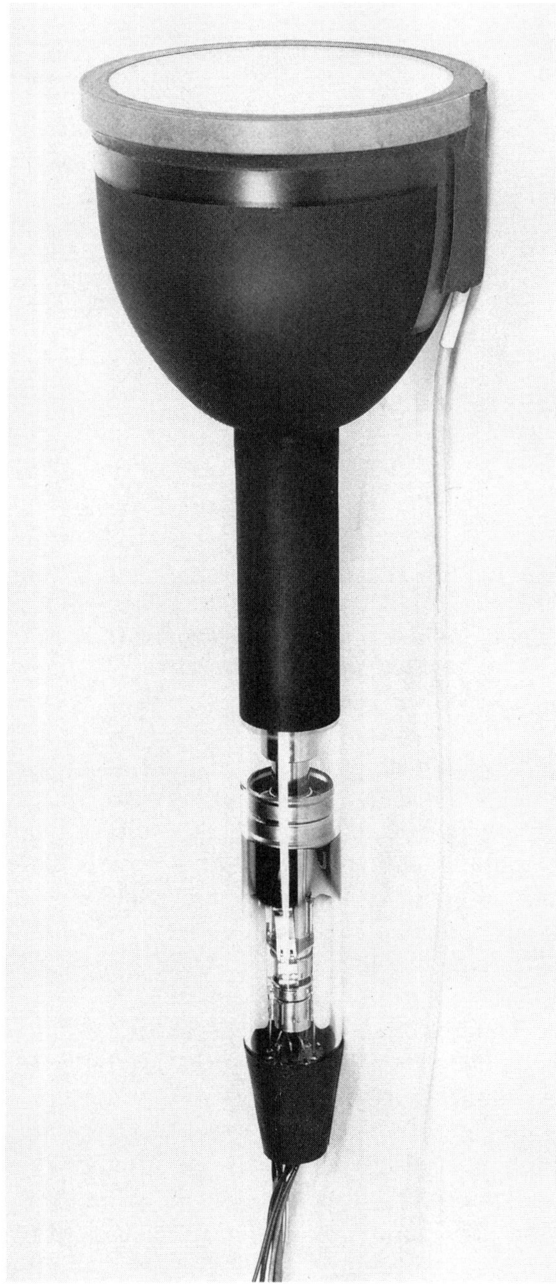

Figure 9.3 Litton L4146 high-resolution, photo-recording CRT. (Courtesy of Litton Electron Tube Division.)

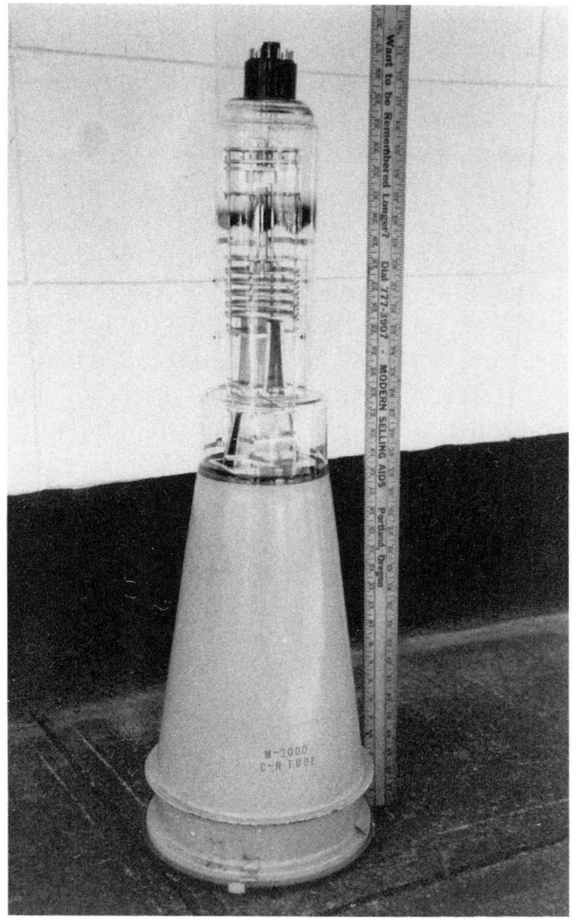

Figure 9.4 Flying-spot film store CRT for computer memory, circa 1958. (Courtesy of Larry Lockwood.)

trically "clean" power supplies and deflection amplifiers were also highly important.

The Litton L4119 was used primarily in Radar Land Mass Simulator systems for the U.S. Air Force. The first system was used to train flight navigators for the Boeing B52 Stratofortress. A five-inch diameter color flying-spot scanner CRT by Litton Industries generated sector scanned, color-coded map and simulated radar signals which were displayed on a conventional monochrome radar indicator. The three signals, which were generated simultaneously, could be selected individually to display radar, map overlay or other information. Special phosphor screens that were a cross between P15 and P24 were used. Particular attention was devoted to producing high-quality phosphor screens without blemishes or contamination as these defects would generate "bogies" or spurious objects on the radar display.[7]

Litton later built similar flying-spot scanner CRTs for Radar Land Mass Simulator systems for other military aircraft. These included a nine-inch diameter tube with special P24 phosphor for the Lockheed C5A and General Dynamics/Grumman F111 and a seven-inch diameter tube with P16 phosphor for the McDonnell F4 Phantom.[7]

Flying-spot scanning was used in the early digital computer years (circa 1950) for high density memory using photographic plates.[8] Some development work continued until about 1958 by Bell Telephone Laboratories on a "flying-spot store" with a rather hefty, special, high-resolution CRT measuring over three feet in length (Figure 9.4). A metal cone and flat faceplate of about 10 inches diameter with very short persistence P16 phosphor were used (the latter permitted high speed access). Precise characteristics were stressed to allow a 256 × 256 data array to be addressed accurately. The flying spot was optically imaged on multiple photographic plates, each with its own photomultiplier tube, to read up to a total of 2.5 megabits of data.[9,10]

General Electric obtained a spot size of eight to nine microns (about 0.0003 inch) in a five-inch microspot tube in 1961.[11] This tube was also intended for flying-spot readout of very high capacity optical computer memory. Several unusual features contributed to the performance including a transparent phosphor screen, a "dispenser" type cathode, a spiral accelerator in the neck section and an electron-optical system analogous to a photographic telephoto lens system. The transparent evaporated phosphor screen was reported to be three times as efficient as that of previous transparent screens.[12–17] Since the tube does not have the usual granular phosphor structure, there is little tendency for spreading of the light emitted from the point of beam impact due to diffusion by the phosphor crystals. The dispenser, or impregnated cathode, is made of a porous tungsten pellet with the electron emitting barium, calcium and aluminum oxides impregnated into it. Barium migrates to the surface when heated and emits electrons. The dispenser cathode is capable of higher temperature and thus higher emission current while having long life due to replenishment of the emitting material from below. An electrostatic prefocus lens and spiral accelerator effectively lengthened the electron-optical path length without unduly increasing overall tube length. Centering coils in the region of the prefocus lens compensated for minor misalignment of the electrodes, maximized beam current and ensured a round spot. A conventional magnetic focus coil after the helical accelerator provided the final focusing of the beam with the beam being large in diameter until it approached the screen to reduce the effect of space-charge spreading of the electrons from mutual repulsion. The deflection angle was only 20 degrees in one axis and two degrees in the other to reduce pattern distortion.

An interesting flying-spot scanner CRT application which existed from the early 1960s to the 1970s was the reading of photographic records on glass plates from "bubble chambers" for nuclear research. The tracks left by particles of nuclear disintegrations on the film emulsion were scanned by the seven- or nine-inch CRTs with P16 phosphor and digitized for computer processing of the information. The total quantity of tubes sold was small, as is usually the case in research applications.[7]

During the early to mid-1960s, Information International developed a forerunner of the optical character recognition (OCR) system using flying-spot scanning techniques. Seven-inch tubes from Ferranti Ltd. were used with Litton later supplying similar tubes with P47 phosphor. P47 was a very short persistence, purplish-blue phosphor composed of yttrium silicate and was registered with EIA by Thomas Electronics in 1972. Its short decay time of about 80 nanoseconds makes it one of the best phosphors for high-speed flying-spot scanners if detection of color information is not required.[7]

In 1970, Sylvania Electric announced two flying-spot scanners for television use. The first, the Sylvania Scanner Color Slide Theater, was a 35 mm carousel-type slide projector modified for flying-spot pickup of the slides and built into a large screen console color television receiver used to display the picture. A five-inch round flying-spot CRT was developed using a new Sylvania proprietary PSP phosphor (now registered as P48). PSP phosphor had greatly enhanced red emission and about ten times shorter persistence which solved the previous limitations of signal-to-noise ratio encountered with color scanners using P24 phosphor. A bi-potential electrostatic focus electron gun and 25,000 volt anode voltage allowed the scanner tube to be operated from the same power supplies as the picture tube in the television receiver portion. A low-cost CRT bulb similar to the 5FP7 radar CRT was utilized.[18]

Sylvania announced a simplified color slide TV scanner for cable television the same year. Similar techniques to those used for the home slide theater were employed. According to an unconfirmed report, Sylvania also announced a color video player for showing 8 mm movies through hotel and motel television receivers during the early 1970s. A 3 × 4-inch rectangular flying-spot scanner CRT was used with PSP phosphor. Reportedly Kodak funded the work but it was unsuccessful. Videotape- and satellite-delivered television movies later took its place with far greater success.[6]

Hazeltine produced a system using a Thomas Electronics CRT with P24 phosphor for determining the printing time for the three-color processing of motion picture film. The negative film scenes were scanned by the flying-spot scanner and reversed electronically for display on a color monitor. The amount of gain in each of the three color channels required to produce a correct color balance of the displayed picture was indicative of the printing exposure time needed for each primary when making prints. Each scene could be individually adjusted for color balance. Similar systems were provided by Kodak for use in printing color slides, although these used a field sequential display with rotating color filter drum.

9.2 KINESCOPE RECORDING CRTS

Great similarity exists between the CRT performance required for kinescope recording, photographic recording and flying-spot scanning. Small spot size, relatively small screen size, high accelerating voltages and phosphor spectral emission that matches a photosensitive device or film rather than the human eye characterize the ideal CRT for all three applications. Often the same tube, differing only in phosphor type, is manufactured for more than one of these applications.

A tube identical to the previously described 5WP15 flying-spot scanner CRT, except for the phosphor, was introduced by RCA in 1948. The 5WP11 was a transcriber kinescope for photographically transcribing television broadcasts directly from a CRT onto motion picture film for later rebroadcast.[19] These films were known as kinescope recordings. P11 phosphor, which is a highly efficient blue-emitting material, was used because of its strong emission at wavelengths of maximum film sensitivity. A related use of the 5WP11 and 5WP15 was the RCA "Ultrafax" facsimile transmission described in the *RCA Review* 1949.[20] Both the 5WP11 and 5WP15 were manufactured by Sylvania Electric and RCA.

As with flying-spot scanner CRTs, 10-inch diameter tubes were used for some kinescope recording. The Sylvania 10NP11 of 1951 used a triode electron gun, magnetic focus and a maximum accelerating voltage of 25,000 volts which was considerably higher than the 10VP15 used for flying-spot scanning.

Kinescope recording was short-lived. The coast-to-coast coaxial cable and microwave networks eliminated the need for shipment of film to other cities during the early 1950s and the videotape recorder of 1955 rapidly displaced most broadcast film use with news reporting and television screenings of Hollywood motion pictures being the last bastions to fall.

9.3 FILM RECORDING CRTS

Many applications now exist for photographically recording information from cathode-ray tube screens. At present film recorder displays are usually a dedicated part of a system rather than a secondary function of displays used for direct viewing of the data. Several approaches are used depending on scanning format, sensitivity of the print medium, space requirements and cost.

Line-scan recording often is used where hard copy of a complete display image is needed. The film motion provides scanning in one axis while the other axis is electrically scanned (Figure 9.5). The scanned line displayed on the CRT is optically imaged on the film. Relatively slow film movement generally is used. Applications include recording side-looking radar images and infrared mapping from aircraft where the film motion corresponds to the aircraft's forward speed to provide a continuous record of the flight path.

Fiber-optic line recording is very similar to line-scan recording except that a fiber-optics CRT faceplate permits direct contact recording and elimination of the optical system. Efficiency of light transfer is greatly enhanced and a more compact recorder is achieved. Major applications of fiber-optic recording include the recording of computer display hard copy and oil well logging of sensors for electrical resistance, sonic returns and induced radiation as the instrumented tool is periodically raised from the hole (Figure 9.6). Color recording with fiber-optic CRTs is possible with screens consisting of three color phosphor stripes (Figure 9.7).

Frame recording of a complete raster image (Figure 9.8) often is used in medical imaging applications such as ultrasound, CAT scans, etc. Although slow scan techniques often are employed, electrical scanning in each axis permits a simpler and more reliable photorecorder than the line-scan approach. High-quality, uniform CRT screens are required to ensure proper diagnosis without artifacts of the display. Phototypesetting requires a similar system.

The earliest film recording with cathode-

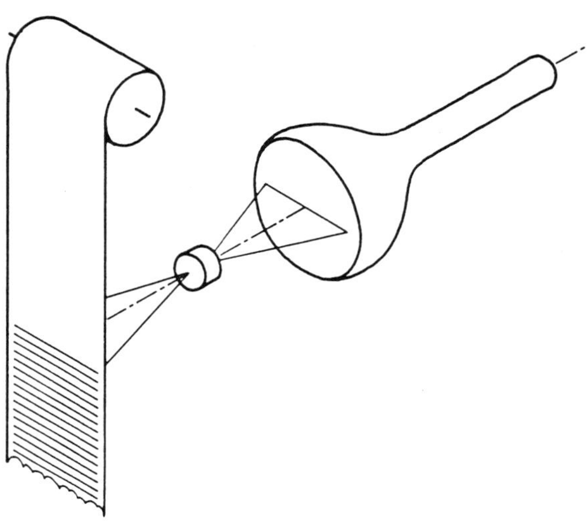

Figure 9.5 Line-scan recording.[30] (Courtesy of Academic Press.)

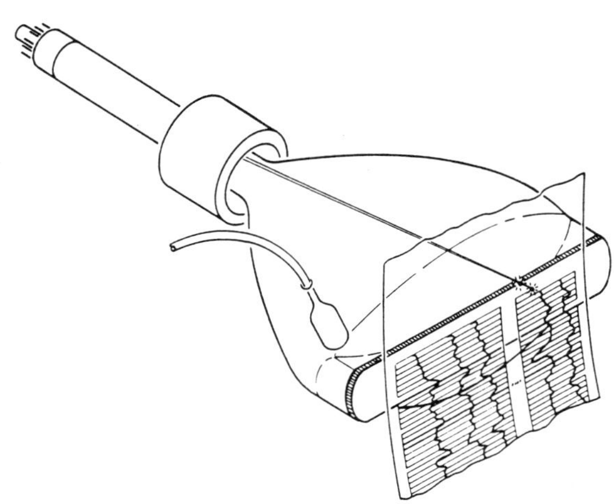

Figure 9.6 Fiber-optic, well-logging record.[30] (Courtesy of Academic Press.)

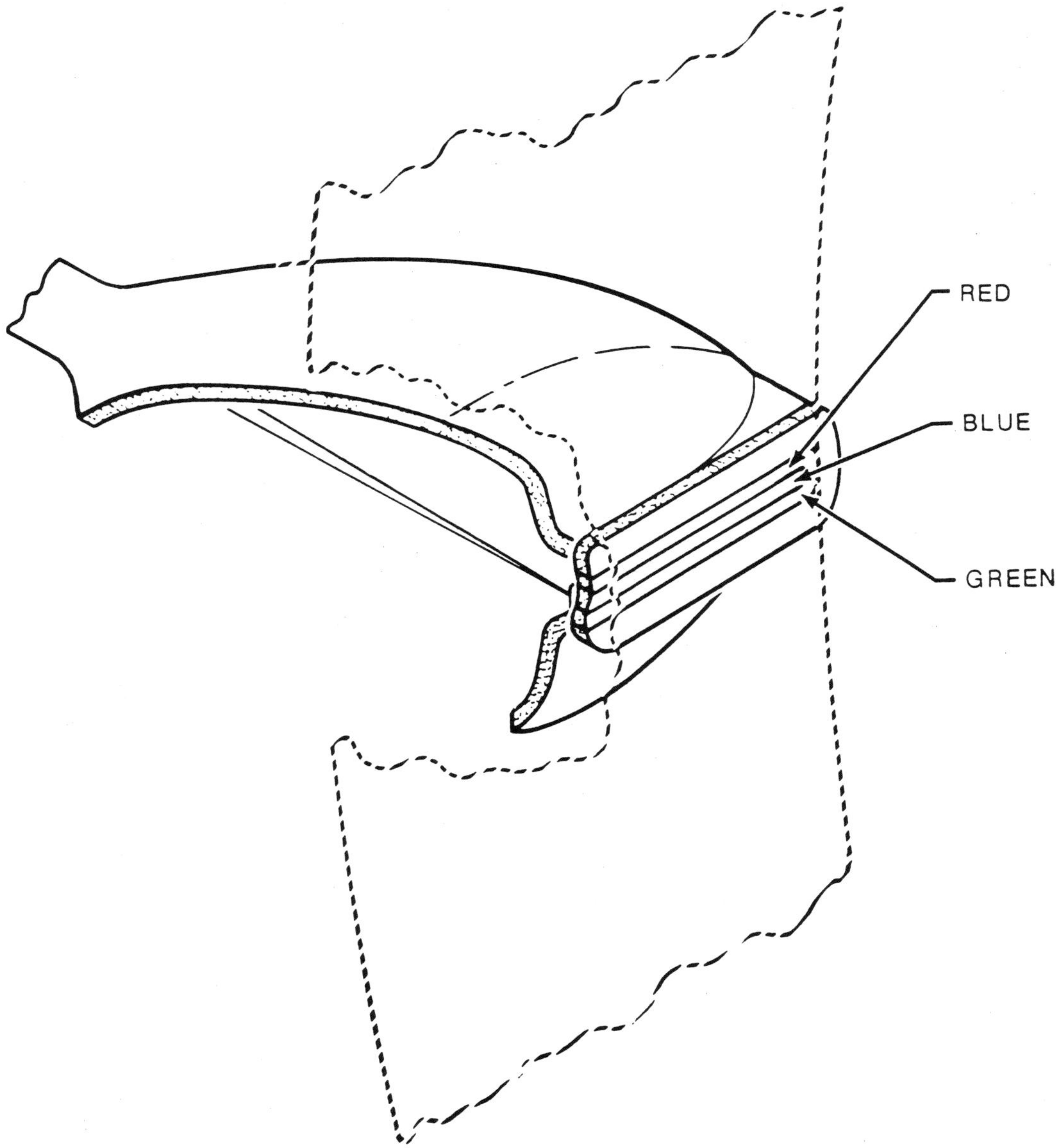

Figure 9.7 Fiber-optic, color-recording CRT.[30] *(Courtesy of Academic Press.)*

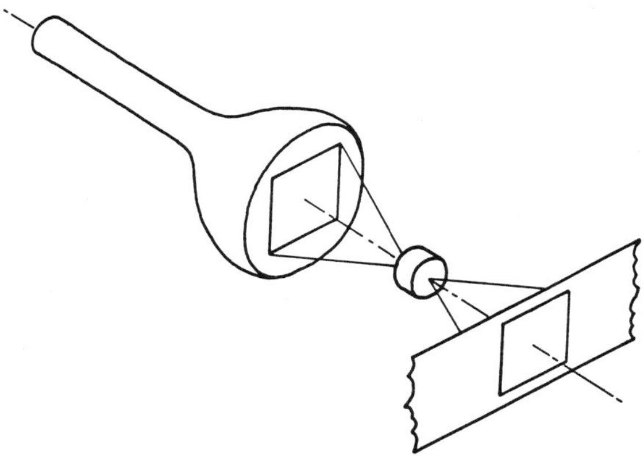

Figure 9.8 Frame recording.[30] (Courtesy of Academic Press.)

rays was accomplished with continuously pumped mechanical monsters which resembled CRTs as we know them only in name and basic electron-optical functioning. They were used for recording high-speed, and high-frequency phenomena by directly exposing the film emulsion with the electron beam rather than using an intermediate phosphor screen. Examples include the oscillographs described in 1920 by Dufour,[21,36,37] and 1929 by R. H. George,[22] as well as several others of that period.[38,39]

Photographic recording from cathode-ray tubes in the mid-1930s to early 1940s merely used standard oscilloscope CRTs made with P5 phosphor and, eventually, P11 phosphor. These short-persistence, blue-emitting materials were an excellent match to the spectral sensitivity characteristics of orthochromatic film emulsion. Short persistence reduced smearing of moving patterns. Usually the subject of the recording was waveforms displayed on an oscilloscope. Examples of CRTs from that period which were produced for photo-recording were the three-inch RCA type 908, the five-inch RCA 907 and five-inch Du Mont 5CP5 and 5RP11. Many other tube types were available with these phosphors and if not cataloged as such, were often available on special order. In 1948, photo-recording branched into kinescope recording for television. During the 1950s, tubes such as the one-inch diameter RCA 1EP11 and National Union 1DP11 (Figure 9.9) were used for some specialized airborne photo-recording applications. Oscillographic recording remained relatively unchanged until the 1980s when digital oscilloscopes made recording of waveforms digitally with inexpensive personal computer printers possible.

Initially photo-recording of television pictures were stills made with ordinary cameras from standard picture tubes. Later, kinescope recording as previously described, was used

Figure 9.9 1DP11 miniature recording CRT. (From the author's collection.)

from 1948 to the late 1950s. By 1960, extremely high resolution CRTs were becoming available from several companies. Notable examples include the five-inch diameter Westinghouse 5CEP- (Figure 9.10) in 1958, the family of CBS-Hytron tubes, the three-inch 3AVP-, 3AWP-, five-inch 5BYP, 5CQP- and seven-inch 7AVP- and several five-inch round Micropix™ tubes by Litton. The 5CEP- was available with P11 or P16 phosphor and was suitable for both photo-recording and flying-spot scanning. It had a spot size of about 0.0015 inch at 20,000 volts accelerating potential. The CBS-Hytron tubes were advertised as being able to resolve up to 2,000 lines per inch which translates to a spot size of about 0.0005 inch which continues as state-of-the-art. Resolution was dependent on the particular tube and phosphor of which P5, P11 and P16 were standard. With the P16 phosphor, the tubes were equally well-suited to flying-spot scanning applications also.

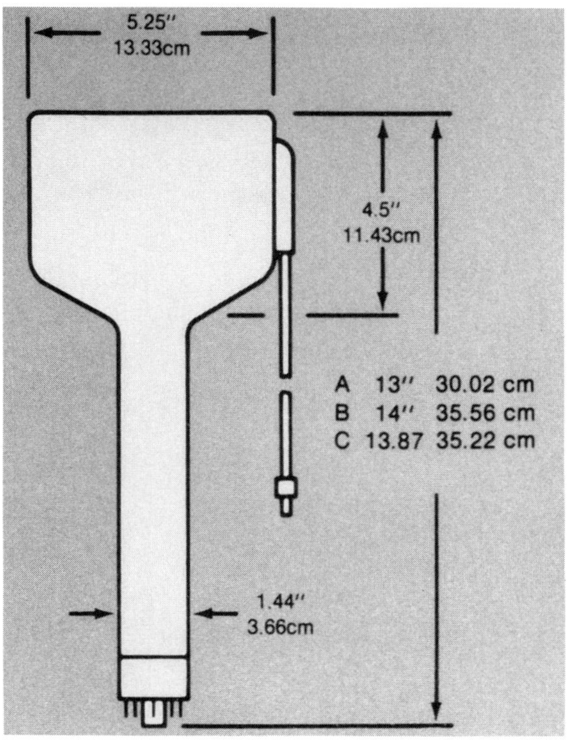

Figure 9.10 Westinghouse 5CEP11 high-resolution CRT. (Courtesy of Imaging and Sensor Technology Corporation.)

The applications, and hence the tubes, became much more diverse after 1960 with the growth of phototypesetting, military reconnaissance, medical imaging, computer graphics, remote sensing, etc., requiring many specialized low-volume cathode-ray tubes with customized characteristics. Areas of photo-recording CRT design improvements since 1960 have included screen quality, phosphors, spot size, ruggedization and life, although generally the changes have not been dramatic once 0.001 inch spot-size was achieved. Cathode-ray tubes with spot size of about 0.001 inch or less were commonly called *microspot* tubes. A typical seven-inch film recorder cathode-ray tube is shown in Figure 9.11.

Many film recorders have been developed for color imaging and graphics in recent years.

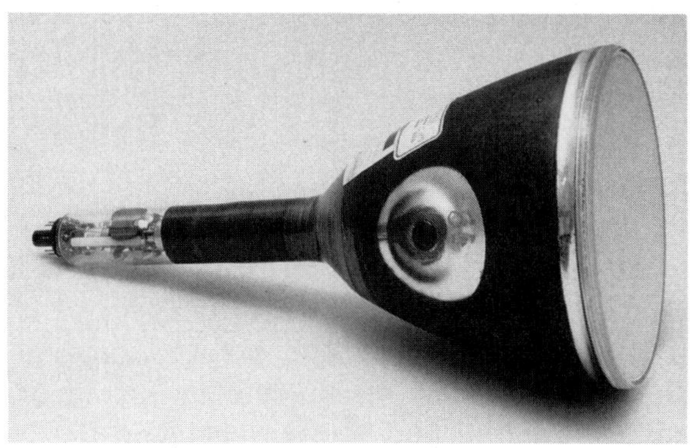

Figure 9.11 Thomas Electronics 7M280 seven-inch film recorder CRT.

Five-inch 2,000-line field sequential systems were built during the 1960s for recording the first color pictures from the Applications Technology Satellite (ATS).[23] P24 phosphor was used, but it required several minutes for exposure of a complete frame, in part because of the phosphor's low red efficiency. Later systems used P48 phosphor for improved red performance. Systems for color graphics using similar CRTs are utilized for making color slides for presentations and even special effects for movies.

Medical imaging systems have depended on CRT displays for many years for photographically recording ultrasound, CAT scan and other diagnostic images. Often these systems have used packaged Original Equipment Manufacturer (OEM) CRT displays by Hewlett-Packard, Tektronix and Kikusui, which were built into the system. Most used electrostatically deflected CRTs similar to those used for low-frequency oscilloscopes. Resolution requirements are typically low compared to microspot CRTs with spot size of about 0.01 inch being adequate in most cases. Screen uniformity and freedom from blemishes are important in medical diagnostics. Raster scanning is used predominantly. Slow vertical scanning rates often are used with exposure times of many seconds being necessary to form one complete raster frame. A typical CRT used is the Tektronix T6340[23] magnetically deflected CRT (Figure 9.12) with its approximately 0.0035 inch spot for systems using NTSC video format. Digital memory has rapidly decreased in cost and is now cheaper than the use of storage tubes and slow-scan systems. Data may be stored as gathered and displayed using standard video or one of the high-resolution formats as needed. Photographic recording will dimin-

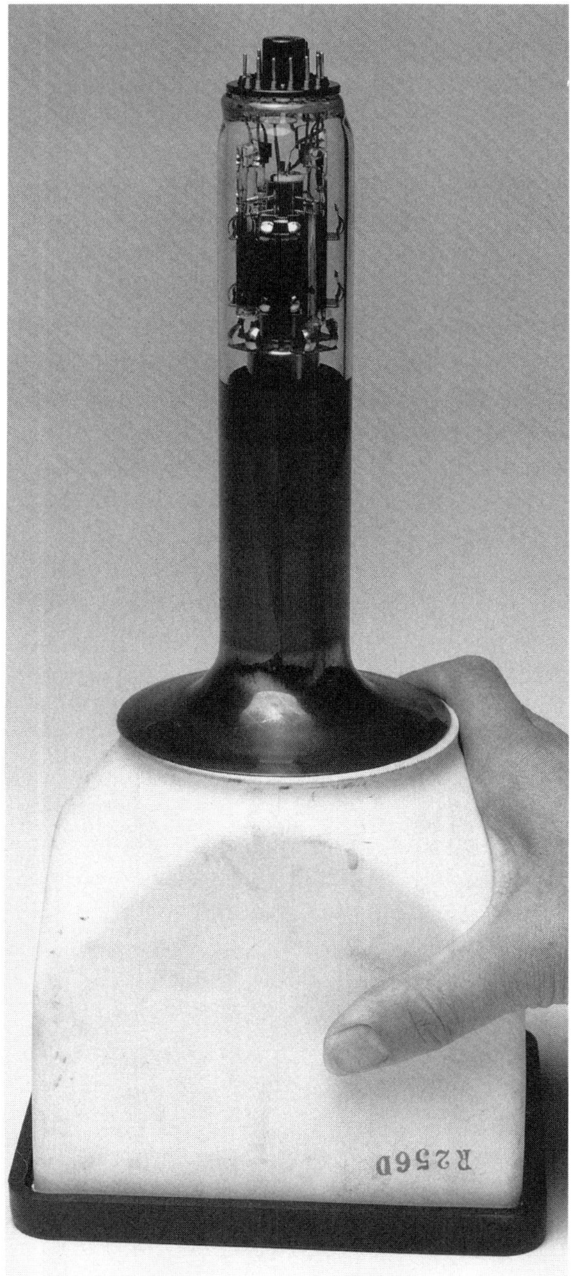

Figure 9.12 Tektronix T6340 raster scan display and photo-recording CRT. (Courtesy of Tektronix Inc.)

ish in importance as digital systems are installed with the capability of recalling images from magnetic or optical storage units at will for display on a large-screen, high-resolution CRT monitor at any of a number of locations, including operating rooms and doctors' offices. These systems are known as *patient archival and communications systems* (PACS).

9.4 FIBER-OPTIC CRTS

A conventional cathode-ray tube screen wastes much of its light output in photographic applications. Light from an unaluminized screen is emitted in all directions with 50 percent being wasted to the rear. Photographic-recording CRTs use aluminized screens which reflect light ordinarily lost to the rear toward the front, thus doubling light output. Still, light is emitted in approximately hemispherical fashion as a "Lambertian" surface. Lambert's Law states that light is emitted from a diffuse surface according to the cosine of the angle. Maximum light is emitted perpendicular to the screen surface, at 45 degrees from perpendicular 70 percent as much is emitted and at 60 degrees 50 percent is available. Since a camera lens intercepts only the light up to perhaps 20 degrees from perpendicular, a considerable amount of the total available light is lost. Fiber-optic faceplates allow much of this otherwise wasted light to be utilized for exposing film in direct contact with no intervening optics, significantly reducing exposure time.

Fiber-optics were the subject of a 1927 British patent by John L. Baird.[24] The first approach to contact printing was reported by R. G. Olden of RCA in 1957 using a thin mica window CRT for exposing Electrofax paper.[25,26] Fiber-optic CRT faceplates became practical in the early 1960s although they have always been expensive to manufacture. Westinghouse appears to be the first in the U.S. to combine fiber-optics and electron tubes (circa 1960).[24] Boeing Aircraft, Honeywell and Seismograph Service Corporation used fiber-optic CRTs during the early 1960s for recording data.[27] In 1966 Fairchild Instruments introduced their Model 977 fiber-optic CRT oscilloscope which was capable of recording transient events of 2,000 cm/μs.[28]

Fiber-optic faceplates are made of many parallel glass fibers pressed into a bundle and heated until just fused into a solid piece with no air space between fibers. Faceplates are sliced from the bundle, then ground and polished for flat surfaces (Figure 9.13). Because of prohibitive costs they usually are made in long narrow strips for line-scan mode rather than using entire CRT faceplates in the "page mode." The length of the strip is equal to the width of a printed page—about 8-1/2 inches. Line-scanning necessitates mechanically moving the photographic film or paper across the fiber-optic strip in the direction of the slow scan, while the electron beam is deflected the length of the fiber-optic strip at the fast scan rate. The combination of electrical fast scan in one axis and the slow mechanical scan in the perpendicular axis forms a raster on the film or paper. The position of the horizontal scanning line the length of the fiber-optic strip may also be slowly shifted a few millimeters back and forth across the fiber's narrow dimension to prevent burning a line into the phosphor. Usually one scan line, modulated by the video signal, is printed at a time. In the case of phototypesetting, one line of type is printed at a time.[29] Larger ar-

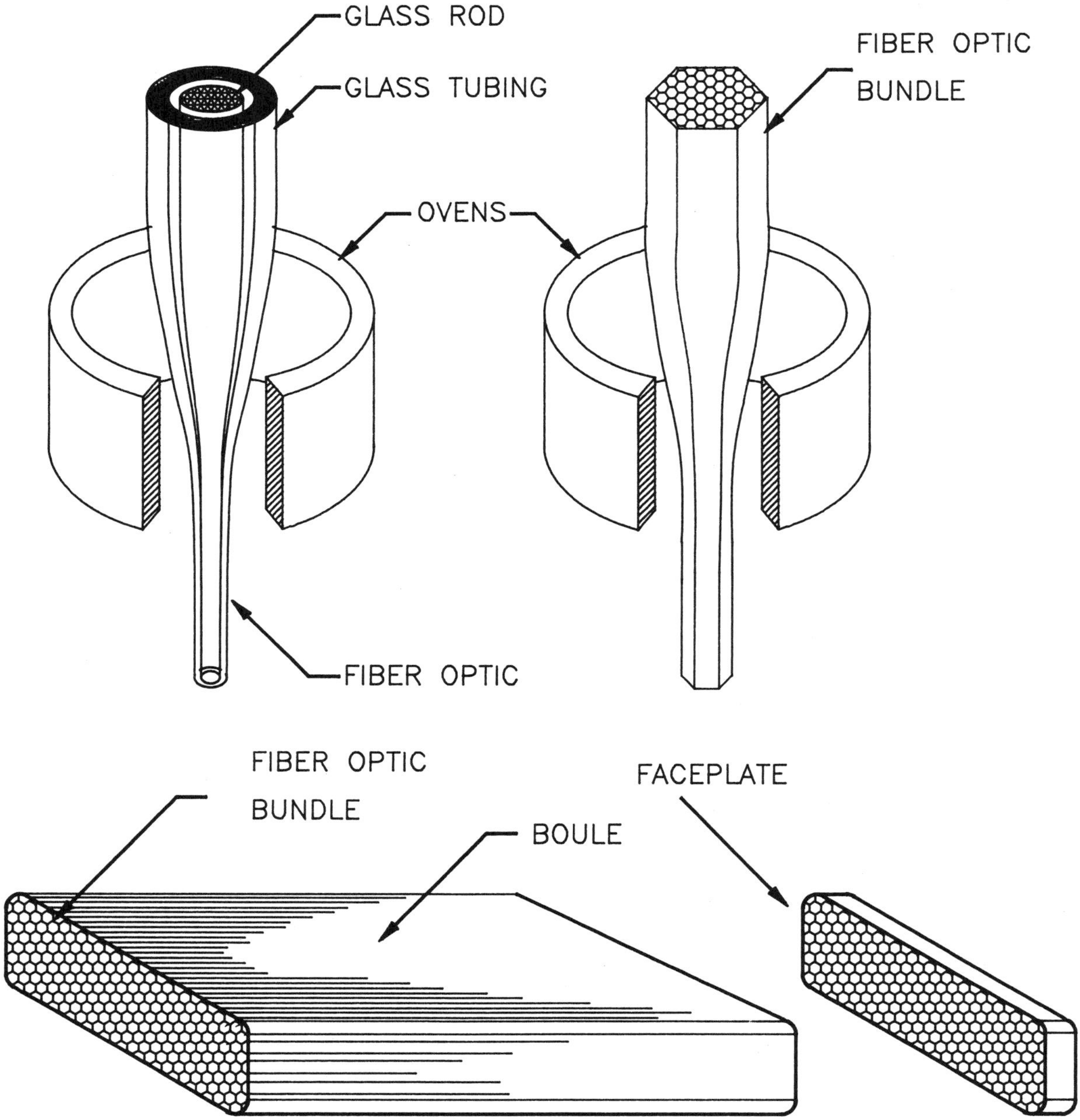

Figure 9.13 Steps in fiber-optic faceplate manufacture. (Courtesy of Jake Brain.)

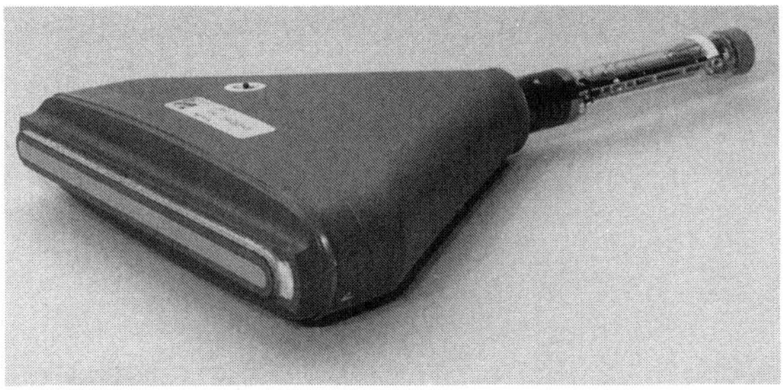

Figure 9.14 Tektronix T4601 fiber-optic, film-recording CRT.

rays of fiber-optics along with ultraviolet-emitting phosphors were evaluated in 1970 by Schlam and Pucilowski, with up to ten-fold improvement in efficiency reported for dry film recording.[30] Line-scan fiber-optics CRTs with prism-shaped fibers were explored by Matsushita in about 1970 for fast facsimile reading using contact flying-spot line scanning. A photomultiplier tube detected the light reflected from the paper being scanned.[40]

Line-scan CRTs are usually, but not always, constructed using highly flattened CRT envelopes as shown in Figure 9.14. The fiber-optic strip is frit-sealed to the CRT envelope to provide a vacuum seal. In order to not impair the resolution capability of the electron gun and phosphor, the diameter of each individual fiber must be no larger than that of the spot on the phosphor. Phosphors suitable for fiber-optic CRTs include P31, P45, P47, P48 and P52. Three parallel color primary phosphor stripes are used for line-scan color printing.[31]

Fiber-optics CRTs are used for computer hard copy, phototypesetting and strip mapping applications. Du Mont, Thompson CSF, Brimar, Thomas Electronics, Westinghouse, Litton and Tektronix are among the manufacturers that have produced or currently produce fiber-optic recording CRTs.

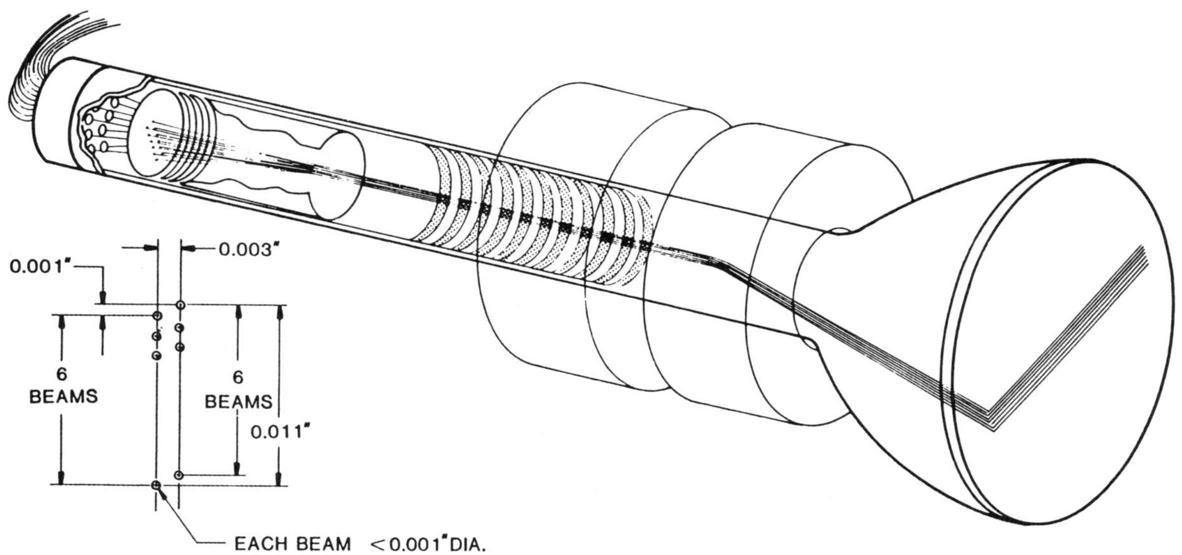

Figure 9.15 Multibeam, film-recording CRT.[30] (Courtesy of Academic Press.)

9.5 MULTIBEAM CRTS

High-resolution, multibeam CRTs for photo-recording applications were developed around 1981.[23,32–34] Use of several closely spaced, vertically stacked electron beams (Figure 9.15) with individual video signals applied and deflected in unison reduce the video amplifier bandwidth inversely to the number of beams. Video amplifiers for an eight-beam tube need only one-eighth of the bandwidth. The problem is that eight separate video amplifiers are required. Fortunately, eight 100 MHz amplifiers are far easier to build than one 800 MHz amplifier. The Litton L4277 is an example of a 12-beam, five-inch flat-faced photo-recording tube. The same technique was later applied to a high-resolution monochrome 19-inch data display CRT as described in Chapter 7.

9.6 PIN TUBES

Pin tubes, being nonviewable cousins to the recording CRT, will be briefly mentioned. During the 1950s, Stanford Research Institute developed the technique for A. B. Dick Co. for the high-speed printing of address labels for magazines and such. Pin tubes resemble fiber-optic, line-scan CRTs in outward appearance. A matrix of fine flush-mounted tungsten wires sealed into the tube's faceplate conducts electrical charges from the electron beam onto a paper strip in the pattern of alphanumeric characters. The paper with its charge patterns passes through a wet xerographic process to deposit toner to the charged areas where it is thermally fused to the paper. Resolution is limited by the number and density of wires used. The pin tube also was applied to high-speed oscillography for petroleum exploration by Seismograph Service Corporation of Tulsa, Oklahoma. Litton was the primary supplier of commercial pin tubes under the trade name Printapix™.[31,35]

10

Flat Cathode-Ray Tubes

10.1 Background

"Picture-on-the-wall" television has been predicted almost since the beginning of television itself. It was not until the 1950s that appreciable progress began to be made at shortening the length of the cathode-ray tube. Initially, this was in the form of wider deflection angles. With the adoption of 90-degree deflection angles in 1953, the tube length finally became less than the screen size. Still, what had been long sought after was a total tube depth of only two to three inches, an order of magnitude less than was attainable by using the classical funnel-shaped cathode-ray tube form.

Many researchers have addressed the flat display need over the years since about 1951 with flat CRTs being only one approach. Vacuum fluorescent displays (VFD), plasma display panels (PDP), electroluminescent displays (EL), liquid-crystal displays (LCD) and light-emitting diodes (LED) have all been intensively explored in recent years. All of these, except the liquid-crystal and light-emitting diode display, have their roots extending to certain aspects of CRT evolution. After all, the vacuum fluorescent display developed by ISE in Japan is essentially a low voltage (50 to 100 volts) cathode-ray tube using matrix addressing instead of conventional electrostatic or electromagnetic deflection. The plasma display, also matrix-addressed, has its origins in the gas-discharge Geissler tube of the nineteenth century, and electroluminescent displays resemble the phosphor screen of the CRT in many ways.

Just as the beginning of the modern CRT was defined at the beginning of this book, we must now decide where the CRT leaves off and its successors begin. It is convenient and generally accepted to exclude matrix-addressed devices and define those as *flat panel displays* as opposed to *flat CRT displays*. Even then, one is faced with additional decisions. What about those displays which are deflection-addressed in one axis and matrix-addressed in the other? It is difficult to draw an actual line between the CRT and flat-panel display. Other researchers may view the distinction differently than presented here.

10.2 Aiken Thin Tube

The Aiken flat CRT, originally dubbed the *thin tube,* is one of the more interesting stories of the trials and tribulations of developing a new technical concept.[1,2] The original concept by W. Ross Aiken harks back to 1951 when he began to address the problem as a self-imposed project while at the University of California Radiation Laboratory at Berkeley. It was apparent that the traditional location of the electron gun at the back end of the tube was ruled out. Instead, Aiken devised a system of electron "mirrors" to electrostatically deflect the electron beam from a side-mounted electron gun to perpendicularly strike the phosphor screen (Figure 10.1). Deflection in both the vertical and horizontal axes was electrostatic.[3,4]

Since funding was unavailable from the university, Aiken obtained a release of the rights to his invention. After the successful construction of a model demonstrating the basic concept, financing of further work was required. Arrangements were made with Kaiser to buy the rights to the invention, file for patents and establish a laboratory initially funded at $30,000 per month. Aiken left the University of California to head-up the research at Kaiser.

After the initial demonstration, the question in Aiken's mind was whether to apply for patent coverage immediately and head off any possibility of someone else independently inventing a similar device, or to delay the patent application until enough details were worked out so that openings would not be left for others to circumvent his patent. Aiken chose the latter course. The patent was filed in 1953.[5] However, interference with a patent application by Dr. Dennis Gabor at the Imperial College of Science and Technology of the University of London was soon found. It seems that Gabor had filed first on a sim-

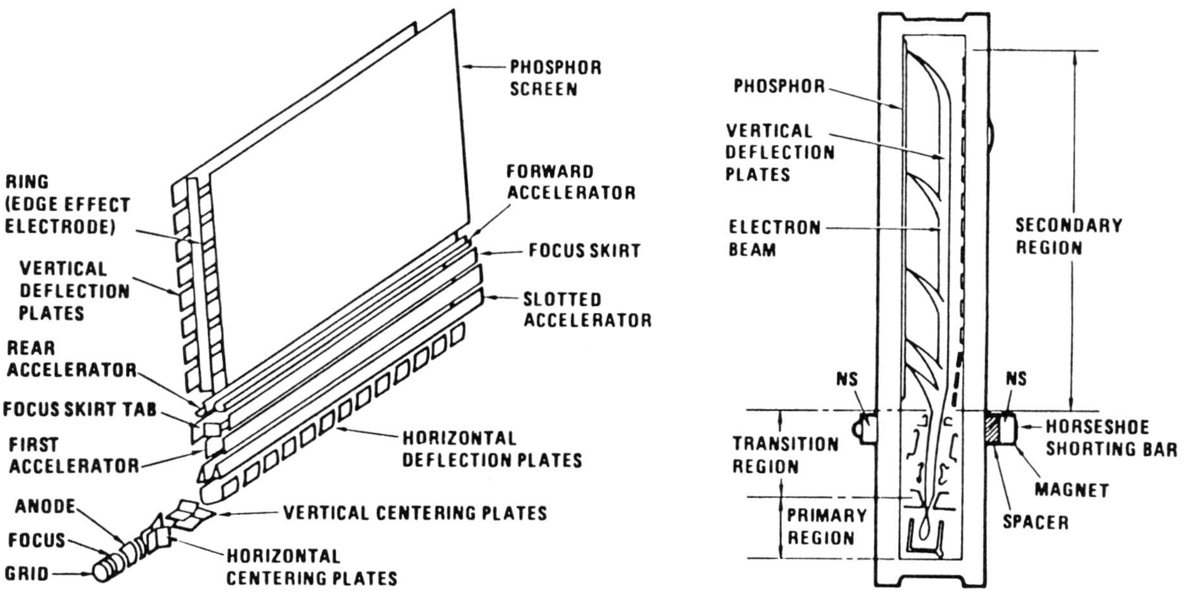

Figure 10.1 Aiken flat CRT construction.[8]

ilar invention in 1952. Aiken had invented it first, but Gabor had filed first. Differences in patent laws between the two countries resulted in Aiken being awarded the patent rights in the United States, where the first inventor has priority, and somewhat limited rights in the United Kingdom where the first to file has priority.

Financial losses at Kaiser precluded further financing for television, but a home was found for the work through the Office of Naval Research. A transparent display was needed that could be mounted in the T2V jet trainer aircraft windshields as a form of heads-up system where the pilot could view the display and what was in front of the aircraft simultaneously. Kaiser continued the development work in a new lab that was funded by the military and was set up in Palo Alto, California. Five- to 17-inch tubes were constructed with effort concentrated in several troublesome areas. The problems included transparency, sufficient brightness for viewing in sunlight and optical flatness, not to mention the necessity for miniaturization of its display computer to the one-cubic-foot volume available in the aircraft. The system was flown at Edwards Air Force Base in 1956,[6] but the work stopped short of production because the anticipated military market was not sufficiently large to justify establishment of a multimillion dollar plant.[1]

Kaiser funded further development work on adapting the thin tube to color for civilian markets. Several approaches were successfully tried including shadow-mask, stripes, phosphor dots, layered phosphor and focus-grid construction. Kaiser dropped additional work in 1961, but Video Color Corporation continued research on it using the Geer three-color screen[7] up to 1965. At that time Twentieth Century Electronics Ltd. began attempts to develop large-screen displays of up to 1.2 × 1.8 meters. Naturally, a tube envelope of that size having adequate strength was a formidable problem and work ceased in 1971.[8] It was several more years before Sinclair, Sony and other manufacturers commercially exploited the features of the flat CRT in its original intended application—television.

10.3 GABOR TUBE

In many respects, the Gabor flat tube was similar to the Aiken thin tube. The major differences were in the position of the electron gun, which was behind the screen in the tube's center, and self-scanning of the beam vertically. The electron beam was directed downward and bent forward and upward by a repeller electrode to strike the screen from the front (Figure 10.2). The self-scanning feature was based on charging of horizontal conductors by secondary emission, which repelled the electron beam progressively downward. Horizontal scanning was accomplished through a pair of electrostatic-deflection plates on either side of the beam. The Gabor tube was proposed for both monochrome and color television displays. The color version used a shadow mask with vertical slits and a vertical array of tricolor phosphor stripes.[9–11] Development work by Dr. Gabor took place during the period of 1952 to 1969.[8] No commercial applications of the Gabor tube are known. Professor Gabor went on to develop the concept of holography for which he was awarded the Nobel Prize in Physics in 1971.

10.4 REFLECTED-BEAM KINESCOPE

Obviously an RCA development, as indicated by the word *kinescope,* the reflected-

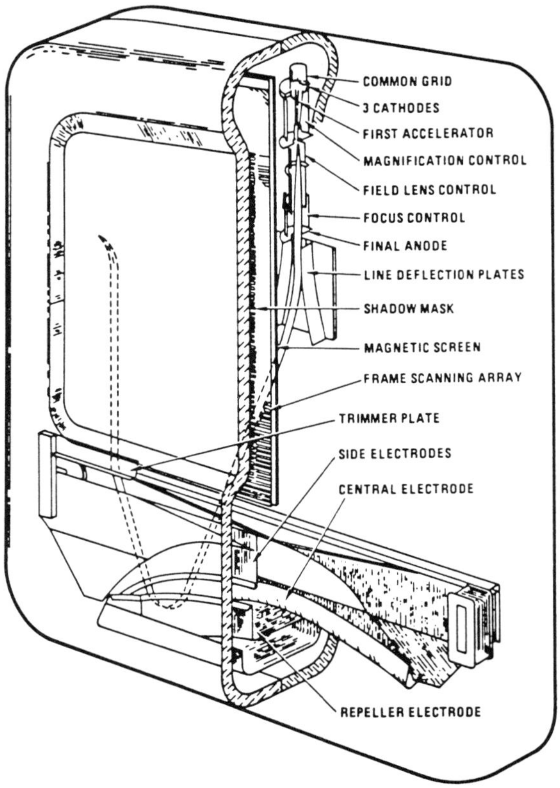

Figure 10.2 Gabor flat CRT construction.[8]

beam kinescope resembled a conventional monochrome picture tube turned inside out. It was described by Law and Ramberg in 1960 for possible use in television or military radar displays.[12] Not truly a flat CRT, it did greatly reduce the tube length for displays of 21 inches or more. A conventional electron gun with 90-degree magnetic deflection was reflected by the curved faceplate having a transparent conductive coating back to the apertured phosphor screen a short distance behind it. This was another device that never achieved commercial success.

10.5 SINCLAIR TUBE

Sinclair Research in the United Kingdom described a flat CRT in 1979 based on the Aiken tube concept. This was a three-inch diagonal CRT developed for a pocket television receiver by T. Krause (of Sinclair) out of his work at Twentieth Century Electronics.[13] Like the Aiken tube, this device used electrostatic deflection in both the vertical and horizontal axes, but was somewhat simplified in construction. The vertical screen height was reduced relative to the width and optically expanded to correct proportions using a Fresnel lens.[14] The construction of the tube was a rectangular cake pan shape with a glass window sealed to the top. The electron gun was mounted in one end and the phosphor screen was at the bottom of the other end. As with most flat CRTs, the beam struck the phosphor from the viewer's side which has higher efficiency than conventional CRTs in which the beam strikes the rear of the phosphor screen and the light must pass through the phosphor to reach the viewer. The receiver was scheduled to be introduced in late 1982,[15] but Sinclair was beaten to the punch by Sony.

10.6 SONY FLAT CRT

In 1982, Sony introduced in Japan what was reported to be the first commercial flat-screen television receiver.[15] The pocket-sized monochrome receiver used a Sony-developed two-inch diagonal flat cathode-ray tube operating at an accelerating voltage of 5,500 volts. Construction was quite similar to the Sinclair tube and it was described by a Sony spokesman as having been developed by careful attention to known design principles, rather than by a dramatic breakthrough. The conventional appearing gun protruded from below the screen (Figure 10.3) and had a longer electron optical path length with less bulk than the Sinclair pan-shaped arrangement. Deflection was electrostatic in the vertical axis and electromagnetic in the horizontal. Perhaps the most unique aspect was the use of internal ferrite bars for vertical deflection plates. These doubled as pole pieces for the external horizontal deflection yoke to reduce power consumption.[16]

This tube has probably been the most successful commercial flat CRT. It and its successor, the 02JM, having a screen curved slightly upward at the top, have sold in relatively large quantities in the Sony FD-200 Watchman TV receiver. The latter tube also was available to OEMs for computer-display applications and featured higher resolution and more conventional electromagnetic deflection in each axis to avoid the requirements for 400 to 5000 volt deflection voltage swings in the vertical axis.[17] An otherwise similar

Figure 10.3 Sony ED15 flat CRT. (From the collection of James Richardson.)

four-inch flat tube was cataloged as the type 04JM. Sony announced the development of a four-inch color version in the same mechanical configuration using dot sequential beam-index techniques in 1986.[18,19]

10.7 OTHER WORK

Many other companies have devoted resources toward flat CRT development. A few of these companies include Philips Research Laboratories, RCA, NEC, Sharp, NHK, Matsushita and Sanyo.[16,20] Announcements and papers (in chronological order) describing some of their work include the following:

Philips–A prototype monochrome nine-inch CRT using a low-voltage beam and channel multiplier was announced in 1982.[21,22] A low-voltage electron beam greatly aided scanning and the U-shape reversal of the beam direction. The channel multiplier overcame the brightness loss resulting from the lower accelerating voltage. Other details of construction were similar to those of the Aiken and Gabor tubes.[8]

Sanyo–The SAN FLAT color CRT was described in 1985 literature[23] which preceded the public announcement of the Sony flat color CRT by about one year. It had a three-inch screen and was of the beam-index type with a physical appearance similar to that of the Sony. Anticipated applications included television, computers, automation systems and automotive dashboard displays.

Matsushita–Planned production of the new four-inch T13-25 and six-inch T20-36 was announced in 1987.[24] Both were monochrome high-resolution (1,000 × 500 pixels) tubes operating at 6,000 volts and 8,000 volts, respectively. Electromagnetic deflection was employed in both axes. The shapes were similar to the "snow shovel" shape of the Sanyo and Sony flat CRTs with slightly inclined phosphor screens.

10.8 THE BEGINNING OF THE END?

Despite predictions set forth since the 1950s[25] that the flat panel replacement for the cathode-ray tube is "just around the corner," we are now entering the 1990s with the CRT at its zenith in terms of its all-pervasiveness and quantities produced annually.[26] To quote David Lanchenbruch, reporter for *TV Digest:*

"My first assignment back in 1950 was an article on the prospects for what we called picture-on-the-wall TV. I reached the conclusion that they were 10 years off, and it amazes me how well this forecast has stood up. They were 10 years off in 1950, and they're still 10 years off today."[27]

The unforeseen element was the continual advances made in CRT technology despite the great strides made in flat panels.[28,29] Yet, the CRT remains comparatively bulky with mostly nothing (just vacuum) inside of it. It has continued to demonstrate superiority in the areas of cost, screen size, brightness, resolution and color fidelity.[31] It now appears that at least the first generation of high-definition television receivers will be dependent on CRTs.

Now one is beginning to see commercial applications of monochrome flat panel displays for laptop and portable computers, avionics, and personal television receivers where space and power are at a premium.[30] This trend should be expected to continue as increasing production volume and technological advances continue to bring cost down and performance continues to improve. Certainly it won't be for lack of effort in any of the display device fields if this prediction fails to materialize.

It is still unclear which of the several competing technologies will eventually prevail. Many manufacturers claim to have the future successor to the CRT with the advances they see just ahead. Possibly different niche markets will exploit various display types for the advantages they alone can provide. A major breakthrough in any of the display technologies (possibly including that of the CRT) or a totally new concept could change the entire picture overnight.

This appears to leave the future of displays sufficiently vague to protect the author no matter what happens. After all, there isn't much future in trying to predict the future. Too often the unexpected happens.

Is the CRT really dead? Will it live on into the twenty-first century? It is hoped that this volume's second edition will provide additional insight based on further hindsight, along with additional information which the author continues to research.

Appendix 1: CRT Manufacturers

Domestic

* Allen B. Du Mont Laboratories–Upper Montclair, Passaic, Clifton, NJ
* American Television Laboratories (ATL)–Chicago, IL
* Arcturus–Newark, NJ

B-Scan (formerly Waterman Products)–Philadelphia, PA

Burle Industries (formerly RCA)–Lancaster, PA

* CBS-Hytron (formerly Hytron)–Salem, MA

Clinton Electronics–Rockford, IL

CRT Scientific Corp.–Van Nuys, CA

Du Mont Electronics (formerly Allen B. Du Mont Labs)–Clifton, NJ

Electronic Tube Corporation (ETC)–Philadelphia, PA

* Farnsworth Television and Radio (formerly Farnsworth Television Inc.)–Philadelphia, PA; Fort Wayne, IN

General Atronics (formerly Electronic Tube Corporation)–Philadelphia, PA

General Electric–Schenectady, NY

* General Electronics†
* General Radio–Cambridge, MA
* Globe Television & Phone Co.–New York, NY
* Hewlett-Packard–Colorado Springs, CO

Hughes Aircraft (formerly Vacuum Tube Products, Thompson CSF)–Carlsbad, CA; Dover, NJ

* Hygrade Sylvania–Emporium, PA
* Hytron–Danvers, MA

Imaging and Sensing Technology Corp. (formerly Westinghouse Electric)–Horseheads, NY

* Ken Rad (subsidiary of General Electric)–Owensboro, KY (Ken-Rad CRTs appeared to have been manufactured by RCA)
* Lansdale Tube Co. (formerly National Union)–Lansdale, PA

Litton Electron Devices–San Carlos, CA; Tempe, AZ

* National Union (formerly National Union Radio Corp.)–Newark, NJ; Lansdale, PA
* National Union Radio Corp.–Newark, NJ
* North American Philips Corp. (Norelco)–Dobbs Ferry, NY

* Philco (formerly Lansdale Tube Co.)–Lansdale, PA
* Radio Corporation of America (RCA)–Harrison, NJ; Lancaster, PA
Rauland Corp. (division of Zenith)–Chicago, IL
Raytheon–Quincy, MA
Richardson Electronics–Chicago, IL
* Sheldon Electric (division of Allied Electric Products)–Irvington, NJ
Stromberg-Carlson–Rochester, NY
Sylvania Electric (formerly Hygrade Sylvania)–Emporium, PA; Seneca Falls, NY
Tektronix–Beaverton, OR
* Tel-O-Tube Corporation of America–East Paterson, NJ
* Telephoto Corporation–New York, NY
Thomas Electronics–Wayne, NJ
Thomson CSF (formerly Du Mont Electronics, now Hughes Aircraft)–Dover, NJ
* Tung Sol Electric–Newark, NJ
* Vacuum Tube Products–Oceanside, CA
* Video Color Corp.†
* Waterman Products–Philadelphia, PA
* Western Electric–New York, NY, Chicago, IL
* Westinghouse Electric–Elmira, NY
* Zetka Laboratories–Clifton, NJ

†Location unknown.

INTERNATIONAL

AEG-Telefunken–Germany
Amperex (Philips)–Netherlands
* Baird Television Ltd.–United Kingdom
* Brimar–United Kingdom
* Cossor Ltd.–London
* Ediswan–United Kingdom
* Electronic Tubes Ltd.–United Kingdom
ELORG–Moscow
* EMI–United Kingdom
English Electric Valve (EEV)–United Kingdom
Ferranti Ltd.–United Kingdom
* General Electric Co. Ltd. (GEC)–United Kingdom
Hitachi–Japan
* Leybold and von Ardenne Oscillograph Co.–Germany
Matsushita (Panasonic)–Japan
* M-O Valve Co. Ltd. (MOV)–United Kingdom
Mullard-Osram–United Kingdom
Philips–Netherlands
* Radio Valve Co. Ltd.–Toronto, Canada
Rank Brimar Ltd. (formerly Thorn EMI Brimar Ltd.)–United Kingdom
Sony–Japan
Sylvania-Thorn–United Kingdom
Thompson CSF–France
Thorn EMI (formerly Brimar)–United Kingdom
Toshiba–Japan

*No longer manufactures CRTs.

Appendix 2: CRT Cross-References

Type	Equivalent	Notes
24-XH	902	Du Mont
34-XH	906, 2501	Du Mont
34-8-HB	2503	Du Mont
54-XH	2505	Du Mont
54-8-H	905, 2507	Du Mont
54-8-HB	2509	Du Mont
54-11-T	2511	Du Mont
94-8-H	2512	Du Mont
94-11-T	2514	Du Mont
114-9-T	2519	Du Mont
906-P1	3AP1	
906-P4	3AP4	
908	3AP5	
1802-P1	5BP1	
1802-P4	5BP4	
1803-P4	12AP4	
1804-P4	9AP4	
1805-P1	5AP1	
1805-P4	5AP4	
1806-P1	3EP1	
1808-P1	3CP1	
1809-P1	9JP1	
1810	VCR 139A	
1811-P1	7CP1	
1811-P4	7CP4	
1812-P7	5FP7	

Type	Equivalent	Notes
1813-P7	7BP7	
1814-P1	2AP1	
1815-P7	12AP7	
2001	913	National Union, similar
2002	902	National Union, similar
2003	906	National Union, similar
2005	54-XH	National Union, similar
2501A3	3AP1	Du Mont, similar
2501C3	3AP5	Du Mont, similar
2501D3	3AP4	Du Mont, similar
2505A5	5MP1	Du Mont
2505C5	5MP5	Du Mont
2505D5	5MP4	Du Mont
2511A5	5LP1	Du Mont
2511B5	5LP2	Du Mont
2511C5	5LP5	Du Mont
2511D5	5LP4	Du Mont
2514A9	9DP1	Du Mont, similar
2514B9	9DP2	Du Mont, similar
2514C9	9DP5	Du Mont, similar
2514D9	9DP4	Du Mont, similar
2519A14	14AP1	Du Mont, similar
2519B14	14AP2	Du Mont, similar
2519C14	14AP5	Du Mont, similar
2519D14	14AP4	Du Mont, similar
2520A20	20AP1	Du Mont, similar
2520B20	20AP2	Du Mont, similar
2520C20	20AP5	Du Mont, similar
2520D20	20AP4	Du Mont, similar
2525A5	5BP1	Du Mont
2525B5	5BP2	Du Mont
2525C5	5BP5	Du Mont
2525D5	5BP4	Du Mont
2529A5	5JP1	Du Mont
2529B5	5JP2	Du Mont
2529C5	5JP5	Du Mont
2529D5	5JP4	Du Mont
2531A14	14AP1	Du Mont
2531B14	14AP2	Du Mont
2531C14	14AP5	Du Mont

Type	Equivalent	Notes
2531D14	14AP4	Du Mont
2532A20	20AP1	Du Mont
2532B20	20AP2	Du Mont
2532C20	20AP5	Du Mont
2532D20	20AP4	Du Mont
2537A3	3GP1	Du Mont
2537C3	3GP5	Du Mont
2537D3	3GP4	Du Mont
B-1014	15DP4	Du Mont, similar
B-1034	12QP4-A	Du Mont, similar
B-1070	5ATP-	Du Mont
B-1217P-	12AWP-	Du Mont
B-2249P-	12AVP-	Du Mont
C-74528	8NP4	RCA
C-74529	8QP4	RCA
C-74537	16AYP4	RCA
CK-1355P19A	7BUP19	Rauland
D13-38	5DMP-	Telefunken
GL-475	5FP7	
K-1170P-	5DTP-	Du Mont
K-1253P-	5BFP-	Du Mont
K-1372P-	12AUP-	Du Mont
K-1435P-	7BTP-	Du Mont
K-1513P-	7BRP-	Du Mont
K-1679P7	5EBP7	Du Mont
K-1813P-	7BSP-	Du Mont
K-1825P16	5DUP16	Du Mont
K-1871P-	5EGP-	Du Mont
K-1904P7	7BFP7	Du Mont
K-1990P-	5DFP-	Du Mont
K-2007P-	12AQP	Du Mont
K-2046P-	5DVP-	Du Mont
K-2076P-	5DJP-	Du Mont
K-2098P-	7BGP-	Du Mont
K-2105P28	12ATP28	Du Mont
K-2109P7	5EAP7	Du Mont
K-2119P-	5DWP-	Du Mont
K-2130P-	5DXP-	Du Mont
K-2139P31	5EFP31	Du Mont
K-2249P-	12AVP-	Du Mont

Type	Equivalent	Notes
KC-2428P19	7BVP19	Fairchild/Du Mont
L-2217P-	5ECP-	Philco/Lansdale
MG13-10	5DHP1	Mullard, Ltd.
MM13-10	5DHP32	Mullard, Ltd.
R-6025	10FP4	Rauland
R-6231	16AVP4	Rauland
SC-2558P-	12AXP-	Sylvania
SC-4437P19	5ERP19	Sylvania
ST-3160	5DLP4	Sylvania
T-32	T316, T3160	Tektronix
T-33	T317, T3170	Tektronix
T-51	T0510, 5BGP-	Tektronix
T-52	T0520, 5CAP-	Tektronix
T-53	T507, T5070	Tektronix
T-54	T0540, 5BHP-	Tektronix
T-55	T0550, 5CBP-	Tektronix
T-56	T536, T5360	Tektronix
T-57	T5511	Tektronix
T-59	T555, T5550	Tektronix
T-60	T502, T5020	Tektronix
T-64	T533, T5330	Tektronix
T-65	T543, T5430, 5ELP-	Tektronix
T-66	T5260	Tektronix
T-310	T3100, 3WP-	Tektronix
T-517	T54P-H, T0541	Tektronix
T-945	T5431	Tektronix
VCR-139A	1810	British
VCR-516	12DP7	British
VT-111	5BP4/1802-P4	Signal Corps Designation
VT-226	3EP1/1806-P1	Signal Corps Designation
VT-248	3CP1-S1	Signal Corps Designation
WX-3956	16ALP7	Westinghouse Electric
Z-4623P7	16AMP7	General Electric
Z-4651P-	7BKP-	General Electric
Z-4652P-	12ASP-	General Electric
Z-4693P7	7BLP7	General Electric
Z-4757P4	5DRP4	General Electric
Z-4819P-	12ARP-	General Electric
Z-4870P7	8SP7	General Electric
Z-4875P4	16ATP4	General Electric

Type	Equivalent	Notes
Z-4887P4	8RP4	General Electric
Z-4938P4	16AWP4	General Electric
ZP-475	5FP7	
ZP-480	5BP1	
ZP-488	12GP7	
ZP-517	5CP1	
ZP-518	5CP4	
ZP-540	3FP7	
ZP-548	12FP7	
ZP-549	9GP7	
ZP-556	4AP10	

Appendix 3: Phosphor Tables

EIA "P" Numbers to Worldwide Type Designation System (WTDS)

P1	GJ	P20	KA	P39	GR
P2	GL	P21	RD	P40	CC
P3	YB	P22	X or XX	P41	YD
P4	WW	P23	WG	P42	GW
P5	BJ	P24	GE	P43	GY
P6	WW	P25	LJ	P44	GX
P7	GM	P26	LC	P45	WB
P8	Obsolete	P27	RE	P46	KG
P9	Obsolete	P28	KE	P47	BH
P10	ZA	P29	SA	P48	KH
P11	BE	P30	Cancelled	P49	VA
P12	LB	P31	GH	P50	VB
P13	RC	P32	GB	P51	VC
P14	YC	P33	LD	P52	BL
P15	GG	P34	ZB	P53	KJ
P16	AA	P35	BG	P54	—
P17	WF	P36	KF	P55	BM
P18	WW	P37	BK	P56	RF
P19	LF	P38	LK	P57	LL

Appendix 4: Commercial Manufacturer Codes

Manufacturer	EIA Mfr. Code	Type Number Prefixes
Du Mont	—	B, K, DC
General Electric	188	GL, ZP
Ken-Rad	188	—
Raytheon	280	CK
RCA	274	C
Sylvania	312	SC, ST
Textronix	—	T
Tung-Sol	322	—
Westinghouse	—	WL

Appendix 5: JAN Manufacturer Codes

The following is a compilation of the Joint Army-Navy (JAN) codes used to identify manufacturers on tubes made on defense contracts beginning in about 1943. These usually were marked on the tube base in the form:

JAN CRC 5CP1

Additional military markings often included an anchor to indicate U.S. Navy acceptance and/or Signal Corps contract number in the form:

SC 753/A

Code	Manufacturer
CAGE	American Television Laboratories–Chicago
CAGX	Waterman Products–Philadelphia, PA
CAHG	Chatham Electronics–Newark, NJ*
CBUP	Thomas Electronics–Wayne, NJ
CDR	General Electronics†
CDU	Allen B. Du Mont Laboratories–Passaic, NJ
CFN	Farnsworth–Fort Wayne, IN
CG	General Electric–Schenectady, NY
CHS	Sylvania Electric–Emporium, PA
CHY	Hytron–Danvers, MA
CKR	Ken Rad (Kentucky Radio)–Owensboro, KY
CNU	National Union–Lansdale, PA
CNY	North American Philips Corp. (Norelco)–Dobbs Ferry, NY
CRC	Radio Corporation of America–Lancaster, PA
CRP	Raytheon–Quincy, MA
CTL	Tung Sol Electric–Newark, NJ
CW	Western Electric–New York, NY; Chicago, IL

*Not a CRT manufacturer. Produced high-voltage rectifiers and gas triodes for CRT applications.

†Location unknown.

Appendix 6: Date Codes

Electronic Industries Association

Example: 74-03	74 indicates year (1974)	
	03 indicates week number	

RCA (World War II)

Example: K3E	First letter indicates year (1943)	Y is 1941
		S is 1942
		K is 1943
		H is 1944
		V is 1945
	Number indicates month	1 is Jan./Feb.
		2 is Mar./Apr.
		3 is May/June
		4 is July/Aug.
		5 is Sept./Oct.
		6 is Nov./Dec.
	Second letter indicates use (optional)	E is probably for Original Equipment Manufacturers (OEM) tube

RCA (1946–1956)

Example: 8-52	8 indicates last digit of year (1948)	
	52 indicates week (last week of year)	
	-indicates OEM tube (Original Equipment Manufacturer)	
	No hyphen indicates replacement tube	

RCA (1956 and later)–Used EIA code

Sylvania (World War II)

Example: B3R	B indicates Month (A is Jan., B is Feb., etc.)	
	3 indicates last digit of year (1943)	
	R probably indicates tube for replacement use (optional)	

Sylvania (Postwar)–Same as RCA Post-war

Appendix 7: Composition of Cathode-Ray Tube Phosphors

Phosphor	Composition
P1	Zinc silicate: manganese
P2	Zinc sulfide: copper
P3	Zinc beryllium silicate: manganese
P4	Zinc sulfide: silver and zinc cadmium sulfide: silver
P5	Calcium tungstate
P6	Similar to P4
P7	Zinc sulfide: silver and zinc cadmium sulfide: copper
P8	The reservation for this phosphor has been cancelled
P9	The reservation for this phosphor has been cancelled
P10	Potassium chloride
P11	Zinc sulfide: silver
P12	Zinc magnesium fluoride: manganese
P13	Magnesium silicate: manganese
P14	Similar to P7
P15	Zinc oxide
P16	Calcium magnesium silicate: cerium
P17	Zinc oxide and zinc cadmium sulfide: copper
P18	Calcium magnesium silicate: titanium and calcium beryllium silicate: manganese
P19	Potassium magnesium fluoride: manganese
P20	Zinc cadmium sulfide: silver
P21	Magnesium fluoride: manganese
P22	Zinc sulfide: silver, zinc silicate: manganese, zinc phosphate: manganese
	Zinc sulfide: silver, zinc cadmium sulfide: silver
	Zinc cadmium sulfide: silver
	Zinc sulfide: silver, zinc cadmium sulfide: silver
	Yttrium vanadate: europium
	Zinc sulfide: silver, zinc cadmium sulfide: silver
	Yttrium oxysulfide: europium
	Zinc sulfide: silver, zinc cadmium sulfide: copper
	Yttrium oxide: europium
	Zinc sulfide: silver, zinc cadmium sulfide: copper

Phosphor	Composition
P22	Yttrium oxysulfide: europium
P23	Similar to P4
P24	Zinc oxide
P25	Calcium silicate: lead: manganese
P26	Same as P19
P27	Zinc phosphate: manganese
P28	Zinc cadmium sulfide: copper
P29	Similar to P2 and P25
P30	The reservation for this phosphor has been cancelled
P31	Zinc sulfide: copper
P32	Calcium magnesium silicate: titanium
	Zinc cadmium sulfide: copper
P33	Magnesium fluoride: manganese
P34	Zinc sulfide: lead: copper
P35	Zinc sulfide selenide: silver
P36	Zinc cadmium sulfide: silver: nickel
P37	Zinc sulfide: silver: nickel
P38	Zinc magnesium fluoride: manganese
P39	Zinc silicate: manganese: arsenic
P40	Zinc sulfide: silver
	Zinc cadmium sulfide: copper
P41	Zinc magnesium fluoride: manganese
	Calcium magnesium silicate: cerium
P42	Zinc sulfide: copper
	Zinc silicate: manganese: arsenic
P43	Gadolinium oxysulfide: terbium
P44	Lanthanum oxysulfide: terbium
P45	Yttrium oxysulfide: terbium
P46	Yttrium aluminate: cerium
P47	Yttrium silicate: cerium
P48	A blend of yttrium aluminate (P46) and yttrium silicate (P47) in the ratio of 70:30
P49	Zinc silicate; yttrium vanadate
P50	Red: Yttrium oxide: europium
	Green: Zinc cadmium sulfide: copper nickel
P51	Red: Yttrium vanadate: europium
	Green: Zinc cadmium sulfide: silver nickel
P52	Zinc silicate: titanium
P53	Yttrium aluminum garnet: terbium
P54	Red: Yttrium oxysulfide: europium
	Green: Zinc sulfide: copper, aluminum
	White: Zinc sulfide: Silver and zinc sulfide: copper, aluminum and yttrium oxysulfide: Europium
P55	Zinc sulfide: silver
P56	Yttrium oxide: europium
P57	Zinc silicate: manganese and magnesium fluoride: manganese

Appendix 8: Cathode-Ray Tube Phosphors

	Date of Registration	
Phosphor	**Registrant**	**Date**
P1	RCA	June 28, 1945
P2	RCA	June 28, 1945
P3	RCA	June 28, 1945
P4	JEDEC	December 18, 1960
P5	RCA	March 21, 1945
P6	GE	September 5, 1946
P7	JEDEC	January 2, 1961
P10	Philco	1945 or before
P11	Du Mont	August 10, 1945
P12	JEDEC	January 2, 1961
P13	Sylvania	June 1945
P14	JEDEC	January 16, 1961
P15	RCA	May 16, 1955
P16	RCA	May 16, 1955
P17	Philco	April 30, 1954
P18	Philco	December 12, 1952
P19	VTP	December 12, 1952
P20	RCA	December 12, 1952
P21	Du Mont	April 3, 1953
P22	RCA	October 17, 1960
	RCA	January 30, 1961
	RCA	February 6, 1961
	Sylvania	June 29, 1964
	RCA	December 1, 1969
	Sylvania	April 13, 1970

Phosphor	Registrant	Date
P23	USR	August 14, 1953
P24	RCA	August 23, 1954
P25	GE	February 7, 1955
P26	CT	December 24, 1956
P27	RCA	August 27, 1956
P28	MOV	April 1, 1957
P29	Litton	September 22, 1958
P31	Mullard	March 7, 1960
P32	Mullard	March 7, 1960
P33	Mullard	March 7, 1960
P34	Ferranti	July 17, 1961
P35	Telefunken	August 12, 1963
P36	Thomas	March 6, 1967
P37	Thomas	March 6, 1967
P38	Thomas	March 6, 1967
P39	Thomas	March 6, 1967
P40	Thomas	March 6, 1967
P41	Sylvania	August 14, 1967
P42	Westinghouse	July 5, 1971
P43	Thomas	June 28, 1971
P44	Thomas	June 28, 1971
P45	Thomas	June 28, 1971
P46	Thomas	July 17, 1972
P47	Thomas	July 17, 1972
P48	Thomas	July 17, 1972
P49	Du Mont	October 21, 1975
P50	Thomas	October 21, 1975
P51	Thomas	October 21, 1975
P52	Thomas	October 21, 1975
P52	Thomas	December 13, 1977
P53	Ferranti	February 5, 1980
P54	Matsu	April 29, 1980
P55	Zenith	October 14, 1980
P56	Zenith	October 14, 1980
P57	Thorn	October 14, 1980

CT	Cinema-Television Ltd.
JEDEC	JT-16 Committee
Du Mont	Du Mont Electron Tubes and Devices Inc.
Ferranti	Ferranti Ltd.

GE	General Electric Co.
Litton	Litton Industries
Matsu	Matsushita Electronics Corp.
MOV	The M-O Valve Co., Ltd.
Mullard	Mullard Ltd.
Philco	Philco-Ford Corp.
RCA	RCA Corp.
Sylvania	GTE Products Corp.
Telefunken	Telefunken
Thomas	Thomas Electronics
Thorn	Thorn Brimar
USR	U.S. Radium Corp.
VTP	Vacuum Tubes Products (Hughes)
Westinghouse	Westinghouse Electric Corp.
Zenith	Zenith Radio Corp.

Appendix 9: WTDS Phosphor Designations

First Letter

A	Reddish-purple, purple.
B	Blue, purplish-blue, greenish-blue.
D	Three color screens where one or more phosphor is significantly different in color from the XX, X, P-22 phosphor for color entertainment.
G,H	Bluish-green, green, yellowish-green.
K	Yellow-green.
L	Orange, yellowish-pink.
M	Custom phosphor.
R	Reddish-orange, red, purplish-red, pink, purplish-pink.
S	Screens intended for two color displays.
V	Multi-color voltage-dependent screens.
W	White.
X	(Single letter) Color entertainment. Data on new color entertainment phosphors will be published in the phosphor registration book on request.

First Letter

X	(Two letter) Tri-color screens intended for data display applications where the colors of the three phosphors vary slightly from color entertainment phosphors.
Y	Greenish-yellow, yellow, orange-yellow.
Z	CRT screens that do not fit in any of the above categories.
M	Custom phosphor controlled by individual manufacturer. See Second Letter For "M-" Phosphor Designation.

In order to avoid confusion, the letters I and O are not used as a second letter designator. The second letter is assigned sequentially.

Second Letter For "M-" Phosphor Designations

A	AEG
B	Mitsubishi
C	Clinton
D	Thomson-CSF

Second Letter for "M-" Phosphor Designations

E	Toshiba
G	GE/RCA
H	Hitachi
M	Matsushita
N	NEG
P	Philips (NV or NAP)
R	Raytheon

Second Letter for "M-" Phosphor Designations

S	Sony
T	Thomas
W	Westinghouse
Z	Zenith-Rauland

Example: A tube with an MM phosphor code would be a tube with a custom phosphor manufactured by Matsushita.

References

Chapter 1

1. Rider, J. F. *The Cathode-Ray Tube at Work,* John F. Rider Publisher Inc., 1935.
2. Rider, J. F. and Uslan, S. D. *Encyclopedia on Cathode-Ray Oscilloscopes and Their Uses,* John F. Rider Publisher Inc., 1950, 1955, 1959.
3. Ruiter, J. H. *Modern Oscilloscopes and Their Uses,* Rinehart, 1949, 1955, 1958.
4. *Cathode-Ray Tubes and Their Associated Circuits,* Department of the Army Technical Manual TM 11-671/TO 16-45-227, September 1951.
5. Soller, T., Starr, M. and Valley, G. *Cathode-Ray Tube Displays,* MIT Radiation Laboratory Series, Vol. 22, McGraw-Hill, 1948.
6. "The Cathode-Ray Tube and Typical Applications," Allen B. Du Mont Laboratories, 1948.
7. *Cathode-Ray Tubes: Getting Down to Basics,* Tektronix, Inc., 1985.
8. Moss, H. "Narrow Angle Electron Guns and Cathode-Ray Tubes," *Adv Electron Electron Phys,* Suppl. 3, Academic Press, 1968.
9. Eisenstein, A. S. "Oxide Coated Cathodes," *Adv Electron,* Vol. 1, pp. 1–64, Academic Press, 1948.
10. Wagener, P. S. *The Oxide-Coated Cathode,* Vols. 1 and 2, Chapman and Hall Ltd., 1951.
11. Zalm, P. "Thermionic Cathodes," *Adv Electron Electron Phys,* Vol. 25, pp. 211–275, Academic Press 1968.
12. Bachman, C. H. and Carnahan, C. W. "Negative-Ion Components in the Cathode-Ray Beam," *Proc IRE,* Vol. 26, No. 5, pp. 529–539, May 1938.
13. Sharpe, J. "The Ion Trap In C. R. Tubes," *Electron Eng,* Vol. 18, No. 226, pp. 385–386, December 1946.
14. Bowie, R. M. "The Negative-Ion Blemish in a Cathode-Ray Tube and Its Elimination," *Proc IRE,* pp. 1482–1486, December 1948.
15. ———. "Cathode Ray Tube," U.S. Patent No. 2,211,613, filed August 14, 1936, issued August 13, 1940.
16. ———. "Cathode Ray Tube and the Like," U.S. Patent No. 2,211,614, filed August 14, 1936, issued August 13, 1940.
17. Pohle, E. and Scott, R. G. "Electron Gun Structure," U.S. Patent No. 2,562,243, filed June 6, 1950, issued July 31, 1951.
18. Branson, H. "Cathode Ray Tube," U.S. Patent No. 2,274,586, issued February 24, 1942.
19. Howes, D. E. "Cathode Ray Oscillograph," U.S. Patent No. 1,810,018, filed May 28, 1924, issued June 16, 1931.
20. Swedlund, L. E. and Saunders, R. "Material-Saving Picture Tube," *Electronics,* pp. 118–120, April 1951.
21. Compton, K. "Electron Guns for CRT Displays," *Inf Disp,* pp. 14–18, June 1989.
22. Swedlund, L. E. "Improved Electron Gun for C-R Tubes," *Electronics,* pp. 122–124, March 1945.
23. "Instrument CRT Captures Picosecond Pulses," *Electr Des,* p. 23, September 14, 1989.
24. *Encyclopaedia Britannica,* Vol. 23, p. 614, 1962.

25. Hill, C. G. A. "Cathode-Ray Tube Phosphors," Soc. Info. Disp. Conf., Lecture Notes, June 1984.
26. Seats, P. "Phosphor Screen Update," Electro-Optical Systems Design, pp. 26–27.
27. Keller, P. A. "Recent Phosphor Screen Registrations and the Worldwide Phosphor Type Designation System," *Proc Soc Info Disp,* Vol. 24, No. 4, pp. 323–328, 1983.
28. McColl, J. R. and Palilla, F. C. "Use of Rare Earths in Television and Cathode-Ray Phosphors," Amer. Chem. Soc. Symposium, pp. 177–193, 1981.
29. Encyclopaedia Britannica, Vol. 14, pp. 484–485, 1962.
30. Leverenz, H. W. *An Introduction to Luminescence in Solids,* John Wiley and Sons, 1950.
31. Larach, S. and Hardy, A. E. "Cathode-Ray Tube Phosphors: Principles and Applications," *Proc IEEE,* Vol. 61, No. 7, pp. 915–926, July 1973.
32. EIA. "Optical Characteristics of Cathode-Ray Tube Screens," JEDEC Publication No. 16, June 1960.
33. ———. "Optical Characteristics of Cathode-Ray Tubes," TEPAC Publication No. 116-B, November 1987.
34. Perkins, T. B. and Kaufmann, H. W. "Luminescent Materials for Cathode-Ray Tubes," *Proc IRE,* Vol. 23, No. 11, pp. 1324–1333, November 1935.
35. Leverenz, H. W. "Problems Concerning the Production of Cathode-Ray Screens," *J Opt Soc Amer,* pp. 396–417, January 1937.
36. Maloff, I. G. and Epstein, D. W. "Screens for Television Tubes," *Electronics,* p. 31, November 1937.
37. "Characteristics of Phosphors for Cathode-Ray Tubes," *Electronics,* p. 31, December 1938.
38. Garlick, G. F. J. "Cathodoluminescence," *Adv Electron,* Vol. 2, pp. 151–184, 1950.
39. McKeag, A. H. and Welch, E. E. "Luminescent Materials and Their Application to Cathode-Ray Tubes–Part 2," *Ind Chem,* pp. 171–178, April 1951.
40. Poole, H. H. *Fundamentals of Display Systems,* pp. 321–348, Spartan Books, 1966.
41. Keller, P. A. "1976 CIE-UCS Chromaticity Diagram with Color Boundaries," *Proc Soc Info Disp,* Vol. 24, No. 4, pp. 317–321, 1983.
42. Law, R. R. "Fluorescent Screen Assembly and Method of Manufacture," U.S. Patent No. 2,233,786, filed November 29, 1939, issued March 4, 1941.
43. ———. "Cathode Ray Tube and Luminescent Screen," U.S. Patent No. 2,303,563, filed May 9, 1941, issued December 1, 1942.
44. Schaefer, V. J. "Cathode-Ray Tube," U.S. Patent No. 2,374,311, filed June 21, 1944, issued April 24, 1945.
45. Bachman, C. H. "Image Contrast in Television," *Gen Elec Rev,* pp. 13–19, September 1945.
46. Epstein, D. W. and Pensak, L. "Improved Cathode-Ray Tubes with Metal-Backed Luminescent Screens," *RCA Rev,* Vol. 7, No. 5, pp. 5–10, March 1946.
47. de Boer, J. H. and Dippel, C. J. "Process of Making a Layer Containing a Fluorescent Material," U.S. Patent No. 1,954,691, filed September 18, 1931, issued April 10, 1934.
48. Williams, F. E. "Some New Aspects of Germanate and Fluoride Phosphors," *J Opt Soc Amer,* Vol. 37, No. 4, pp. 302–307, April 1947.
49. Studer, F. J., Cusano, D. A. and Young, A. H. "Transparent Luminescent Films," *J Opt Soc Amer,* Vol. 41, p. 559, August 1951.
50. Cusano, D. A. and Studer, F. J. "Transparent Luminescent Screens," *J Opt Soc Amer,* Vol. 42, p. 878, November 1952.
51. Studer, F. J. and Cusano, D. A. "Transparent Phosphor Coatings," *J Opt Soc Amer,* Vol. 45, No. 7, July 1955.
52. "Transparent Phosphors Improve CRT," *Electronics,* pp. 92–94, February 28, 1958.
53. Steier, H. P., Kelar, J., Lattimer, C. T. and Faulkner, R. D. "Development of a Large Metal Kinescope for Television," *RCA Rev,* pp. 43–58, March 1949.
54. Rose, A. S. and Turnbull, J. C. "The Evaluation of Chromium-Iron Alloys for Metal Kinescope Cones," *RCA Rev,* pp. 593–599, December 1949.
55. Brain, J. "From Cups to CAD: A History of Glass with CRTs in Mind," *Inf Disp,* March 1990.
56. Smith, C. P., Morrell, A. M. and Demmy, R. C. "Design and Development of the 21CYP22 21-Inch Color Picture Tube," *RCA Rev,* pp. 334–348, September 1958.
57. Schneider, R. K. "Making Glass for Television

Picture Tubes," in *RCA Picture Tubes,* pp. 89–95, RCA 1980.

58. Wilbanks, W. H., Johnson, R. H., Wang, C. H. and Erzen, F. C. "Electron Beam Display Device," U.S. Patent No. 3,207,936, filed August 21, 1961, issued September 21, 1965.
59. Bramall, D. H. "Ceramic Cathode Ray Tubes," *SID Dig,* pp. 68–69, 1970.
60. Robinder, R. C., Bates, D. J., Green, P. J., Lewen, G. D. and Rath, D. R. "A High-Brightness Shadow-Mask Color CRT for Cockpit Displays," *SID Dig,* Vol. 14, pp. 72–73, 1983.
61. Sheffield, B. "Recent Developments in German Cathode Ray Tubes," *Electronics,* p. 60, November 1937.
62. Lempert, I. E. and Feldt, R. "The 5RP Multiband Tube: An Intensifier-Type Cathode-Ray Tube for High-Voltage Operation," *Proc IRE,* pp. 432–440, July 1946.
63. Seats, P. "The Evolutionary CRT," Soc. Inf. Disp. Conf., Lecture Notes, 1987.
64. Mathew, R. L. "Without Glass, It's Not a Tube," *Inf Disp,* pp. 18–21, June 1988.
65. Palac, K., Dietch, L., Prazak, C. and Chiodi, W. "The Perfectly Flat Tension Mask Color CRT," *IEEE Intl. Conf. Cons. Elec.,* pp. 82–83, 1986.
66. Strauss, P., Prazak, C. J. and Lee, S. D. "Construction of a Flat Tension Mask Color CRT," *SID Dig,* pp. 324–326, 1986.
67. Dietch, L., Palac, K. and Chiodi, W. "Performance of High-Resolution Flat Tension Mask Color CRTs," *SID Dig,* pp. 322–323, 1986.
68. Izumitani, T., Nakagawa, K., Yamashita, T. and Matsu-ura, T. "Glass Filters for Contrast Enhancement in CRT Displays," *SID Dig,* pp. 172–173, 1980.
69. Weidert, F. "Protective Glass," U.S. Patent No. 2,219,122, filed May 21, 1937, issued October 22, 1940.
70. Schulz, P. and Janus, G. "Television Receiver Comprising an Optical Filter," U.S. Patent No. 3,078,343, filed August 8, 1956, issued February 19, 1963.
71. Ban, G., Hersh, H. N. and Kaplan, S. "Enhancing the Brightness Capability of Color-TV Receivers Through the Use of a Spectrally Selective Tinted Glass," *J Electrochem Soc,* pp. 531–534, April 1969.
72. "High-Resolution Improvements Make Life Easier for Color CRT Users," *Inf Disp,* p. 18, January 1983.
73. von Ardenne, M. *Cathode-Ray Tubes,* Sir Issac Pitman and Sons, London, 1939 (translation from the German edition of 1933).
74. Kohl, W. H. *Materials and Techniques for Electron Tubes,* Reinhold, 1960.
75. Weingartner, H. C. and Kennedy, S. W. "Modern Vacuum Pumps in Electronics Manufacturing," *Adv Electron,* Vol. 5, pp. 239–244, Academic Press, 1953.
76. Welch, D. *Spectrum Analyzer Measurements,* Tektronix, Inc., 1969 (out of print).
77. "Polar Coordinate Oscillograph," *Electronics,* p. 46, December 1936.
78. Moran, R. B. "Spiral Sweep Oscilloscope Timer," *Electronics,* pp. 120–123, October 1947.
79. Jaski, T. "Better Yet, Use a Spiral," *Radio-Elect,* p. 88, November 1959.
80. *Reference Data for Radio Engineers,* 4th ed., International Telephone and Telegraph Corp., 1956.
81. Sherr, S. *Electronic Displays,* John Wiley and Sons, 1979.
82. *Information Display Concepts,* Tektronix, Inc., 1968 (out of print).
83. Bingley, F. J. "A Half Century of Television Reception," *Proc IRE,* pp. 799–804, May 1962.
84. Fink, D. G. "Perspectives on Television: The Role Played by the Two NTSCs in Preparing Television Service for the American Public," *Proc IEEE,* Vol. 64, No. 9, pp. 1322–1331, September 1976.
85. ———. *Principles of Television Engineering,* McGraw-Hill, 1940.
86. ———. "Television Broadcasting Practice in America–1927 to 1944," *JIEE,* Vol. 92, No. 19, pp. 145–164, September 1945.
87. ———. *Television Engineering,* 2nd ed., McGraw-Hill, 1952.
88. Kiver, M. S. *Television Simplified,* Van Nostrand, 1946.
89. Zworykin, V. K. and Morton, G. A. *Television,* John Wiley and Sons, 1940, 1954.
90. Du Mont, A. B. "Method and System for Television Communication," U.S. Patent No. 2,186,634, filed January 26, 1938, issued January 9, 1940.
91. "Television Without Sync Signals," *Electronics,* pp. 33–34, 68, March 1938.

92. "Universal Television System," Radio World, p. 26, April 1938.
93. Keller, P. A. "Cathode-Ray Displays for Medical Imaging" *J Digital Imaging,* Vol. 3, No. 1, pp. 15–25, February 1990.
94. Secrest, J. D. *Electronic Industries Association: The First Fifty Years,* EIA, 1974.
95. Federal Trade Commission Trade Regulation Rule "Deceptive Advertising as to Sizes of Viewable Pictures Shown by Television Receiving Sets," pp. 1–4, effective July 1, 1966.
96. EIA. "Worldwide Type Designation System for TV Picture Tubes and Monitor Tubes," TEPAC Publication No. 106-B, June 1988.
97. ———. "Industrial Cathode-Ray Tube Test Methods" TEPAC Publication No. 105 and Supplements, various dates.
98. ———. "Glossary of Cathode-Ray Tube Terms and Definitions," TEPAC Publication No. 192, September 1984.
99. Harris, F. K. "A New Cathode-Ray Oscillograph and Its Application to the Study of Power Loss in Dielectric Materials," *Bureau of Standards J Res,* RP 636, Vol. 12, No. 1, pp. 87–102, January 1934.
100. Epstein, D. W. "Electron Optical System of Two Cylinders as Applied to Cathode-Ray Tubes," *Proc IRE,* Vol. 24, No. 8, pp. 1095–1139, August 1936.
101. Nottingham, W. B. "Electrical and Luminescent Properties of Willemite Under Electron Bombardment," *J Appl Phys,* Vol. 8, pp. 762–778, November 1937.
102. Martin, S. T. and Headrick, L. B. "Light Output and Secondary Emission Characteristics of Luminescent Materials," *J Appl Phys,* Vol. 10, pp. 116–127, February 1939.
103. Perkins, P. "The Design and Development of a Non-Compressing Post-Deflection Acceleration Method for Cathode-Ray Tubes," Tektronix Internal Report, p. 5, June 1967.

CHAPTER 2

1. Harvey, E. N. *A History of Luminescence from the Earliest Times to 1900,* American Philosophical Society, Chap. 4, 1957.
2. *Encyclopaedia Britannica,* Vol. 14, pp. 484–484d, 1962.
3. Hill, C. G. A. "Cathode Ray Tube Phosphors," SID Conf. Lecture Notes, pp. 6.2-2–6.2-20, June 8, 1984.
4. Harvey, E. N. *A History of Luminescence from the Earliest Times to 1900,* American Philosophical Society, Chap. 5, 1957.
5. Shiers, G. "Ferdinand Braun and the Cathode-Ray Tube," *Sci Amer,* Vol. 230, No. 3, pp. 92–101, March 1974.
6. *Encyclopaedia Britannica,* Vol. 14, p. 103, 1962.
7. Harvey, E. N. *A History of Luminescence from the Earliest Times to 1900,* American Philosophical Society, Chap. 7, 1957.
8. Tyne, G. F. J. *Saga of the Vacuum Tube,* Chap. 1, Howard W. Sams & Co., 1977.
9. *Encyclopaedia Britannica,* Vol. 21, p. 181, 1962.
10. Harrison, S. W. and Dickenson, W. A. "A Short History of the Cathode-Ray Tube in Television, Part 1," *Sylvania Technol,* Vol. 5, No. 1, pp. 2–6, January 1952.
11. Abramson, A. *The History of Television,* McFarland, 1987.
12. MacGregor-Morris, J. T. and Henley, J. A. *Cathode Ray Oscillography, Volume II,* Instruments Publishing Co: Pittsburgh, 1936.
13. Thomson, J. J. "Cathode-Rays," Presented before the Royal Institution April 30, 1897, reprinted in *Proc Roy Inst,* Vol. 15, pp. 419–432, 1897.
14. Ruiter, J. H. *Modern Oscilloscopes and Their Uses,* Rinehart, 1949, 1955, 1958.
15. *Encyclopaedia Britannica,* Vol. 9, pp. 70–71, 1962.
16. *The Cathode-Ray Tube and Typical Applications,* Allen B. Du Mont Laboratories, 1948.
17. Magie, W. F. *A Source Book in Physics,* Harvard University Press, 1935, 1963.
18. Tannas, L. E. *Flat Panel Displays and CRTs,* Van Nostrand, 1985.
19. *Encyclopaedia Britannica,* Vol. 11, p. 610, 1962.
20. Tyne, G. F. J. *Saga of the Vacuum Tube,* Chap. 2, Howard W. Sams & Co., 1977.
21. *Encyclopaedia Britannica,* Vol. 6, p. 747, 1962.
22. Mezger, G. R. *The Hot-Cathode Low-Voltage Cathode-Ray Tube,* Allen B. Du Mont Laboratories, Inc., 1939.
23. *Encyclopaedia Britannica,* Vol. 13, pp. 909–910, 1962.

24. Ryan, H. J. "The Cathode Ray Alternating Current Wave Indicator," *Amer Inst Elec Eng,* Vol. 22, pp. 539–552, 1903.
25. *Encyclopaedia Britannica,* Vol. 17, p. 539, 1962.
26. Perrin, J. "Nouvelles proprietes des rayons cathodiques," *Acad Sci Comptes Rendus,* Vol. 121, pp. 1130–1134, 1895.
27. *Encyclopaedia Britannica,* Vol. 19, p. 527, 1962.
28. ———. Vol. 23, pp. 842–858.
29. Glasser, O. *Dr. W. C. Röntgen,* 2nd ed., Charles C. Thomas, 1958.
30. Thomson, J. J. "Cathode-Rays," *Philosophical* magazine, Vol. 44, Series 5, p. 293, 1897.
31. *Encyclopaedia Britannica,* Vol. 17, pp. 869–876, 1962.
32. ———. Vol. 22, p. 152.
33. ———. Vol. 8, pp. 303–308.
34. Susskind, C. "Ferdinand Braun: Forgotten Forefather," *Adv Electron Electron Phys,* Vol. 50, pp. 241–260, Academic Press, 1980.
35. Kurylo, F. and Susskind, C. *Ferdinand Braun,* MIT Press, 1981.
36. *Encyclopaedia Britannica,* Vol. 4, p. 47, 1962.
37. Stinchfield, J. M. "Cathode Ray Tubes and Their Application," *Electr Eng,* pp. 1608–1615, December 1934.
38. Atherton, W. A. "Pioneers: Karl Ferdinand Braun (1850–1918): Inventor of the Oscilloscope," *Electron World* and *Wireless World,* pp. 12–14, January 1990.
39. Braun, K. F. Ueber ein Verfahren zur Demonstration und zum Studiumdes zeitlichen Verlaufes variabler Strome, *Ann Phys Chem,* pp. 552–559, February 1897.
40. Zenneck, J. "Eine Methode zur Demonstration und Photographic von Stromcurven," *Ann Phys,* Vol. 69, pp. 838–853, 1899.
41. Tyne, G. F. J. *Saga of the Vacuum Tube,* Chap. 3, Howard W. Sams & Co., 1977.
42. ———. Chap. 4.
43. ———. Chap. 5.
44. Tressler, M. E. "The Cathode Ray Tube and Its Applications," Gen *Elec Rev,* Vol. 18, pp. 816–820, 1915.
45. Bingley, F. J. "A Half Century of Television Reception," *Proc IRE,* pp. 799–804, May 1962.
46. McGee, J. D. "Distant Electronic Vision," *Proc IRE,* pp. 596–608, June 1950.
47. *Encyclopaedia Britannica,* Vol. 21, pp. 910a–913, 1962.
48. Atherton, W. A. "Pioneers—Vladimir Kosma Zworykin," *Electron World* and *Wireless World,* Vol. 93, No. 1620, pp. 1019–1020, October 1987.
49. Rosing, B. "Art of Electric Telescopy," U.S. Patent No. 1,161,734, filed April 5, 1911, issued November 23, 1915.
50. Nipkow, P. Deutsches Reichs Patent No. 30,105, January 6, 1884.
51. Dufour, A. "Oscillographe Cathodique," *J Phys Radium,* Vol. 1, p. 147–160, 1920.
52. ———. "Oscillographe Cathodique Pour L'Étude des Basses, Moyennes et Hautes Fréquences," *L'Onde Électr,* Vol. 1, pp. 699–715, 1922.
53. ———. "Oscillographe Cathodique Pour L'Étude des Basses, Moyennes et Hautes Fréquences," *L'Onde Électr,* Vol. 2, p. 19–42, 1923.
54. George, R. H. "A New Type of Hot Cathode Oscillograph," *J AIEE,* Vol. 48, pp. 534–538, July 1929.
55. Fagen, M. D. *A History of Engineering and Science in the Bell System: The Early Years (1875–1925),* Bell Laboratories, 1975.
56. Rider, J. F. and Uslan, S. D. *Encyclopedia on Cathode-Ray Oscilloscopes and Their Uses,* John F. Rider Publisher Inc., 1950, 1955, 1959.
57. Johnson, J. B. "A Low Voltage Cathode-Ray Oscillograph," *Phys Rev,* Vol. 17, pp. 420–421, 1921.
58. ———. "A Low Voltage Cathode-Ray Oscillograph," *J Opt Soc Amer Rev Sci Inst,* Vol. 6, pp. 701–712.
59. ———. "A Low Voltage Cathode-Ray Oscillograph," *Elec Comm,* Vol. 1, pp. 57–62, November 1922.
60. ———. "A Low Voltage Cathode-Ray Oscillograph," *Bell Syst Tech J,* Vol. 1, pp. 142–151, November 1922.
61. ———. "Vacuum Tube," U.S. Patent No. 1,565,855, filed November 20, 1920, issued December 15, 1925.
62. ———. "Electric Discharge Device," U.S. Patent No. 1,603,284, filed November 24, 1922, issued October 19, 1926.
63. ———. "Some Uses of the Cathode-Ray Oscillograph," *Bell Lab Rec,* Vol. 2, No. 2, pp. 57–59, April 1925.

64. ———. "Electric Discharge Device," U.S. Patent No. 1,632,080, filed December 27, 1921, issued June 14, 1927.
65. Smits, F. M. *A History of Engineering and Science in the Bell System: Electronics Technology (1925–1975),* AT&T Bell Laboratories, 1985.
66. Van der Bijl, H. J. "Vacuum Tube and Method of Operating the Same," U.S. Patent No. 1,565,873, filed August 10, 1920, issued December 15, 1925.
67. Jones, L. T. and Tasker, H. G. "A Thermionic Braun Tube with Electrostatic Focusing," *J Opt Soc Amer Rev Sci Inst,* pp. 471–478, October 1924.
68. Rogowski, W. and Grosser, W. "Kathode-Ray Oscillograph," Arch *Electrotech,* Vol. 15, pp. 377–384, December 5, 1925.
69. News Release, Franklin Institute, October 16, 1957.
70. Johnson, J. B. "A Movable-Screen Cathode Ray Tube," *Bell Lab Rec,* Vol. 26, No. 5, pp. 219–221, May 1948.
71. Takayanagi, K. "Experiments on Television," *J IEE* (Japan), No. 482, pp. 932–942, September 1928.
72. Zworykin, V. K. "Television System," U.S. Patent No. 1,691,324, filed July 13, 1925, issued November 13, 1928.
73. ———. "Television System," U.S. Patent No. 2,141,059, filed December 29, 1923, issued December 20, 1938.
74. ———. "Television with Cathode-Ray Tube for Receiver," *Radio Eng,* pp. 38–41, December 1929.
75. ———. "Improvements in Cathode-Ray Tube Design," *Electronics,* pp. 188–190, November 1931.
76. ———. "Cathode Ray Apparatus," U.S. Patent No. 1,988,469, filed June 30, 1931, issued January 22, 1935.
77. ———. "Television System," U.S. Patent No. 2,084,364, filed December 24, 1930, issued June 22, 1937.
78. ———. "Vacuum Tube," U.S. Patent No. 2,109,245, filed November 16, 1929, issued February 22, 1938.
79. Zworykin, V. K. and Painter, W. H. "Development of the Projection Kinescope," *Proc IRE,* Vol. 25, no. 8, pp. 937–953, August 1937.
80. Zworykin, V. K. and Morton, G. A. *Television,* John Wiley and Sons, 1940, 1954.
81. Morgan, J. *Electronics in the West,* Chap. 7, National Press Books, 1967.
82. Everson, G. *The Story of Television: The Life of Philo T. Farnsworth,* Arno Press, 1974.
83. Farnsworth, P. T. "Synchronizing System," U.S. Patent No. 1,844,949, filed April 25, 1928, issued February 16, 1932.
84. ———. "Thermionic Oscillograph" U.S. Patent No. 2,099,846, filed June 14, 1930, issued November 23, 1937.
85. ———. "Electrical Discharge Apparatus," U.S. Patent No. 1,986,330, filed April 17, 1928, issued January 1, 1935.
86. von Ardenne, M. (translated by G. S. McGregor), *Cathode-Ray Tubes,* Pitman and Sons, London, 1939.
87. ———. (translated by O. S. Puckle), *Television Reception,* D. Van Nostrand Co., Inc., New York, 1936.
88. Dornberg, J. "East Germany's Technology Baron," *Technol Illus,* pp. 46–51, July 1983.
89. von Ardenne, M. "Investigations on Gas-Filled Cathode Ray Tubes," *Proc IRE,* Vol. 20, No. 8, pp. 1310–1323, August 1932.
90. Batcher, R. R. "Commercial Cathode-Ray Tubes," *Instruments,* p. 267, October 1937.
91. Thiessen, A. *A History of the General Radio Company,* General Radio 1965.
92. von Ardenne, M. "A Double Beam Cathode-Ray Tube," *Electronics,* pp. 31–33, October 1936.
93. Sheffield, B. "Recent Developments in German Cathode-Ray Tubes," *Electronics,* p. 60, November 1937.
94. Turmail, R. L. "The Du Mont Legacy Lives On, Reflected in the Modern CRT," *Electron Des,* Vol. 24, pp. 126–127, November 22, 1973.
95. Settel, I. and Lass, W. *A Pictorial History of Television,* pp. 38–39, Grosset and Dunlap, 1969.
96. *Broadcasting* magazine, p. 89, June 15, 1981.
97. Henney, K. "Engineering Trends in Receiving Sets at Atlantic City," *Electronics,* p. 177, July 1930.
98. "An Electronic Motor," *Electronics,* p. 251, August 1930.
99. Du Mont, A. B. "Electron Turbine," U.S. Patent No. 1,999,407, filed January 31, 1930, issued April 30, 1935.

100. ———. "Synchronous Electron Motor," U.S. Patent No. 2,067,382, filed January 31, 1930, issued January 12, 1937.
101. ———. "Current Generator and Converter," U.S. Patent No. 2,082,327, filed March 25, 1935, issued June 1, 1937.
102. ———."Commutating Device," U.S. Patent No. 2,085,576, filed March 25, 1935, issued June 29, 1937.
103. *Time, Tubes and Television,* Allen B. Du Mont Laboratories, 1944.
104. Du Mont, A. B. "Cathode Ray Instrument for Measuring Electrical Quantities," U.S. Patent No. 1,960,333, filed October 7, 1932, issued May 29, 1934.
105. ———. "Cathode Ray Tube," U.S. Patent No. 2,087,280, filed November 5, 1934, issued July 20, 1937.
106. ———. "Cathode Ray Tube," U.S. Patent No. 2,163,256, filed September 18, 1934, issued June 20, 1939.
107. ———. "Cathode Ray Tube," U.S. Patent No. 2,186,635, filed November 7, 1938, issued January 9, 1940.
108. ———. "An Investigation of Various Electrode Structures of Cathode-Ray Tubes Suitable for Television Reception," *Proc IRE,* Vol. 20, pp. 1863–1877, December 1932.
109. Goldsmith, T. T. Interview, February 9, 1989.
110. ———. "Television Receivers Using Electrostatic Deflection," *Electronics,* pp. 16–19, 89, June 1940.
111. *Pioneering the Cathode-Ray and Television Arts,* Du Mont Laboratories, 1941.
112. Du Mont, A. B. and Goldsmith, T. T. "Television Broadcast Coverage," *Proc IRE,* pp. 192–205, April 1944.
113. Du Mont, A. B. "Method and System for Television Communication," U.S. Patent No. 2,186,634, filed January 26, 1938, issued January 9, 1940.
114. ———. "Cathode-Ray Tube Control Device for Television Scanning Apparatus," U.S. Patent No. 2,229,556, filed July 10, 1939, issued January 21, 1941.
115. Hess, G. N. *An Historical Study of the Du Mont Television Network,* Doctoral Dissertation, June 1960 (published by Arno Press, 1979).
116. Television/Radio Age, WNEW-TV 40th Anniversary supplement, pp. A3–A38, 1984.
117. Du Mont, A. B. "The Cathautograph," *Electronics,* p. 7, January 1933.
118. ———. "Teleautograph," U.S. Patent No. 2,000,014, filed May 31, 1932, issued May 7, 1935.
119. ———. "Cathautograph," U.S. Patent No. 2,185,705, filed November 30, 1935, issued January 2, 1940.
120. ———. "Voltmeter for Vacuum Tubes," U.S. Patent No. 2,014,106, filed September 19, 1932, issued September 10, 1935.
121. ———. "Cathode Ray Device," U.S. Patent No. 2,098,231, filed May 28, 1932, issued November 9, 1937.
122. ———. "Method and System for Television Communication," U.S. Patent No. 2,098,231, filed January 11, 1938, issued May 9, 1939.
123. ———. "Elimination of Pattern Distortion in Cathode-Ray Tubes," *Electronics,* pp. 16–17, January 1935.
124. ———. "System for Color Television Receivers," U.S. Patent No. 2,337,980, filed April 26, 1941, issued December 28, 1943.
125. Patterson, D. O. "Allen B. Du Mont Laboratories, Inc. Part I," *Radio Age,* Vol. 16, No. 5, pp. 1–2, 5, May 1989.
126. ———. "Allen B. Du Mont Laboratories, Inc. Part II," *Radio Age,* Vol. 16, No. 6, pp. 1–5, June 1989.
127. ———. "Allen B. Du Mont Laboratories, Inc. Part III," *Radio Age,* Vol. 16, No. 8, pp. 1–3, August 1989.
128. ———. "Allen B. Du Mont Laboratories, Inc. Part IV," *Radio Age,* Vol. 16, No. 10, pp. 8–10, October 1989.
129. ———. "Allen B. Du Mont Laboratories, Inc. Part V," *Radio Age,* Vol. 16, No. 12, pp. 1–4, December 1989.
130. McArthur, T. and Waddell, P. *The Secret Life of John Logie Baird,* Hutchenson, London, 1986.
131. Marchant, F. W. "A New Baird Colour-Television System," *Telev Shortwave World,* pp. 541–542, September 1939.
132. "New Television Development," *Electrician,* p. 359, December 26, 1941.
133. Baird, J. L. "Stereoscopic Television," *Electron Eng,* pp. 620–621, February 1942.

134. "Stereoscopic Colour Television," *Wireless World,* pp. 31–32, February 1942.
135. "J. L. Baird's Improved Colour Television," *Electron Eng,* p. 327, January 1943.
136. "Colour Television Development," *Wireless World,* p. 41, February 1943.
137. Parr, G. "Review of Progress in Colour Television," *J R Telev Soc,* Vol. 3, No. 10, pp. 251–256, 1943.
138. "Electronic Colour Television," *Electron Eng,* pp. 140–141, September 1944.
139. "J. L. Baird's 'Telechrome,'" *J R Telev Soc,* Vol. 4, pp. 58–59, September 1944.
140. "Baird 'Telechrome,'" *Wireless World,* Vol. 50, pp. 316–317, October 1944.
141. "New Baird Tube Gives Television in Color," *Electronics,* pp. 190–198, October 1944.
142. Dickenson, W. A. and Harrison, S. W. "A Short History of the Cathode-Ray Tube in Television, Part 2," *Sylvania Technol,* Vol. 5, no. 2, pp. 30–33, April 1952.
143. Seats, P. "The Evolutionary CRT," SID Conf. Lecture Notes, p. 7.29, 1987.
144. *Encyclopaedia Britannica,* Vol. 15, p. 500, 1962.
145. Patterson, D. "Du Mont: The Final Chapter," Part VII, *Radio Age,* Vol. 15, No. 6, pp. 1–4, June 1990.
146. Millikan, R. A. "A New Modification of the Cloud Method of Determining the Elementary Electrical Charge and the Most Probable Value of that Charge," *Philos Mag and J Sci,* 6th Series, Vol. 19, No. 110, pp. 209–229, February 1910.
147. Sterling, G. E. "Spies Use Radio—The Radio Intelligence Division in WWII," *AWA Rev,* Vol. 5, pp. 63–107, 1990.
148. von Ardenne, M. "An Experimental Television Receiver Using a Cathode-Ray Tube," *Proc IRE,* Vol. 23, No. 3, pp. 409–424, March 1936.
149. von Ardenne, M. "Der Elektronen-mikrooszillograph," Hochfrequenztechnik und Elektroakustik, Vol. 54, No. 6, pp. 181–188, December 1939.
150. Patterson, D. O., "Du Mont Gets Ready for Color," Part VI, *Radio Age,* Vol. 15, No. 3, pp. 1–4, March 1990.

CHAPTER 3

1. *Encyclopaedia Britannica,* Vol. 11, p. 525, 1962.
2. ———. Vol. 18, pp. 872–874, 1962.
3. Zahl, H. A. *Radar Spelled Backwards,* Vantage Press, 1972.
4. Ridenour, L. N. *Radar System Engineering,* MIT Radiation Laboratory Series, Vol. 1, McGraw-Hill, 1947.
5. Marconi, D. *My Father Marconi,* pp. 244–245, McGraw-Hill, 1962.
6. Marconi, G. "Radio Telegraphy," *Proc IRE,* p. 215, August 1922.
7. Page, R. M. *The Origin of Radar,* Anchor Books, Garden City, New York, 1962.
8. Skolnik, M. I. *Radar Applications,* pp. 1–29, IEEE Press, 1987.
9. Fink, D. G. "The SCR-268 Radar," *Electronics,* Vol. 18, pp. 100–109, September 1945.
10. Fink, D. G. "Radar Warfare," *Electronics,* Vol. 18, pp. 92–97, October 1945.
11. "Report on Wartime Electronic Developments," *Electronics,* Vol. 18, pp. 92–93, November 1945.
12. Fink, D. G. "The SCR-584 Radar, Part 1," *Electronics,* Vol. 18, pp. 104–109, November 1945.
13. ———. "The SCR-584 Radar, Part 2," *Electronics,* Vol. 18, pp. 104–109, December 1945.
14. ———. "The SCR-584 Radar, Part 3," *Electronics,* Vol. 19, pp. 110–117, February 1946.
15. ———. "Radar Specifications," *Electronics,* Vol. 18, pp. 116–119, November 1945.
16. "Radar Stories Are Released by U.S. and Great Britain," *Electronics,* pp. 274–282, June 1943.
17. Berkner, L. V. "Naval Airborne Radar," *Proc IRE,* pp. 671–706, September 1946.
18. Guerlac, H. E. *Radar in World War II, Vols. 1–2,* American Institute of Physics, 1987.
19. Dickenson, W. A. "A Short History of the Cathode-Ray Tube in Television, Part II," *Sylvania Technol,* Vol. 5, No. 2, pp. 30–33, April 1952.
20. Sollar, T., et al. *Cathode-Ray Tube Displays,* MIT Radiation Laboratory Series, Vol. 22, McGraw-Hill 1948.
21. Helgeson, D. "VT-Numbered Tubes and Early U.S. Army Radar," *AWA J,* Vol. 30, No. 3, November 1989.
22. Swedlund, L. E. "Improved Electron Gun for C-R Tubes," *Electronics,* pp. 122–124, March 1945.
23. *Reference Data for Radio Engineers,* 4th ed., International Telephone and Telegraph Corp., pp. 806–807, 1956.
24. "Polar Coordinate Oscillograph," *Electronics,* p. 46, December 1936.

25. Puckle, O. S. "Cathode-Ray Tubes," *Wireless World,* pp. 275–278, March 23, 1939.
26. Rider, J. F. and Uslan, S. D. *Encyclopedia on Cathode-Ray Oscilloscopes and Their Uses,* John F. Rider Publisher Inc., 1950, 1955, 1959.
27. Jesty, L. C., Moss, H. and Puleston, R. "War-Time Developments in Cathode-Ray Tubes for Radar," *J IEE,* Vol. 93-3A, pp. 149–166, 1946.
28. *Cathode-Ray Tubes and Their Associated Circuits,* Department of the Army Technical Manual TM 11-671/TO 16-45-227, September 1951.
29. Hohlbauch, R. E. Interview on December 27, 1988.
30. Leverenz, H. W. An Introduction to Luminescence in Solids, John Wiley and Sons, 1950.
31. Leverenz, H. W. "Luminescence and Tenebrescence as Applied to Radar," *RCA Rev,* Vol. 7, No. 2, pp. 199–239, June 1946.
32. Nozick, S. "Recent Dark-Trace Tube Developments," *IRE Trans Elec Dev,* pp. 66–70, January 1956.
33. Rottgardt, K. H. J. and Heller, G. "Rectangular Dark-Trace Tube," *Electronics,* pp. 208–216, June 1, 1957.
34. Brain, J. Interview of December 4, 1989.
35. Pensak, L. "The Graphechon—A Picture Storage Tube," *RCA Rev,* pp. 59–73, March 1949.
36. McNaney, J. T. "Data Displaying Cathode-Ray Tube," *Electronics,* pp. 174–186, August 1952.
37. Knoll, M., Rudnick, P. and Hook, H. O. "Viewing Storage Tube with Halftone Storage," *RCA Rev,* Vol. 14, pp. 492–501, December 1953.
38. Hook, H. O., Knoll, M. and Stone, R. P. "Viewing Storage Tubes for Large Displays," *RCA Rev,* pp. 503–514, December 1956.
39. George, D. E. "The Wamoscope: A Microwave Display Device," *Sylvania Technol,* Vol. 10, pp. 5–7, January 1957.
40. Skolnik, M. I. *Introduction to Radar Systems,* p. 394, McGraw-Hill, 1962.
41. Koller, L. R. and Williams, F. E. "Production of Colored Images," U.S. Patent No. 2,590,018, filed October 24, 1950, issued March 18, 1952.
42. Hallett, J. and Rhodes, C. "Beam-Penetration Cathode-Ray Tube Display," *Proc SID,* Vol. 10, No. 4, pp. 9–14, Fall 1969.
43. Martin, A. "Multicolor and Multipersistence Penetration CRTs," *SID J,* March/April 1974.
44. Bartlett, N. J. "The Penetron Colour Display," *Elec Eng,* pp. 57–62, April 1984.
45. Kell, R. D. "Luminescent Screen Comprising Phosphor Cores Luminescent in First Color and Phosphor Coatings Luminescent in Second Color," U.S. Patent No. 3,714,490, filed August 24, 1964, issued January 30, 1973.
46. Galves, J. P. "High Brightness Voltage Sensitive Screens," Electrochemical Society, paper given in Las Vegas, Nevada, 1975.
47. Trimmier, R. and Knutson, G. "High Contrast Thin-Film Color Penetron," *SID Dig,* pp. 90–91, 1981.
48. Robinder, R. C. and Fogelson, M. "Two-Color Display Using Dual-Neck Flat-Face Cathode Ray Tube," E-O Systems Design Conf., pp. 332–345, 1970.
49. Rawdin, E., Schwartz, J. and Smilowitz, S. "Multigun CRT Color Display," *SID Dig,* pp. 94–95, 1981.
50. Seats, P. "The Evolutionary CRT," SID Conf. Lecture Notes, p. 7.16, 1987.
51. Hunt, V. R. "The Next-Generation Air Traffic Control Display," *SID Digest,* pp. 20–22, 1986.
52. "New Generation Color Displays Planned for Flight Control System," *Inf Disp,* pp. 18–20, September 1986.
53. "Zenith Zaps FAA Award," *Elect Eng Times,* p. 1, March 27, 1989.
54. Sanders, F. H. "Radar Development in Canada," *Proc IRE,* pp. 195–200, February 1947.
55. Strebeigh, F. "How England Hung the 'Curtain' that Held Hitler at Bay," *Smithsonian Mag,* Vol. 21, No. 4, pp. 120–129, July 1990.
56. Burns, R. W. "Aspects of UK Air Defence from 1914 to 1935: Some Unpublished Admiralty Contributions," *IEE Proc,* Vol. 136, Part A, No. 6, November 1989.
57. Knoll, M., Hook, H. O. and Stone, R. P. "Characteristics of a Transmission Control Viewing Storage Tube with Halftone Display," *Proc. IRE,* pp. 1496–1504, October 1954.
58. "Wave-Modulated Oscilloscope for TV," *Electronics,* p. 232, October 1956.
59. Briskin, H. "Typical Applications of the Wamoscope," *Sylvania Technol,* Vol. 10, pp. 8–9, January 1957.

CHAPTER 4

1. Batcher, R. R. "Commercial Cathode-Ray Tubes," *Instruments,* p. 267, October 1937.

2. Ruiter, J. H. *Modern Oscilloscopes and Their Uses,* p. 22, Rinehart, 1958.
3. Rider, J. F. *The Cathode-Ray Tube at Work,* John F. Rider Publisher Inc., 1935.
4. von Ardenne, M. *Cathode Ray Tubes,* Pitman and Sons, London, 1939.
5. Bedell, F. and Reich, H. J. "The Oscilloscope: A Stabilized Cathode-Ray Oscillograph with Linear Time-Axis," *J AIEE,* Vol. 46, p. 563, June 1927.
6. Metcalf, G. F. "A New Cathode-Ray Oscillograph Tube," *Electronics,* pp. 158–159, May 1932.
7. Kuehni, H. P. "A Low-Voltage Cathode-Ray Oscillograph with Fluorescent Viewing Screen," *Gen Electr Rev,* Vol. 37, No. 9, pp. 429–432, September 1934.
8. Sutherlin, L. and Harcher, A. J. "Cathode-Ray Tubes," *Electr J,* Vol. 29, No. 8, pp. 388–389, August 1932.
9. Appleton, V. *Tom Swift and His Television Detector,* p. 54, Whitman, 1933.
10. JEDEC Registration Files, Elec. Ind. Assoc., Washington, D.C.
11. "The 902—A New 2-inch Cathode-Ray Tube," *Radio World,* pp. 36–40, May 1938.
12. "Television Components," *Electronics,* p. 60, May 1939.
13. "New Small Cathode-Ray Tube," *Radio World,* pp. 46–50, December 1936.
14. Stocker, A. C. "Oscillograph for Television Development," *Proc IRE,* Vol. 25, No. 8, pp. 1012–1033, August 1937.
15. Du Mont advertisement, *Electronics,* p. 46, June 1937.
16. "New Type Numbers for Du Mont Cathode-Ray Tubes," *Du Mont Oscillographer,* Vol. 5, No. 10, pp. 1–3, December 1942.
17. Rider, J. F. and Uslan, S. D. *Encyclopedia on Cathode-Ray Oscilloscopes and Their Uses,* Chap. 1, John F. Rider Publisher Inc., 1959.
18. Swedlund, L. E. "Improved Electron Gun for C-R Tubes," *Electronics,* pp. 122–124, March 1945.
19. Sollar, T., et al. *Cathode-Ray Tube Displays,* MIT Radiation Laboratory Series, Vol. 22, pp. 61–62, 1948.
20. Feldt, R. "High Speed Photographic Recording and Projection Oscillography with the New Du Mont Type 5RP Multi-Band Tube," *Du Mont Oscillographer,* pp. 1–4, July–August 1945.
21. ———. "Projection C-R-O Tube," Electronic Industries, September 1945.
22. Lempert, I. and Feldt, R. "The 5RP Multiband Tube: An Intensifier-Type Cathode-Ray Tube for High-Voltage Operation," *Proc IRE,* pp. 432–440, July 1946.
23. *Catalog, Du Mont Cathode-Ray Equipment,* Allen B. Du Mont Laboratories, p. 140, April 1949.
24. Hoagland, K. A. "Cathode-Ray Tube for Wideband Oscillography," *Communications,* May 1949.
25. von Ardenne, M. "A Double Beam Cathode-Ray Tube," *Electronics,* pp. 31–33, October 1936.
26. Fleming-Williams, B. C. "The Double Beam Cathode-Ray Oscillograph," *Telev Short Wave World,* Vol. 12, p. 457, August 1939.
27. Rider, J. F. and Uslan, S. D. *Encyclopedia on Cathode-Ray Oscilloscopes and Their Uses,* Chap. 6, John F. Rider Publisher Inc., 1950.
28. ———. Chapter 4.
29. Advertisement, *Electronics,* p. 37, January 1938.
30. Smits, F. M. *A History of Engineering and Science in the Bell System: Electronics Technology (1925–1975),* AT&T Bell Laboratories, pp. 142–145, 1985.
31. Advertisement, *Electronics,* p. 36, September 1945.
32. Grossbohlin, H. W. and Hoagland, K. A. "Improved Instrument Cathode-Ray Design," *Proc Natl Elec Conf,* Vol. 9, pp. 414–421, February 1954.
33. Day, J. E. "Recent Developments in the Cathode-Ray Oscilloscope," *Advances in Electronics and Electron Physics,* L. Marton, Ed., pp. 239–298, Academic Press, 1958.
34. "A New 50 MC Oscilloscope Based on an Advanced CRT," *Hewlett-Packard J,* pp. 1–8, April 1962.
35. "A New DC-50 MHz Transistorized Oscilloscope of Basic Instrumentation Character," *Hewlett-Packard J,* pp. 1–12, August 1966.
36. Hollmann, H. E. "Ultra-High-Frequency Oscillography," *Proc IRE,* pp. 213–219, May 1940.
37. Haeff, A. V. "Device for and Method of Controlling High-Frequency Currents," U.S. Patent No. 2,064,469, filed October 23, 1936, issued December 15, 1936.
38. Pierce, J. R. "Travelling-Wave Oscilloscope," *Electronics,* pp. 97–99, November 1949.
39. Lee, G. M. "A Three-Beam Oscillograph for Recording at Frequencies Up to 10,000 Megacy-

cles," *Proc IRE,* Vol. 34, pp. 121W-127W, March 1946.
40. Owaki, K., Terahata, S., Hada, T. and Nakamura, T. "The Traveling-Wave Cathode-Ray Tube," *Proc IRE,* Vol. 38, pp. 1172–1180, October 1950.
41. Smith, S. T., Talbot, R. V. and Smith, C. H. "Cathode-Ray Tube for Recording High-Speed Transients," *Proc IRE,* Vol. 40, pp. 297–302, March 1952.
42. Germeshausen, K. J., Goldberg, S. and McDonald, D. F. "A High-Sensibility Cathode-Ray Tube for Millimicrosecond Transients," IRE Trans. on electron devices ED-4, p. 152–158, April 1957.
43. *Cathode-Ray Tubes,* Circuit Concept Series, Tektronix, Inc., pp. 27–36, 1967 (out of print).
44. *Cathode-Ray Tubes: Getting Down to Basics,* Tektronix, Inc., 1985.
45. *Cathode-Ray Tubes,* Circuit Concept Series, Tektronix, Inc., pp. 30–35, 1967 (out of print).
46. Janko, B. "A New High-Speed CRT," *Proc SID,* Vol. 20, No. 2, 1979.
47. "A Parallax-Free No-Glare CRT for H-P Oscilloscopes," *Hewlett-Packard J,* pp. 1–3, July/August 1961.
48. Wilbanks, W. H., Johnson, R. H., Wang, C. H. and Erzen, F. C. "Electron Beam Display Device," U.S. Patent No. 3,207,936, filed August 21, 1961, issued September 21, 1965.
49. Wilbanks, W. H. "Lead-in Seal for Evacuated Envelope of an Electron Discharge Device," U.S. Patent No. 3,275,878, filed February 27, 1963, issued September 27, 1966.
50. Hawes, J. D. and Nelson, G. "Mesh PDA Lenses in Cathode-Ray Tubes," *Proc SID,* Vol. 19, No. 1, pp. 17–21, 1978.
51. Perkins, P. "The Design and Development of a Non-Compressing Post-Deflection Acceleration Method for Cathode-Ray Tubes," Master's Thesis, Oregon State University, 1968.
52. Gass, G. "CRT-Design Review Mesh and Frame-Grid Characteristics," Tektronix Service Scope, pp. 1–3, October 1963.
53. Grein, W., Russell, M. and Sverdrup, J. "The Gamble That Paid Off: A Solid State 50 MHz Oscilloscope," *Electronics,* pp. 95–99, July 25, 1966.
54. Chaffee, D. "A Fast-Writing, High Frequency Cathode Ray Tube," *H-P J,* pp. 9–10, January 1970.
55. Pettit, J. "A DC-to-VHF Oscilloscope," *H-P J,* pp. 2–8, January 1970.
56. Russell, M. E. "Factors in Designing a Large-Screen, Wideband CRT," *H-P J,* pp. 10–11, December 1967.
57. Odenthal C. "A Box-Shaped Scan Expansion Lens for an Oscilloscope CRT," *SID Dig,* pp. 134–135, 1977.
58. Odenthal, C. J. "The Scan Magnification Lens Used in the Microchannel Plate CRT," Electro-Optical Systems Design, pp. 27–34, August 1979.
59. ———. "Box-Shaped Scan Expansion Lens for Cathode Ray Tube," U.S. Patent No. 4,142,128, filed April 18, 1977, issued February 27, 1979.
60. Klemperer, *Electron Optics,* pp. 295–296, Cambridge University Press, 1953.
61. El-Kareh, A. B. "High-Speed Oscilloscope with Electron-Optical Magnification Using Four-Pole Lenses," *Rev Sci Inst,* Vol. 32, No. 4, pp. 421–424, April 1961.
62. Zeppenfeld, K. "Oscilloscope Tube with Magnetic Correction and Scan Magnification," *Elec Compon Appl,* Vol. 4, No. 3, pp. 185–188, May 1982.
63. Janko, B. "Cathode-Ray Tube Having an Electron Lens System Including a Meshless Scan Expansion Post-Deflection Acceleration System," U.S. Patent No. 4,137,479, filed January 6, 1977, issued January 30, 1979.
64. ———. "Cathode Ray Tube Having an Electron Lens System Including a Meshless Scan Expansion Post Deflection Acceleration Lens," U.S. Patent No. 4,188,563, filed August 31, 1978, issued February 12, 1980.
65. Franzen, N. "A Quadrupole Scan-Expansion Accelerating Lens System for Oscilloscopes," *SID Dig,* pp. 120–121, 1983.
66. Janko, B., Franzen, N. R. and Bostwick, M. A. "Accelerating and Scan Expansion Electron Lens Means for a Cathode Ray Tube," U.S. Patent No. 4,623,819, filed August 12, 1985, issued November 18, 1986.
67. Janko, B., Franzen, N. R. and Sonneborn, J. "CRT Architecture with Meshless Scan Expansion," *SID Dig,* pp. 118–119, 1983.
68. Franzen, N. and Janko, B. "Meshless Scan Expansion CRT Improves Trace Quality," *Electron Imaging,* pp. 48–54, August 1983.

69. Hawken, K. W., Franzen, N. and Sonneborn, J. H. "A Novel High-Performance CRT with Meshless Scan Expansion," Japan Display '83, pp. 136–139, 1983.
70. Sonneborn, J. H. and Hawken, K. W. "Design of a Microchannel Plate CRT for a General Purpose Oscilloscope," *SID Dig,* pp. 240–243, 1986.
71. Janko, B. "CRT for Displaying Nanosecond Transients," *SID Dig,* pp. 68–69, 1979.
72. Bristol, R. and Hillman, A. K. "Portable Analog Scope Bares Signals Hidden Even from Digital Storage Units," *Electron Des,* pp. 125–130, March 20, 1986.
73. Kazan, B. and Knoll, M. *Electronic Image Storage,* p. 1, Academic Press, 1968.
74. Austin, L. and Starke, H. "Uber die Reflexion der Kathodenstrahlen und eine Damit Verbundene Neue Erscheinung Sekundarer Emission," *Ann. Physik,* Vol. 9, pp. 272–292, 1902.
75. Haeff, A. V. "A Memory Tube," *Electronics,* pp. 80–83, September 1947.
76. Smith, S. T. and Brown, H. E. "Direct-Viewing Memory Tube," *Proc IRE,* pp. 1167–1171, September 1953.
77. "Memotron Direct Display Cathode-Ray Storage Tube," Technical Data, Hughes Products, Undated.
78. Lee, M. M. *Winning with People: The First Forty Years of Tektronix,* Tektronix, Inc., 1986.
79. Anderson, R. H. "Storage Tube and Electrical Readout Apparatus and Method for Such Tube," U.S. Patent No. 3,214,516, filed December 19, 1962, issued October 26, 1965.
80. ———. "Tube Having Target with Store and/or Non-Store Sections," U.S. Patent No. 3,214,631, filed August 6, 1962, issued October 26, 1965.
81. ———. "Principles of Direct-Viewing Storage Tubes," Tektronix, Inc., 1961.
82. ———. "A Simplified Direct-Viewing Storage Tube," *IEEE Trans on Elec Dev,* Vol. ED-14, No. 12, pp. 838–844, December 1967.
83. Kazan, B. and Knoll, M. *Electronic Image Storage,* pp. 225–230, Academic Press, 1968.
84. Curtin, C. "Recent Advances in Direct-View Storage Tubes," *SID Dig,* pp. 132–133, 1977.
85. DeVere, C. *Storage Cathode-Ray Tube and Circuits,* Circuit Concept Series, Tektronix, Inc., 1971 (out of print).
86. Kolar, R. H. "Variable Persistence Increases Oscilloscope's Versatility," *Electronics,* pp. 66–70, November 29, 1965.
87. Donaldson, C. A. and Gustafson, C. A. "Easier and Brighter Display of High-Frequency Signals," *Hewlett-Packard J,* pp. 10–15, May 1968.
88. Harrison, V., Carnahan, P. and Haley, J. F. "Variable-Persistence Storage CRT Achieves High Writing Speed," *Electronics,* pp. 110–114, August 3, 1978.
89. Edens, G. "A Bistable Storage Tube Scan Converter," *SID Dig,* pp. 60–61, 1970.
90. Sang, E. "The Signal-Generation Mechanism in Bistable Storage-Scan Converters," *SID Dig,* pp. 104–105, 1973.
91. Loty, C. "A 7 GHz CRT for Realtime Digital Oscilloscopy," *SID Dig,* pp. 126–127, 1983.
92. Campbell, D. E. "A Review of Single Transient Oscillographic Recorders with Gigahertz Bandwidth," *IEEE Trans on Nucl Sci,* Vol. NS-30, No. 1, pp. 267–271, February 1983.
93. Sollar, T., et al. *Cathode-Ray Tube Displays,* MIT Radiation Laboratory Series, Vol. 22, pp. 73–75, 1948.
94. "Three New Instruments, Three Kinds of Storage," *Tekscope,* Vol. 4, No. 4, pp. 1–6, July 1972.
95. Woodhead, A. W. and Eschard, G. "Microchannel Plates and their Applications," *Acta Electron,* Vol. 14, No. 2, pp. 181–200, 1971.
96. Janko, B. "Space-Charge Effects in a CRT with Post-Deflection Acceleration," *IEEE Trans on Elec Dev,* Vol. ED-21, No. 3, pp. 184–189, March 1974.

CHAPTER 5

1. Zworykin, V. K. "Vacuum Tube," U.S. Patent No. 2,109,245, filed November 16, 1929, issued February 22, 1938.
2. ———. "Television with Cathode-Ray Tube for Receiver," *Radio Eng,* pp. 38–41, December 1929.
3. Sutherlin, L. and Harcher, A. J. "Cathode-Ray Tubes," *Electr J,* Vol. 29, No. 8, pp. 388–389, August 1932.
4. Zworykin, V. K. "Improvements in Cathode-Ray Tube Design," *Electronics,* pp. 188–190, November 1931.
5. ———. "Television System," U.S. Patent No.

2,084,364, filed December 24, 1930, issued June 22, 1937.
6. ———. "Cathode-Ray Apparatus," U.S. Patent No. 1,988,469, filed June 30, 1931, issued January 22, 1935.
7. *RCA Cathode-Ray Tubes and Allied Types,* RCA, Technical Series TS-2, 1935.
8. "Television Tubes," *Electronics,* p. 71, November 1937.
9. Burnap, R. S. "Television Cathode-Ray Tubes for the Amateur," *RCA Rev,* pp. 297–302, January 1938.
10. Wilder, M. P. "Television in Europe," *Electronics,* pp. 13–15, September 1937.
11. "Television European Style," *Electronics,* p. 32, March 1937.
12. McArthur, T. and Waddell, P. *The Secret Life of John Logie Baird,* Hutchinson, London, 1986.
13. Goldsmith, T. T. Interview, February 9, 1989.
14. Fink, D. G. "Television Broadcasting Practice in America—1927 to 1944," *J IEE,* Vol. 92, No. 19, pp. 145–164, September 1945.
15. ———. "Perspectives on Television: The Role Played by the Two NTSCs in Preparing Television Service for the American Public," *Proc IEEE,* Vol. 64, No. 9, pp. 1322–1331, September 1976.
16. Lavin, H. "NTSC Part 1: How the NTSC Standard Developed," Electro-Optical Systems Design, pp. 64–65, June 1980.
17. Bingley, F. J. "A Half Century of Television Reception," *Proc IRE,* pp. 799–804, May 1962.
18. Marshall, C. J. and Katz, L. "Television Equipment for Guided Missiles," *Proc IRE,* pp. 375–401, June 1946.
19. Dickenson, W. A. and Harrison, S. W. "A Short History of the Cathode-Ray Tube in Television, Part II," *Sylvania Techn,* Vol. 5, No. 2, pp. 30–33, April 1952.
20. Wright, A. "Television Receivers," *RCA Rev,* pp. 5–28, March 1947.
21. "Television Receivers in Mass Production," *Electronics,* pp. 86–91, June 1947.
22. Bachman, C. H. "Image Contrast in Television," *GE Rev,* pp. 13–19, September 1945.
23. Epstein, E. W. and Pensak, L. "Improved Cathode-Ray Tubes with Metal-Backed Luminescent Screens," *RCA Rev,* Vol. 7, No. 1, pp. 5–10, March 1946.
24. Gulliland, A. A. "Television in Germany," *Telev Short-Wave World,* pp. 538–539, September 1939.
25. Brown, W. J. "British Vision Receivers—I," *Electronics,* pp. 26–29, December 1939.
26. ———. "British Vision Receivers—II," *Electronics,* pp. 26–29, March 1940.
27. Zworykin, V. K. and Painter, W. H. "Development of the Projection Kinescope," *Proc IRE,* pp. 311–325, August 1937.
28. Law, R. R. "High Current Electron Gun for Projection Kinescopes," *Proc IRE,* Vol. 25, pp. 954–976, August 1937.
29. Epstein, D. W. and Malof, I. G. "Projection Television," *J SMPE,* Vol. 44, No. 6, pp. 443–455, June 1945.
30. "New Large Screen RCA Television Receiver," *Radio News,* pp. 98, 100, June 1945.
31. Bradley, W. E. and Traub, E. "A New Television Projection System," *Electronics,* pp. 84–89, September 1947.
32. Lytel, A. *TV Picture Projection and Enlargement,* John F. Rider Publisher Inc., 1949.
33. Rinia, H., de Gier, J. and van Alphen, P. M. "Home Projection Television: Part 1," *Proc IRE,* Vol. 36, No. 3, pp. 395–400, March 1948.
34. Boyle, H. G. and Doll, E. B. "Compact Projection Television System," *Electronics,* pp. 72–77, April 1948.
35. Lytel, A. "Projection Television," Radio Service Dealer, pp. 13, October 1948.
36. Robinson, D. M. "The Supersonic Light Control and Its Application to Television with Special Reference to the Scophony Television Receiver," *Proc IRE,* pp. 483–486, August 1939.
37. Sieger, J. "The Design and Development of Television Receivers Using the Scophony Optical System," *Proc IRE,* pp. 487–492, August 1939.
38. Maloff, I. G. and Epstein, D. W. "Resumé of the Technical Aspects of RCA Theater Television," *RCA Rev,* Vol. 6, No. 1, pp. 5–11, July 1941.
39. Patrick, N. W. "Developments in Cathode-Ray Tubes for the Schmidt Projector," Proc. IEEE Conf. on Disp. Dev., pp. 109–115, October 1972.
40. Baumann, E. "The Fischer Large-Screen Projection System," *J SMPTE,* pp. 344–356, April 1953.
41. Chafaris, G. J. "Display Technology—Today and Tomorrow: Part I—Light Valve Techniques," *Proc SID,* Vol. 10, No. 1, pp. 3–8, Winter 1969.

42. Donal, J. S. "Cathode-Ray Control of Television Light Valves," *Proc IRE,* pp. 195–208, May 1943.
43. ———. "A Type of Light Value for Television Reproduction," *Proc IRE,* pp. 208–214, May 1943.
44. van Raalte, J. A. "A New Light Value for Television Projection," *J SMPTE,* Vol. 80, pp. 474–476, June 1971.
45. Good, W. E. "Recent Advances in the Single-Gun Color Television Light-Valve System," *SID Dig,* pp. 24–25, 1975.
46. Jacobson, A. D., Grinberg, J., Bleha, W. P., Miller, L. J., Frass, L. M. and Boswell, D. D. "A New Television Projection Light Valve," *SID Dig,* pp. 26–27, 1975.
47. Haven, D. "Electron-Beam Addressed Liquid-Crystal Light Valve," *IEEE Trans Elec Dev,* Vol. ED-30, No. 5, pp. 489–492, May 1983.
48. Haven, D., Whitlow, D. E. and Jones, T. "Full-Color Electron-Beam-Addressed Light-Valve Projector," *SID Dig,* pp. 372–374, 1986.
49. Buzak, T. S., Vatne, R. S., Nelson, G. A. and Whitlow, D. E. "Electron-Beam-Addressed Light Valve Operating Modes and Performance," *SID Dig,* pp. 64–67, 1987.
50. Felgenblatt, R. I. "Projection Display Technology," *J Imaging Technol,* Vol. 13, No. 4, pp. 108–115, August 1987.
51. Swedlund, L. E. and Thierfelder, C. W. "Projection Kinescope 7NP4 for Theater Television," *J SMPTE,* Vol. 56, pp. 332–342, March 1951.
52. Steier, H. P. et al. "Development of a Large Metal Kinescope for Television," *RCA Rev,* pp. 43–58, March 1949.
53. Rose, A. S. and Turnbull, J. C. "The Evaluation of Chromium-Iron Alloys for Metal Kinescope Cones," *RCA Rev,* pp. 593–599, December 1949.
54. Swedlund, L. E. and Steier, H. P. "Short 16-inch Metal-Cone Kinescope Development," Tele-Tech, p. 40, August 1950.
55. Swedlund, L. E. and Saunders, R. "Material-Saving Picture Tube," *Electronics,* pp. 119–120, April 1951.
56. McQueen, J. Interview, June 1989.
57. "German Television Receiver," *Electronics,* p. 50, October 1939.
58. Dickenson, W. A. "Monochrome Picture Tubes," *IEEE Trans Bcst Tel Recvrs,* Vol. BTR-13, No. 3, pp. 44–48, November 1967.
59. "Characteristics—Television Picture Tubes," Sylvania Electric, Publication ET-1925, April 1967.
60. "Picture Tube Product Guide," RCA, Publication PIX-300H, September 1974.
61. Cunningham, T. M. and Trim, M. E. "Evolution of TV Picture Tubes at RCA," *RCA Picture Tubes,* pp. 10–13, RCA, 1980.
62. "The 8YP4 as a 'Universal' Check Tube," *Sylvania News,* Vol. 27, No. 7, November/December 1960.
63. Beers, G. L., Schade, O. H. and Shelby, R. E. "The RCA Portable Television Pickup Equipment," *Proc IRE,* pp. 450–458, October 1940.
64. Trainer, M. A. "Orthicon Portable Television Equipment," *Proc IRE,* pp. 15–19, January 1942.
65. Beers, G. L. "The Focusing View-Finder Problem in Television Cameras," *Proc IRE,* pp. 100–106, March 1943.
66. *Pioneering the Cathode-Ray and Television Arts,* Du Mont Labs, September 1941.
67. Campbell, R. L., Kessler, R. E., Rutherford, R. E. and Landsburg, K. U. "Mobile Television Equipment," *Proc IRE,* Vol. 30, No. 1, pp. 1–7, January 1942.
68. Fink, D. G. *Television Engineering,* pp. 574–577, McGraw-Hill, 1952.
69. "A New TV Waveform Oscilloscope for Precision Measurement on Video Test Signals," *H-P J,* Vol. 17, No. 6, pp. 1–6, February 1966.
70. Jensen, A. G. "The Davisson Cathode Ray Television Tube using Deflection Modulation," *Bell System Tech J,* Vol. 30, No. 4, pp. 855–866, October 1951.
71. Darrow, K. K. "The Scientific Work of C. J. Davisson," *Bell System Tech J,* Vol. 30, No. 4, pp. 786–797, October 1951.
72. Kelly, M. J. "Dr. C. J. Davisson," *Bell System Tech J,* Vol. 30, No. 4, pp. 778–785, October 1951.
73. Calbick, C. J. U.S. Patent No. 2,312,206.
74. "Rochester 1937," *Electronics,* p. 13, December 1937.
75. Labin, E. "The Eidophor Method for Theater Television," *J SMPTE,* Vol. 54, pp. 393–406, April 1950.
76. Dornberg, J. "East Germany's Technology Baron," *Technol Illustr,* pp. 46–51, July 1983.
77. "Radio Progress During 1946," *Proc IRE,* Vol. 35, No. 4, pp. 414–415, April 1947.

78. Doyle, R. J. and Glenn, W. E. "Lumatron: A High-Resolution Storage and Projection Display Device," *IEEE Trans Elec Dev,* Vol. ED. 18, No. 9, pp. 739–747, September 1971.
79. Salvo, C. J. "Solid State Light Value," *IEEE Trans Elec Dev,* Vol. ED. 18, No. 9, pp. 748–755, September 1971.

CHAPTER 6

1. Herold, E. W. "A History of Color Television Displays," *Proc IRE,* Vol. 64, No. 9, pp. 1331–1338, September 1976.
2. Goldmark, P. C., Dyer, J. N., Piore, E. R. and Hollywood, J. M. "Color Television—Part I," *Proc IRE,* Vol. 30, pp. 162–182, April 1942, also Part II, Vol. 31, September 1943.
3. Zworykin, V. K. "Television System," U.S. Patent No. 1,691,324, filed July 13, 1925, issued November 13, 1928.
4. Parr, G. "Review of Progress in Colour Television," *R Telev Soc J,* Vol. 3, No. 10, pp. 251–256, 1943.
5. Marchant, F. W. "A New Baird Colour-Television System," *Television Short Wave World,* Vol. 12, pp. 541–542, September 1939.
6. "New Television Development," *Electrician,* p. 359, December 26, 1941.
7. Baird, J. L. "Stereoscopic Television," *Electron Eng,* pp. 620–621, February 1942.
8. "Stereoscopic Colour Television," *Wireless World,* pp. 31–32, February 1942.
9. "J. L. Baird's Improved Colour Television," *Electronic Engineering,* Vol. 15, p. 327, January 1943.
10. "Colour Television Development," *Wireless World,* Vol. 49, No. 2, p. 41, February 1943.
11. "J. L. Baird's Telechrome," *R Televi Soc J,* Vol. 4, pp. 58–59, September 1944.
12. "Electronic Colour Television," *Electron Eng,* pp. 140–141, September 1944.
13. "Baird 'Telechrome,'" *Wireless World,* Vol. 50, pp. 316–317, October 1944.
14. "New Baird Tube Gives Television in Color," *Electronics,* pp. 191, 194, 198, October 1944.
15. Kiver, M. S. *Television Simplified,* Chapter 12, Van Nostrand, 1946.
16. Fink, D. G. *Television Engineering,* Chapter 8, McGraw-Hill, 1952.
17. Fink, D. G. "Perspectives on Television: The Role Played by the Two NTSC's in Preparing Television Service for the American Public," *Proc IRE,* Vol. 64, No. 9, pp. 1322–1331, September 1976.
18. Goldsmith, T. T. Interview, February 9, 1989.
19. "Simultaneous All-Electronic Color Television a Progress Report, *RCA Rev,* Vol. 7, No. 4, pp. 459–468, December 1946.
20. Wendt, K. R., Fredendall, G. L. and Schroeder, A. C. "Part III—Radio Frequency and Reproducing Equipment," *Proc IRE,* pp. 871–875, September 1947.
21. "A Six-Megacycle Compatible High-Definition Color Television System," *RCA Rev,* pp. 504–524, December 1949.
22. Geer, C. W. "Color Television Device," U.S. Patent No. 2,480,848, filed July 11, 1944, issued September 6, 1949.
23. "Experimental C-R Tubes for Television," *Electronics,* pp. 112–115, March 1947.
24. "No Longer a Black-and-White Issue, 1950," *Wall St J,* p. B1, July 19, 1989.
25. Law, H. B. "The Shadow-Mask Color Picture Tube: How it Began," *RCA Picture Tubes,* pp. 3–9, RCA, 1980.
26. Herold, E. W. "Methods Suitable for Television Color Kinescopes," *RCA Rev,* Vol. 12, No. 3, pp. 445–465, September 1951.
27. ———. "Methods Suitable for Television Color Kinescopes," *Proc IRE,* Vol. 39, pp. 1177–1185, October 1951.
28. Law, R. R. "A One-Gun Shadow-Mask Color Kinescope," *RCA Rev,* Vol. 12, No. 3, pp. 487–502, September 1951.
29. ———. "A One-Gun Shadow-Mask Color Kinescope," *Proc IRE,* Vol. 39, pp. 1194–1201, October 1951.
30. Law, H. B. "A Three-Gun Shadow-Mask Color Kinescope," *RCA Rev,* Vol. 12, No. 3, pp. 466–486, September 1951.
31. Freedman, N. S. and McLaughlin, K. M. "Phosphor-screen Application in Color Kinescopes," *RCA Rev,* Vol. 12, No. 3, pp. 568–582, September 1951.
32. Moodey, H. C. and Van Ormer, D. D. "Three-Beam Guns for Color Kinescopes," *RCA Rev,* Vol. 12, No. 3, pp. 583–592, September 1951.

33. Barnes, B. E. and Faulkner, R. D. "Mechanical Design of Aperture-Mask Tri-Color Kinescopes," *RCA Rev,* Vol. 12, No. 3, pp. 593–602, September 1951.
34. Van Ormer, D. D. and Ballard, D. C. "Effects of Screen Tolerances on Operating Characteristics of Aperture-Mask Tri-Color Kinescopes," *RCA Rev,* Vol. 12, No. 3, pp. 603–611, September 1951.
35. Friend, A. W. "Deflection and Convergence in Color Kinescopes," *RCA Rev,* Vol. 12, No. 3, pp. 612–644, September 1951.
36. Law, H. B. "A Three-Gun Shadow-Mask Color Kinescope," *Proc IRE,* Vol. 39, No. 10, pp. 1186–1194, October 1951.
37. Freedman, N. S. and McLaughlin, K. M. "Phosphor-Screen Application in Color Kinescopes," *Proc IRE,* Vol. 39, pp. 1230–1236, October 1951.
38. Moodey, H. C. and Van Ormer, D. D. "Three-Beam Guns for Color Kinescopes," *Proc IRE,* Vol. 39, pp. 1236–1240, October 1951.
39. Barnes, B. E. and Faulkner, R. D. "Mechanical Design of Aperture-Mask Tri-Color Kinescopes," *Proc IRE,* Vol. 39, pp. 1241–1245, October 1951.
40. Van Ormer, D. D. and Ballard, D. C. "Effects of Screen Tolerances on Operating Characteristics of Aperture-Mask Tri-Color Kinescopes," *Proc IRE,* Vol. 39, pp. 1245–1249, October 1951.
41. Friend, A. W. "Deflection and Convergence in Color Kinescopes," *Proc IRE,* Vol. 39, pp. 1249–1264, October 1951.
42. Forgue, S. V. "A Grid-Controlled Color Kinescope," *Proc IRE,* Vol. 39, pp. 1212–1218, October 1951.
43. Forgue, S. V. "A Grid-Controlled Color Kinescope," *RCA Rev,* Vol. 12, No. 3, pp. 583–592, September 1951.
44. Bond, D. S., Nicoll, F. H. and Moore, D. G. "Development and Operation of a Line-Screen Color Kinescope," *RCA Rev,* Vol. 12, No. 3, pp. 542–567, September 1951.
45. Bond, D. S., Nicoll, F. H. and Moore, D. G. "Development and Operation of a Line-Screen Color Kinescope," *Proc IRE,* pp. 1218–1221, October 1951.
46. Weimer, P. K. and Rynn, N. "A 45-Degree Reflection-Type Color Kinescope," *RCA Rev,* Vol. 12, No. 3, pp. 503–526, September 1951.
47. Weimer, P. K. and Rynn, N. "A 45-Degree Reflection-Type Color Kinescope," *Proc IRE,* p. 1201, October 1951.
48. Herold, E. W. "History and Development of the Color Picture Tube," *Proc SID,* Vol. 15, No. 4, pp. 141–149, 1974.
49. "Constructing the Tricolor Picture Tube," *Electronics,* pp. 86–88, May 1951.
50. Grimes, M. J., Grimj, A. C. and Wilhelm, J. F. "Improvements in the RCA Three-Beam Shadow-Mask Color Kinescope," *Proc IRE,* pp. 315–326, January 1954.
51. Fyler, N. F., Rowe, W. E. and Cain, C. W. "The CBS-Colortron: A Color Picture Tube of Advanced Design," *Proc IRE,* pp. 326–334, January 1954.
52. Cunningham, T. M. and Trim, M. E. "Evolution of TV Picture Tubes at RCA," *RCA Picture Tubes,* pp. 10–13, RCA, 1980.
53. Seelen, H. R., Moody, H. C., VanOrmer, D. D. and Morrell, A. M. "Development of a 21-inch Metal-Envelope Color Kinescope," *RCA Rev,* pp. 122–139, March 1955.
54. Janes, R. B., Headrick, H. B. and Evans, J. "Recent Improvements in the 21AXP22 Color Kinescope," *RCA Rev,* pp. 143–167, June 1956.
55. Lob, C. G. "General Electric Post Acceleration Color Tube," *IRE Conv Rec,* Vol. 4, Part 3, pp. 114–117, 1956.
56. Hergenrother, R. C. "Design of Lens-Mask Three-Gun Color Television Tubes," *Proc IRE,* Vol. 43, pp. 943–951, August 1955.
57. Amdursky, M. E., Pohl, R. G. and Szegho, C. S. "A New High-Efficiency Parallax Mask Color Tube," Vol. 43, pp. 936–943, August 1955.
58. Fyler, N., Cain, C. and Hambleton, P. "The Unipotential Mask-Focusing Colortron," *IRE Conv Rec,* Vol. 4, Part 3, pp. 122–127, 1956.
59. Ramberg, E. G., Law, H. B, Allwine, H. S., Darling, D. C., Henderson, C. W. and Rosenthal, H. "Focusing Grill Color Kinescopes," *IRE Conv Rec,* Vol. 4, Part 3, pp. 128–134, 1956.
60. Smith, C. P., Morrell, A. M. and Demmy, R. C. "Design and Development of the 21CYP22 21-Inch Glass Color Picture Tube," *RCA Rev,* pp. 334–348, September 1958.
61. Larach, S. and Hardy, A. E. "Cathode-Ray-Tube Phosphors: Principles and Applications," *Proc*

IEEE, Vol. 61, No. 7, pp. 915–926, July 1973.

62. Hardy, A. E. "Major Developments in Phosphors and Screen Application Techniques for Cathode-Ray Tubes," *RCA Engin,* Vol. 25, No. 2, pp. 12–18, August/September 1979.
63. "New Lawrence Tube Shown," *Electronics,* p. 146, November 1951.
64. "Phosphor-Strip Tricolor Tubes," *Electronics,* pp. 89–91, December 1951.
65. Dressler, R. "The PDF Chromatron—A Single or Multi-Gun Tri-Color Cathode-Ray Tube," *Proc IRE,* pp. 851–858, July 1953.
66. "(Agreement) Speeds Single Gun Color Sets," *Electronics,* pp. 8, 10, November 1956.
67. Lyons, N. *The Sony Vision,* Chapter 8, Crown Publishers, Inc., 1976.
68. Clapp, R. G., Creamer, E. M., Moulton, S. W., Partin, M. E. and Bryan, J. S. "A New Beam-Indexing Color Television Display System," *Proc IRE,* pp. 1108–1114, September 1956.
69. Payne, D., Colgate, H., Moulton, S., Comeau, C. and Kelley, D. "Recent Improvements in the Apple Beam-Indexing Color Tube," *IRE Natl Conv Rec,* Part 3, pp. 238–241, 1957.
70. Bloomsburgh, R. A., Hopengarten, A., Moore, R. C. and Wilson, H. H. "An Advanced Color Television Receiver Using a Beam Indexing Picture Tube," *IRE Natl Conv Rec,* Part 3, pp. 243–249, 1957.
71. Barnett, G. F., Bingley, F. J., Parsons, S. L., Pratt, G. W. and Sadowsky, M. "A Beam-Indexing Color Picture Tube—The Apple Tube," *Proc IRE,* pp. 1115–1119, September 1956.
72. Ohkoshi, A., Tohyama, T., Yukawa, T. and Tohyama, A. "A New 30V Beam-index Color Cathode Ray Tube," *IEEE Conf on Cons Elec,* pp. 1–11, June 4, 1981.
73. "Banana-Tube Colour-Television Display," *Wireless World,* pp. 351–352, July 1961.
74. Schagen, P. "The Banana-Tube Display System," *Proc IEE* (London), Vol. 108, Part B, No. 42, pp. 577–586, November 1961.
75. Eastwell, B. A. and Schagen, P. "Development of the Banana Tube," *Proc IEE* (London), Vol. 108, Part B, No. 42, pp. 587–595, November 1961.
76. Schagen, P. "Banana-Tube Color-Television Display System," *Electronics,* pp. 44–46, January 26, 1962.
77. Morrell, A. M., Law, H. B., Ramberg, E. G. and Herold, E. W., *Color Television Picture Tubes: Supplement 1 Advances in Image Pickup and Display,* Academic Press, 1974.
78. Tannas, L. E. *Flat Panel Displays and CRTs,* Chapter 7, Van Nostrand, 1985.
79. Technical Notebook, "Zebra Colour Television Display Tube," *Wireless World,* p. 95, February 1962.
80. Morrell, A. M. and Hardy, A. E. "Development of the RCA 25-Inch 90-Degree Rectangular Color Picture Tube," *IEEE Trans Bcst TV Recvrs,* BTR 10, pp. 15–22, 1964.
81. Morrell, A. M. "Development of the RCA Family of 90-Degree Rectangular Color Picture Tubes," *IEEE Trans Bcst TV Recvrs,* BTR 11, pp. 90–95, 1965.
82. Federal Trade Commission Trade Regulation Rule "Deceptive Advertising As to Sizes of Viewable Pictures Shown by Television Receiving Sets," pp. 1–4, effective July 1, 1966.
83. "Optical Characteristics of Cathode-Ray Tube Screens," EIA JEDEC Publication No. 16-C, November 1975.
84. Fiore, J. P. and Kaplan, S. H. "The Second Generation Color Tube Providing More than Twice the Brightness and Improved Contrast," *IEEE Trans Bcst TV Recvrs,* Vol. BTR-15, No. 3, pp. 267–276, October 1969.
85. Kaplan, S. H. and Lange, H. "Chroma-Filter Color Tube: A High-Contrast, Negative Guard Band Color Tube Using Selective Color Filters," *SID Dig,* pp. 70–71, 1975.
86. Masterton, W. D. and Barbin, R. L. "Designing Out the Problems of Wide-Angle Color TV Tubes," *Electronics,* Vol. 44, No. 9, pp. 60–64, April 26, 1971.
87. Thierfelder, C. W. RCA Large-Screen Narrow-Neck 110° Color Television System," *IEEE Trans Bcst TV Recvrs,* Vol. BTR-17, pp. 141–147, August 1971.
88. Kanai, H., Yamzaki, E. and Yoshikazu, W. "A New High-Performance Color Picture Tube: The Mask-Focusing Color Tube," *IEEE Trans on Elec Dev,* Vol. ED-23, No. 1, pp. 45–49, January 1976.
89. Verweel, J. "Magnetic Post-Focusing in Shadowmasks for Color Picture Tubes," *SID Dig,* pp. 48–49, 1980.

90. ———. "Magnetic Focusing for CTV Tube Masks," *SID Dig*, pp. 54–55, 1982.
91. Fastenau, R. H. J. and Koorneef, J. "Colour C.R.T. with a Magnetic Post-Focusing Shadow Mask," *Eurodisplay '84 Proc*, Paris, pp. 93–96, September 18–20, 1984.
92. Hockings, E. F., Bloom, S. and Catanese, C. A. "A Focus Mask for Color Picture Tubes Using Dipolar and Quadrupolar Lensing," *IEEE Trans on Elec Dev*, Vol. ED-30, No. 1, pp. 912–919, August 1983.
93. Hockings, E. F., Bloom, S. and Tamutus, D. J. "A High-Transmission Focus Mask for Color Picture Tubes," *RCA Rev*, Vol. 44, pp. 371–383 September 1983.
94. Takano, Y. *Advances in Image Pickup and Display, Volume 6, Recent Advances in Color Picture Tubes*, pp. 80–81, Academic Press, Inc., 1983.
95. Braestrup, C. B. and Mooney, R. T. "X-Ray Emission from Television Sets," *Science*, pp. 1071–1074, October 1959.
96. *Electronic Industries Association: The First Fifty Years*, EIA, pp. 120, 142, 186, 1974.
97. Richardson, J. R. "Generation, Characteristics and Shielding of X-Radiation Produced in High Voltage Rectifier and Shunt Regulator Tubes in Television Sets," *IEEE Trans Bcst TV Recvrs*, BTR-17, No. 4, pp. 232–238, November 1971.
98. "X-Radiation Characteristics of Future RCA Color Picture Tubes," RCA Color Picture Tube Engineering Note, September 1969.
99. Sheldon, J. L. "Progress in the Development of X-Ray Absorbing Glasses for Television Picture Tubes," Intl. Symp. on Radiological Protection Problems Associated with Parasitic X-Ray Emission from Electronic Products, Toulouse, 3–6, XI, 1970.
100. Yoshida, S., Ohkoshi, A. and Miyaoka, S. "The 'Trinitron'—A New Color Tube," *IEEE Trans Bcst TV Recvrs*, Vol. BTR-14, pp. 19–27, July 1968.
101. Yoshida, S., Ohkoshi, A. and Miyaoka, S. "A Wide-Deflection Angle (114 Degree) Trinitron Color-Picture Tube," *IEEE Trans Bcst TV Recvrs*, Vol. BTR-19, No. 4, pp. 231–238, November 1973.
102. ———. "25-V Inch 114-Degree Trinitron Color Picture Tube and Associated New Developments," *IEEE Trans Bcst TV Recvrs*, Vol. BTR-20, No. 3, pp. 193–200, August 1974.
103. Yoshida, S. "A New 30V-Inch Screen Size Color Picture Tube," *IEEE Trans Cons Electron*, Vol. CE-23, No. 1, pp. 1–7, February 1977.
104. Yamada, Y., Ashiya, R., Kikuchi, M. and Saita, K. "A New High-Resolution Trinitron Color Picture Tube for Display Application," *IEEE Trans Cons Electron*, Vol. CE-26, pp. 466–472, August 1980.
105. Nakamura, M., Makino, T., Soematsu, S. and Saito, T. "A 45-inch Flat and-Square Trinitron Color CRT," *SID Dig*, pp. 386–388, 1988.
106. Barbin, R. L. and Hughes, R. H. "A New Color Picture Tube System for Portable TV Receivers," *IEEE Trans Bcst TV Recvrs*, Vol. BTR-18, No. 3, pp. 193–200, August 1972.
107. Branton, T. W., Godfrey, R. H., Masterton, W. D., Morrell, A. M., Trim, M. E. and Weber, D. L. "Recent Trends in Color Picture Tube Design," *RCA Picture Tubes*, pp. 14–21, RCA, 1980.
108. Morrell, A. M. "Color Picture Tube Design Trends," *SID Dig*, pp. 162–163, 1980.
109. ———. "Color Picture Tube Design Trends," *Proc SID*, Vol. 22, No. 1, 1981.
110. Trond, S. S. "Filter Phosphors," *RCA Picture Tubes*, pp. 3–9, RCA, 1980.
111. Hamano, E., Okada, H. and Koshigoe, S. "Mini-Neck Color Picture Tube," *Toshiba Rev*, 1980.
112. Morrell, A. M. "An Overview of the RCA COTY-29 Tube System," Presented at IEEE Intl. Conf. on Consumer Elect. (Chicago IL) RCA Reprint ST-5125, pp. 1–7, 1982.
113. Ashizaki, S., Konosu, O. and Suzuki, Y. "43-Inch Direct-View Color CRT," *IEEE Tokyo Sec*, No. 25, pp. 58–60, 1986.
114. Ashizaki, S., Suzuki, Y., Konosu, O. and Adachi, O. "43-inch Direct View Color CRT," *Proc Jpn Disp '86*, pp. 226–229, 1986.
115. Kitani, T., Aya, M. and Suzuki, Y. "Pocketable Color Display," *Jpn Disp '83*, pp. 124–126, 1983.
116. Hergenrother, G. R. and Kloss, H. E. "Projection Television Tube System," U.S. Patent No. 3,969,648, filed September 3, 1974, issued July 13, 1976.
117. ———. "Projection Television Tube System and

Method of Producing Same," U.S. Patent No. 4,024,579, filed June 4, 1976, issued May 17, 1977.
118. Kloss, H. E. "Projection Television Tube and Process for Forming Same," U.S. Patent No. 4,365,183, filed February 17, 1981, issued December 21, 1982.
119. ———. "Process for Making Envelope for Projection Television Tube," U.S. Patent No. 4,433,962, filed September 7, 1982, issued February 28, 1984.
120. Patrick, N. W. "Developments in Cathode-Ray Tubes for the Schmidt Projector," *Proc IEEE Conf on Disp Dev* RCA Reprint ST-4889, pp. 109–115, October 1972.
121. Gilmore, C. P. "Giant Screen Color Television," *Electronics* (Australia), pp. 10–16, December 1979.
122. Todd, L. "Projection Display Devices," *SID Conf Lec Notes,* Vol. 2, pp. 12-2–12-26, May 13, 1983.
123. Kikuchi, M., Kobayashi, K., Chiba, T. and Fujii, Y. "A New Coolant Sealed CRT for Projection TV," *IEEE Trans Cons Elec,* Vol. CE-27, No. 3, pp. 478–485, August 1981.
124. Hockenbrock, R. and Rowe, W. A. "A Self-Converging Three-Tube Projection System," *Proc SID,* Vol. 23, No. 3, pp. 135–140, 1982.
125. New Product Announcement, "40-Inch Color CRT," *Electron Compon News,* November 1984.
126. Good, W. E. "Projection Television," *Proc SID,* Vol. 17, pp. 3–7, 1976.
127. Kaplan, S. H. "The History of Color Picture Tubes and Some Future Projections," *SMPTE J,* pp 396–400, May 1990.

CHAPTER 7

1. Computer Issue, *Proc IRE,* Vol. 41, No. 10, October 1953.
2. Williams, F. C. and Kilburn, T. "A Storage System for Use with Binary-Digital Computing Machines," *Proc IEE* (London), Vol. 95, pp. 183–202, December 1948.
3. Williams, F. C. and Kilburn, T. "A Storage System for Use with Binary-Digital Computing Machines," *Proc IEE,* Part III, pp. 81–100, March 1949.
4. Eckert, J. P. "A Survey of Digital Computer Memory Systems," *Proc IRE,* Vol. 71, No. 10, pp. 1393–1406, October 1953.
5. Eckert, J. P., Lukoff, H. and Smoliar, G. "A Dynamically Regenerated Electrostatic Memory System," *Proc IRE,* pp. 498–510, May 1950.
6. Greenwald, S., Haueter, R. C. and Alexander, S. N. "SEAC," *Proc IRE,* Vol. 71, No. 10, pp. 1300–1313, October 1953.
7. Huskey, H. D., Thorensen, R., Ambrosio, B. F. and Yowell, E. C. "The SWAC-Design Features and Operating Experience," *Proc IRE,* Vol. 41, No. 10, pp. 1294–1299, October 1953.
8. Mutter, W. E. "Improved Cathode-Ray Tube for Application in Williams Memory System," *Electr Eng,* Vol. 71, pp. 352–356, April 1952.
9. Frizzell, C. E. "Engineering Description of the IBM Type 701 Computer," *Proc IRE,* Vol. 41, No. 10, pp. 1275–1287, October 1953.
10. Rider, J. F. and Uslan, S. D. *Encyclopedia on Cathode-Ray Oscilloscopes and Their Uses, Second Edition,* pp. 12-8–12-11, John F. Rider Publisher Inc., 1959.
11. King, G. W., Brown, G. W. and Ridenour, L. N. "Photographic Techniques for Information Storage," *Proc IRE,* Vol. 41, No. 10, pp. 1421–1428, October 1953.
12. Hoover, C. W., Staehler, R. E. and Ketchledge, R. W. "Fundamental Concepts in the Design of the Flying Spot Store," *Bell Sys Tech J,* pp. 1161–1194, September 1958.
13. Cooper, H. G. "Design of a High-Resolution Electrostatic Cathode Ray Tube for the Flying Spot Store," *Bell Sys Tech J,* pp. 723–759, May 1961.
14. Rider, J. F. and Uslan, S. D. *Encyclopedia on Cathode-Ray Oscilloscopes and Their Uses, Second Edition,* pp. 12-14–12-15, John F. Rider Publisher Inc., 1959.
15. McNaney, J. T. "Data Displaying Cathode-Ray Tube," *Electronics,* pp. 174–186, August 1952.
16. McNaney, J. T. "The Type C19K Charactron Tube and Its Application to Air Surveillance Systems," *IRE Convention Record,* New York, 1955.
17. Corpew, C. R. "An Improved Charactron® Shaped Beam Tube," *Proc SID,* Vol. 12, No. 1, pp. 22–30, 1971.
18. Haflinger, D. J. "A New CRT for Alphanumeric Computer Terminals," *SID Dig,* pp. 126–127, 1976.

19. Hills, D. S. "Multi-Beam Cathode-Ray-Tube Displays," *SID Dig,* pp. 74–75, 1972.
20. Rider, J. F. and Uslan, S. D. *Encyclopedia on Cathode-Ray Oscilloscopes and Their Uses, Second Edition,* pp. 12-15–12-16, John F. Rider, Publisher Inc., 1959.
21. Winningstad, C. N. "The Simplified Direct-View Bistable Storage Tube in Computer-Output Applications," *Proc SID,* Vol. 8, pp. 129–136, 1967.
22. Mootry, F. B. and Frankland, R. A. "Recent Developments in Direct-View Storage Tubes," *SID Dig,* pp. 58–59, 1970.
23. Frankland, R. A. "Developments in Simplified Direct View Storage," *SID Dig,* pp. 82–83, 1972.
24. Curtin, C., Hutcheon, J., Mason, W. and McTeague, J. "A Large-Screen Display for Bistable Storage of Up to 17,000 Characters," *SID Dig,* pp. 98–99, 1973.
25. DeVey, B. "Large-Screen Direct-View Storage Monitor," *SID Dig,* pp. 116–117, 1978.
26. Woody, T. "Improved Phosphor Life in Direct-View Storage CRTs," *SID Dig,* pp. 118–119, 1978.
27. Virgin, L., Langford, D. and Penn, C. "Write-Thru on Direct-View Bistable Storage Tubes," *IEDM Dig,* pp. 268–271, December 1978.
28. Woody, T. "A Color Write-Through Direct-View Storage Tube," *SID Dig,* pp. 98–99, 1981.
29. Tam, T. and Bloom, S. "CRT Studies Show Growing Preference for White Phosphor Displays," *Comput Tech Rev,* pp. 91–93, Fall 1987.
30. Murch, G. M., Virgin, L. and Beaton, R. J. "Resolution and Addressability: How Much is Enough?" *Proc SID,* Vol. 26, No. 4, pp. 305–308, 1985.
31. Murch, G. M. and Beaton, R. J. "Matching Display Resolution and Addressability to Human Visual Capacity," *Displays,* Vol. 9, No. 1, pp. 23–26, January 1988.
32. Virgin, L. "Understanding and Evaluating a Computer Graphics Display," *Tekniques* (Tektronix, Inc.), Vol. 10, No. 4, pp. 34–38, 1987.
33. EIA, "Relating Display Resolution and Addressability," *TEPAC Engin Bull,* TEB-27, pp. 1–7, August 1988.
34. "Cutting Pixels to 4 mils Hikes CRTs Resolution," *Electron Des,* p. 39–40, May 26, 1983.
35. Odenthal, C. J. and McKibben, B. A. "CRT Astigmatism Correction Apparatus with Stored Correction Values," U.S. Patent No. 4,672,276, filed May 29, 1984, issued June 9, 1987.
36. Odenthal, C. J. "A 19-in. Very-High-Resolution Display CRT," *SID Dig,* pp. 258–261, 1984.
37. Infante, C., Denham, D. and McKibben, B. A. "A 230 MHz Bandwidth High-Resolution Monitor," *SID Dig,* pp. 124–125, 1983.
38. Carpenter, M. E. and Say, D. L. "A CRT for Realtime Display of Slow-Scan Information," *SID Dig,* pp. 94–95, 1973.
39. Quick, R. G. and Marshall, F. J. "Multibeam CRT for High-Speed Information Recording," *SID Dig,* pp. 122–123, 1981.
40. Wurtz, J. E. "Multibeam CRTs Get the Picture," *Electro-Optical Sys Des,* pp. 53–58, September 1982.
41. Odenthal, C. J. and Quick, R. G. "A Gatling-Gun Multibeam CRT," *SID Dig,* pp. 53–55, 1986.
42. Quick, R. G. "Multibeam CRT for Direct View Display of Photographic Quality Medical Images," *SPIE Conf Proc,* paper 536-05, February 1985.
43. Sisneros, T. E., Faeth, P. A., Davis, J. A. and Hilborn, E. H. "Current Sensitive, Single-Gun Color CRT," *Inf Disp,* pp. 33–37, April 1970.
44. Avella, F. J. "Phosphor Screens for Current-Sensitive Single-Gun Polychromatic Cathode-Ray Tubes," *IEEE Trans on Electron Dev,* Vol. ED-18, No. 9, pp. 719–724, September 1971.
45. Sisneros, T. E. "Evaluation of Current-Sensitive Color CRT Screens," *IEEE Trans on Electron Dev,* Vol. ED-18, No. 9, pp. 798–800, September 1971.
46. Ohkoshi, A., Takeuchi, O., Kambayashi, K. and Yukawa, T. "Current-Sensitive Multi-Color CRT Display," International Display Research Conference, pp. 64–65.
47. "Current-Sensitive CRT Combines High Resolution with Some Color," *Electron Imaging,* p. 18, January 1985.
48. "Color Comes to Monochrome CRTs," *Mach Des,* pp. 25–28, January 24, 1985.
49. Nakamura, H. "High-Resolution Color Picture Tube," *Jpn Electron Engin,* pp. 44–48, May 1974.
50. Barbin, R. L., Simpson, T. F. and Marks, B. G. "A Color-Data-Display CRT: A Product Whose Time Has Come," *RCA Engin,* Vol. 27, No. 4, July/August 1982.

51. Chen, H. Y. "An In-Line Gun for High Resolution Color Display," *SID Symposium,* San Diego, CA, pp. 1–7, 1982.
52. Yoshida, H. and Yamazaki, E. "High-Resolution Color Display Tubes for Display Terminals," *Hitachi Rev,* Vol. 32, No. 1, pp. 33–36, 1983.
53. Yoshida, H. and Yamazaki, E. "Superhigh-Resolution Color Display Tubes," *Hitachi Rev,* Vol. 32, No. 4, pp. 199–202, 1983.
54. Kobayashi, K., Okamoto, H., Koba, H., Hamono, E. and Ikegaki, M. "A 17 in. Full Square, Superhigh-Resolution CRT," *Jpn Disp* '86, pp. 222–225, 1986.
55. Williams, G., Gallaro, A. V., Bates, D. J., Nelson, G. A. and Robinder, R. C. "A High-Resolution CRT for a Monitor with Auto-Convergence Features," *SID Jpn 83,* pp. 16–18, October 1983.
56. Denham, D. and Meyer, W. "A High Resolution Color Display with Autoconvergence," *Jpn Disp,* pp. 12–14, 1983.
57. "Automatic Convergence 'Paints' Unprecedented Color Display," *Electron Prod,* pp. 17–18, May 12, 1983.
58. Denham, D., Singer, M. and Meyer, B., "Autoconvergence Enhances High-Resolution of 4115B Color Display," *Tekniques* (Tektronix, Inc.), Vol. 7, No. 3, pp. 21–24, Fall 1983.
59. Denham, D., Singer, M., Meyer, B. and Murch, J. "Autoconvergence Enhances High Resolution Display," *Electron Imaging,* pp. 30–34, January 1984.
60. Yamada, Y., Ashiya, R., Kikuchi, M. and Saita, K. "A New High-Resolution Trinitron Color Picture Tube for Display Application," *IEEE Trans Consum Electron,* Vol. CE-26, pp. 466–472, August 1980.
61. Sudo, M., Ashiya, R., Uba, T., Murata, A. and Amano, Y. "High-Resolution 20V-inch × 20V-inch Trinitron and Monochrome CRT," *SID Dig,* pp. 338–341, 1986.
62. Ohno, H., Amano, Y. and Inouye, H. "Super-High-Resolution Gun with Complex Prefocus Lens for 2K × 2K Color Display," SID *Dig,* pp. 45–48, 1989.
63. Dietch, L. and Palac, K. "Tension Mask Color Cathode Ray Tube," U.S. Patent No. 4,593,225, filed August 31, 1984, issued June 3, 1986.
64. Palac, K. "Tension Mask Mounting Structure," U.S. Patent No. 4,614,892, filed August 31, 1984, issued September 30, 1986.
65. Dietch, L., Palac, K. and Chiodi, W. "Performance of High-Resolution Flat Tension Mask Color CRTs," *SID Dig,* pp. 322–323, 1986.
66. Strauss, P., Prazak, C. J. and Lee, S. D. "Construction of a Flat Tension Mask Color CRT," *SID Dig,* pp. 324–326, 1986.
67. Palac, K., Dietch, L., Prazak, C. and Chiodi, W. "The Perfectly Flat Tension Mask Color CRT," *IEEE Intl Conf Cons Elec,* pp. 82–83, 1986.
68. Robinder, R. C. et al, "A High-Brightness Shadow-Mask Color CRT for Cockpit Displays," *SID Dig,* pp. 72–73, 1983.
69. Kawamura, T., Tomita, Y. and Kumada, M. "A Flat-Screen High-Resolution 18-in. Viewable-Diagonal Color CRT," *SID Dig,* pp. 264–265, 1989.
70. Vatne, R., Johnson, P. A. and Bos, P. J. "A LC/CRT Field-Sequential Color Display," *SID Dig,* pp. 28–29, 1983.
71. Bos, P. J., Johnson, P. A. and Koehler-Beran, K. R. "A Liquid-Crystal Optical-Switching Device (π Cell)," *SID Dig,* pp. 30–31, 1983.
72. Bos, P. J. and Johnson, P. A. "Field Sequential Color Display System Using Optical Retardation," U.S. Patent No. 4,582,396, filed May 9, 1983, issued April 15, 1986.
73. Bos, P. J., Thomas, B. and Vatne, R. "A Full-Color Field-Sequential Color Display," *Eurodisp '84 Proc,* pp. 7–9 September 18–20, 1984.
74. Bos, P. J. "Stereoscopic Imaging System with Passive Viewing Apparatus," U.S. Patent No. 4,719,507, filed April 26, 1985, issued January 12, 1988.
75. Murch, G. M. "Human Factors and Flat Panels Challenge the CRT," *Inf Disp,* pp. 8–11, March 1987.

CHAPTER 8

1. Ostrolenk, S. "The Cathode-Ray Aircraft Compass," *Electronics,* pp. 12–15, August 1936.
2. Untitled *Du Mont Oscillographer,* Vol. 1, No. 2, p. 4, April 1937.
3. Sanders, F. H. "Radar Development in Canada," *Proc IRE,* pp. 195–200, February 1947.
4. Berkner, L. V. "Naval Airborne Radar," *Proc IRE,* Vol. 34, No. 9, pp. 671–706, September 1946.

5. Barnes, C. B. "Radar for Carrier-Based Planes," *Electronics,* pp. 100–107, October 1946.
6. Soller, T., Starr, M. A. and Valley, G. E. *Cathode-Ray Tube Displays,* McGraw-Hill, 1948.
7. *Reference Data for Radio Engineers,* 4th ed., pp. 806–807, International Telephone and Telegraph Corp., 1956.
8. Jesty, L. C., Moss, H. and Puleston, R. "War-Time Developments in Cathode-Ray Tubes for Radar," *J IEE* (London), Vol. 93-3A, pp. 149–166, 1946.
9. Bradfield, G., Bartlett, J. B. and Stewart-Watson, D. "A Survey of Cathode-Ray Tube Problems in Service Applications with Special Reference to Radar," *J IEE* (London), Vol. 93-3A, pp. 128–148, 1946.
10. Seats, P. "CRT Improvements Planned and On-going," *Proc SID,* Vol. 19, No. 4, pp. 191–192, 1978.
11. Buhler, F. "CRT Displays for the Military," *Avionics,* pp. 14–17, May 1988.
12. Dowd, C. "Color CRT Applications in a High Performance Military Aircraft," *Proc SID,* Vol. 20, No. 1, pp. 29–32, 1979.
13. Gard, J. H. "A Somewhat Vignetted History of the HUD," *Proc SID,* Vol. 19, No. 4, pp. 163–168, 1978.
14. Seats, P. "Space Shuttle CRT Displays," *Inf Disp,* pp. 4–6, March 1984.
15. Haeff, A. V. "A Memory Tube," *Electronics,* pp. 80–83, September 1947.
16. Knoll, M., Rudnick, P. and Hook, H. "Viewing Storage Tube with Halftone Display," *RCA Rev,* pp. 492–501, December 1953.
17. Knoll, M., Hook, H. O. and Stone, R. P. "Characteristics of a Transmission Control Viewing Storage Tube with Halftone Display," *Proc IRE,* pp. 1496–1504, October 1954.
18. Knoll, M. and Kazan, B. *Advances in Electronics and Electron Physics,* Vol. 8, pp. 447–501, Academic Press, 1956.
19. Hook, H. O., Knoll, M. and Stone, R. P. "Viewing Storage Tube for Large Displays," *RCA Rev,* pp. 503–514, December 1956.
20. Kazan, B. and Knoll, M. *Electronic Image Storage,* pp. 240–241, Academic Press, 1968.
21. Damon, P. P. "The Multimode Tonotron," *SID Proc,* pp. 177–190, February 1965.
22. Sandberg, J. W. "Operation of Hughes Tonotron and Multi-Mode Tonotron Direct-View Storage Tubes," Hughes Application Note 91-11-001, pp. 1–13, Revised November 1978.
23. Wheeler, P. R. "Cathode-Ray Tube Gunsight," U.S. Patent No. 2,459,206, filed December 19, 1945, issued January 18, 1949.
24. Flint, E. E. "Aircraft Landing Instrument Having an Optical Sight," U.S. Patent No. 2,517,779, filed October 3, 1946, issued August 8, 1950.
25. Seats, P. "The Cathode-Ray Tube—A Review of Current Technology and Future Trends," *IEEE Trans Electron Devices,* Vol. ED-18, No. 9, pp. 679–684, September 1971.
26. EIA, "Optical Characteristics of Cathode Ray Tube Screens," TEPAC Publication, No. 116, December 1980.
27. Seymour, R. S. "Phosphors for Heads-Up Display Cathode-Ray Tube," *Displays,* pp. 123–130, July 1988.
28. Kalmanash, M. H. "Miniature High Contrast Color Display Indicator," *SID Dig,* pp. 92–93, 1981.
29. Hayashi, K., Nakamura, K. and Matsushita, M. "Development of a Shadow-Mask Color Picture Tube for Cockpit Display," *IEEE Display Research Conf,* pp. 120–121, 1980.
30. Ikegaki, M., Nakanishi, M. and Tsuruzoe, H. "High Resolution Color CRTs for Avionics Display," *Toshiba Rev,* No. 142, pp. 24–28, November/December 1982.
31. Hockenbrock, R. "New Technology Advances for Brighter Color CRT Displays," *Proc SPIE,* Vol. 1117, pp. 219–226, 1989.
32. Robinder, R. C., Bates, D. J., Green, P. J., Lewen, G. D. and Rath, D. R. "A High Brightness Shadow-Mask Color CRT for Cockpit Displays," *SID Dig,* pp. 72–73, 1983.
33. "Hi-Res Color CRT Meets Stringent Avionic Requirements," *Inf Disp,* pp. 12, 14, September 1986.
34. Kolbeck, R., Schwertfeger, W. and Selfert, H. "Improvements in Filter, Phosphor and Mask Technologies for Avionics Color CRTs," *SID Dig,* pp. 182–184, 1987.
35. Martin, A. "Ruggedized Color CRT Assemblies," *Inf Disp,* pp. 10–13, September 1988.
36. Schwartz, J. W. "Beam Index Technology," *Proc SID,* Vol. 20, No. 2, pp. 45–53, 1979.
37. Barten, P. G. J. "CRT: Present and Future," *SID Conf Lec Notes,* pp. 3-1–3-44, May 1989.

38. MacWhirter, I. "Beam Indexing Tubes (part 1)," *Wireless World,* pp. 2–7, January 1961.
39. ———. "Beam Indexing Tubes (part 2)," *Wireless World,* pp. 92–98, February 1961.
40. Turner J. A. "A New Beam Indexing Color Display Tube Using Direct Pick-Up of the Electron Beam—The Turner Tube," *IEEE/SID Biennial Disp Conf Rec,* pp. 83–87, October 1976.
41. Turner, J. A. "An Electron Beam Indexing Colour Display System," *Displays,* pp. 155–158, October 1979.
42. Saito, S., Tanaka, Y., Yamamoto, M. and Iida, K. "A 6.3 × 6.3 in. Avionics Beam Index CRT," *SID Dig,* pp. 178–181, 1987.
43. Yano, T., Fujita, K., Irie, O. and Matsuo, T. "A 5-in. Beam-Index Color Monitor for Automotive Applications," *SID Dig,* pp. 389–392, 1988.
44. Catalog. "Color Display Beam Index CRTs," Thomson Tubes Electroniques, Publication TTE155A, 1989.
45. Forstmann, R. "Beam Index Display—The Answer to Modern Cockpit Design," *IEEE,* pp. 220–224, 1987.
46. Woodcock, S. and Leyland, J. D. "High-Resolution CRTs and Their Application to Helmet-Mounted Displays," *Proc SID,* Vol. 20, No. 2, pp. 105–109, 1979.
47. Wurtz, J. E. "Applications of Miniature CRT Displays," *Inf Disp,* pp. 16–18, September 1987.
48. Seats, P. "The Evolutionary CRT," *SID Conf Lec Notes,* p. 7.23, May 1987.
49. Sanford, E. and Seats, P. "Miniature CRT Improvement Program Study," A. F. Aerospace Medical Res. Lab. Report AFAMRL-TR-83-075, September 1983.
50. Wurtz, J. E. "Miniature CRTs Meet High Performance Specs," *Inf Disp,* pp. 22–23, February 1986.
51. Seymour, R. S. "HUD CRT Phosphors Exhibit Distinctive Preferential Qualities," *Inf Disp,* pp. 24–30, April 1986.
52. Gritz, E. E. "High-Brightness High-Resolution Miniature CRTs," *SID Dig,* pp. 340–341, 1984.
53. Shmulovich, J. and Koclan, D. F. "Thin-Film Phosphors for Miniature CRTs Used in Helmet-Mounted Displays," *SID Dig,* pp. 200–202, 1989.
54. Levy, U. and Yaffe, H. H. "Saturation and Heating of Miniature CRT Single-Crystal Phosphor Screens," *SID Dig,* pp. 336–339, 1984.

CHAPTER 9

1. Abramson, A. *The History of Television, 1880 to 1941,* p. 34, McFarland and Co., 1987.
2. Zworykin, V. K. *Photoelectricity,* p. 368, John Wiley & Sons, 1949.
3. Fink, D. G. *Principles of Television Engineering,* p. 90, McGraw-Hill, 1940.
4. ———. *Television Engineering,* pp. 91–95, McGraw-Hill, 1952.
5. Rider, J. F. and Uslan, S. D. *Encyclopedia on Cathode-Ray Oscilloscopes and Their Uses,* pp. 176, 839–840, John F. Rider Publisher Inc., 1950.
6. Neal, C. B. "Simplified Color Slide TV Camera System for Cable Television," *SMPTE Proc Symp Cable Telev,* New York City, pp. 76–81, October 5–6, 1970.
7. Wurtz, J. Interview, August 4, 1989.
8. King, G. W., Brown, G. W. and Ridenour, L. N. "Photographic Techniques for Information Storage," *Proc IRE,* Vol. 41, No. 10, pp. 1421–1428, October 1953.
9. Hoover, C. W., Staehler, R. E. and Ketchledge, R. W. "Fundamental Concepts in the Design of the Flying Spot Store," *Bell Sys Tech J,* pp. 1161–1194, September 1958.
10. Cooper, H. G. "Design of a High-Resolution Electrostatic Cathode Ray Tube for the Flying Spot Store," *Bell Sys Tech J,* pp. 723–759, May 1961.
11. Schlesinger, K. "A Microspot Tube with Very High Resolution," *IRE Trans Electron Devices,* pp. 281–288, May 1962.
12. de Boer, J. H. and Dippel, C. J. "Process of Making a Layer Containing a Fluorescent Material," U.S. Patent No. 1,954,691, filed September 18, 1931, issued April 10, 1934.
13. Williams, F. E. "Some New Aspects of Germanate and Fluoride Phosphors," *J Opt Soc Amer,* Vol. 37, No. 4, pp. 302–307, April 1947.
14. Studer, F. J., Cusano, D. A. and Young, A. H. "Transparent Luminescent Films," *J Opt Soc Amer,* Vol. 41, No. 8, p. 559, August 1951.
15. Cusano, D. A. and Studer, F. J. "Transparent Luminescent Screens," *J Opt Soc Amer,* Vol. 42, No. 11, p. 878, November 1952.
16. Studer, F. J. and Cusano, D. A. "Transparent Phosphor Coatings," *J Opt Soc Amer,* Vol. 45, No. 7, pp. 493–497, July 1955.

17. "Transparent Phosphors Improve CRT," *Electronics,* pp. 92–93, February 28, 1958.
18. Neal, C. B. and Smithgall, H. E. "Integrated Flying-Spot Color Slide Scanner and Television Receiver," *IEEE Trans Broadcast Telev Receivers,* Vol. BTR16, pp. 56–62, February 1970.
19. Fraser, R. M. "Motion Picture Photography of Television Images," *RCA Rev,* Vol. 9, No. 2, pp. 202–217, June 1948.
20. Bond, D. S. and Duke, V. J. "Ultrafax," *RCA Rev,* pp. 99–115, March 1949.
21. Dufour, A. "Oscillographe Cathodique," *J Phys Radium,* pp. 147–160, 1920.
22. George, R. H. "A New Type of Hot Cathode Oscillograph," *J AIEE,* Vol. 48, pp. 534–538, July 1929.
23. Wurtz, J. "The Evolution of CRT Color Film Recording Technology," *Comput Graphics World,* pp. 25, 26, 28, 32, 34, 96, February 1985.
24. Stow, R. L. "Fiber Optics and Their Application to Electronic Tubes," Westinghouse Electric Reprint ET-1258, 1960.
25. Martin, A. *Advances in Electronics and Electron Physics,* ed. P. W. Hawkes, Vol. 67, pp. 302–304, Academic Press, 1986.
26. Olden, R. G. "A Thin-Window Cathode-Ray Tube for High-Speed Printing with 'Electrofax,'" *RCA Rev,* Vol. 18, 1957.
27. Wurtz, J. "The Forgotten Fiber-Optic Application," *Electro-Opt Syst Des,* pp. 40–45, November 1979.
28. Katzman, F. L. "Improving Ultra-Fast Transient Recording Using Fibre-Optics Cathode-Ray Tubes," *Elect Inst Dig,* pp. 41–47, October 1966.
29. Durbeck, R. C. and Sherr, S. *Output Hardcopy Devices,* pp. 66–74, Academic Press, 1988.
30. Schlam, E. and Pucilowski, J. J. "High Efficiency Cathode-Ray Tubes for Dry-Process Recording," *IEEE Trans on Electron Devices,* Vol. ED-18, No. 9, pp. 713–718, September 1971.
31. Seats, P. "The Evolutionary CRT," *SID Conf Lec Notes,* pp. 7.19–7.20, 1987.
32. Quick, R. G. and Marshall, F. J. "The Multi-Beam CRT for High-Speed Information Recording," *SID Dig,* pp. 122–123, 1981.
33. Marshall, F. J. and Quick, R. G. "A Micro-Array Multi-Beam Gun for Hard Copy Printout Cathode-Ray Tubes," *Eurodisplay,* September 1981.
34. Wurtz, J. E. "Multi-Beam CRTs Get the Picture," *Electro-Opt Syst Des,* pp. 53–58, September 1982.
35. Durbeck, R. C. and Sherr, S. *Output Hardcopy Devices,* pp. 63–64, Academic Press, 1988.
36. Dufour, A. "Oscillographe Cathodique pour l'Étude des Basses, Moyennes et Hautes Fréquences," *L'onde Electrique,* Vol. 1, pp. 699–715, 1922.
37. Dufour, A. "Oscillographe Cathodique pour l'Étude des Basses, Moyennes et Hautes Fréquences," *L'onde Electrique,* Vol. 1, pp. 638–663, 1922.
38. Mac Gregor-Morris, J. T. "Measurements in Electrical Engineering by Means of Cathode Rays," *J IEE,* Vol. 63, pp. 1056–1107, November 1925.
39. ———, and Henley, J. A. *Cathode Ray Oscillography,* Vol. II, Chapter 4, Instruments Publishing Co. Pittsburgh, 1936.
40. Miyazaki, E. and Tomii, K. "Prism Fiber Reading Tube," *IEEE Trans Electron Devices,* Vol. ED-18, No. 10, pp. 872–878, October 1971.

CHAPTER 10

1. Aiken, W. R. "History of the Kaiser-Aiken Thin Cathode Ray Tube," *IEEE Trans Electron Devices,* Vol. ED-31, No. 11, pp. 1605–1608, November 1984.
2. Marshall, F. J. "Commentary on the Aiken Thin CRT," *IEEE Trans Electron Devices,* Vol. ED-31, No. 11, pp. 1608–1609, November 1984.
3. Aiken, W. R. "A Thin Cathode-Ray Tube," *Proc IRE,* Vol. 45, No. 12, pp. 1599–1604, December 1957.
4. Aiken, W. R. "Development of the Thin Cathode-Ray Tube," *J SMPTE,* Vol. 67, No. 7, pp. 452–455, July 1958.
5. Aiken, W. R. "Cathode Ray Tube," U.S. Patent No. 2,795,731, filed December 4, 1953, issued June 11, 1957.
6. Gard, J. H. "A Somewhat Vignetted History of the HUD," *Proc SID,* Vol. 19, No. 4, pp. 163–168, 1978.
7. Geer, W. "Color in the Thin-Tube," *SID Proc,* pp. 278–285, February 1964.
8. Tannas, L. E. *Flat-Panel Displays and CRTs,* Chapter 7, Van Nostrand Reinhold, 1985.
9. Gabor, D., Stuart, P. R. and Kalman, G. "A New Cathode-Ray Tube for Monochrome and Colour Television," *Proc IEE* (London), Vol. 105, part B, pp. 581–606, May 1958.

10. Commentary. "The Gabor Television Tube," *Research,* pp. 209–210, June 1958.
11. Gabor, D., Tothill, H. A. W. and Smith-Whittington, J. E. "A Fully Electrostatic, Flat, Thin Television Tube," *IEE Proc* (London), Vol. 115, No. 4, pp. 467–478, April 1968.
12. Law, H. B. and Ramberg, E. G. "The Reflected-Beam Kinescope," *Proc IRE,* pp. 1409–1417, August 1960.
13. Smith, K. "CRT Slims Down for Pocket and Projection TVs," *Electronics,* pp. 67–68, July 19, 1979.
14. Sinclair, C. "Small Flat Cathode Ray Tube," *SID Dig,* pp. 138–139, 1981.
15. Cohen, C. "Sony's Pocket TV Slims Down CRT Technology," *Electronics,* pp. 81–82, February 10, 1982.
16. Ohkoshi, A., Sato, H., Nakano, T. N. and Hatanaka, M. "A Compact Flat Cathode Ray Tube," *IEEE Trans Cons Elect,* Vol. CE-28, No. 3, pp. 431–435, August 1982.
17. "Sony Engineers Develop High-Resolution, Flat CRT," *Electr Eng Times,* pp. 51, 66, July 18, 1983.
18. Matsuzaki, A., Ikegami, H., Hatanaka, M., Kobayashi, H. and Wakita, S. "A Color Flat CRT and Its Application," *IEEE Intl Conf Cons Elec,* pp. 76–77, June 4, 1986.
19. ———. "A Color Flat CRT and its Application," *IEEE Trans Cons Elec,* Vol. CE-32, No. 3, pp. 194–201, August 1986.
20. "Flat Screen CRTs Add Color and Resolution," *Electron Imag,* p. 18, April 1985.
21. Smith, K. "Electrons Make U Turn in Flat CRT," *Electronics,* pp. 81–82, October 20, 1982.
22. Lamport, D. L., Woodhead, A. W., Washington, D. and Overall, C. D. "Flat Deflection System for a Channel Multiplier CRT," *IEE Proc* (London), Vol. 131, part 1, No. 1, pp. 10–12, February 1984.
23. Yamano, M., Hinotani, K., Hayama, H., Kishimoto, S., Sugishita, S. and Matsudaira, M. "A Color Flat Cathode Tube," *IEEE Trans Cons Elect,* Vol. CE-31, No. 3, pp. 163–173, August 1985.
24. "Monochrome Flat Tubes with 1000-Line Resolution," *JEE,* p. 20, 1987.
25. Schwartz, J. W. "Twenty-Five Years Without Panel TV," *SID Dig,* pp. 124–125, 1976.
26. Wurtz, J. E. "The Not-So-Amazing Survival of the CRT," *Inf Disp,* pp. 5–6, 18, September 1989.
27. Lieberman, D. "Colorful Future for Flat Panels," *Electron Eng Times,* p. 71, January 8, 1990.
28. Werner, K. I. Editorial, *Inf Disp,* p. 5, June 1988.
29. Infante, C. "CRTs-Present and Future," *Inf Disp,* pp. 8–11, December 1988.
30. Lieberman, D. "The CRT: Still King of the Displays" *Elecr Eng Times,* pp. 40–45, July 16, 1990.
31. Kaplan, S. H. "The History of Color Picture Tubes and Some Future Projections," *SMPTE J,* pp. 396–400, May 1990.

About the Author

Peter Keller has been fascinated with the cathode-ray tube and its many applications since 1949 when, at the age of 12, he constructed his first oscilloscope using a war surplus 2AP1 CRT. This was the first of many oscilloscopes built, bought and traded in subsequent years. A career in electronics followed with Keller holding positions at Stanford University, General Precision Laboratories, and Associated Universities for Research in Astronomy before he joined Tektronix (in 1963) to further pursue his interest in cathode-ray tubes. At Tektronix he specialized in the measurement of cathode-ray tube and phosphor performance.

Peter Keller is the chairman of the Society for Information Display Committee on Definitions and Standards, and the Electronic Industries Association Committee on Cathode-Ray Display Devices. He has authored a number of papers and articles on cathode-ray tubes and measurements. Not surprisingly, he is a collector of early CRTs and a number of these were used for this book's illustrations. Many CRT developments have occurred during Keller's lifetime and he has witnessed many of them firsthand. During this time, the cathode-ray tube has evolved from its specialized applications and use in early commercial television to a daily presence in almost everyone's lives in the form of computer displays and television.

Main Index

Name Index